THE THEORY OF
MATRICES

BY

F. R. GANTMACHER

VOLUME TWO

CHELSEA PUBLISHING COMPANY
NEW YORK, N. Y.

LIBRARY OF CONGRESS CATALOG CARD NO. 59-11779
INTERNATIONAL STANDARD BOOK NO. 0-8284-0133-0

THE PRESENT WORK, PUBLISHED IN TWO VOLUMES, IS AN ENGLISH TRANSLATION, BY K. A. HIRSCH, OF THE RUSSIAN-LANGUAGE BOOK TEORIYA MATRITS BY F. R. GANTMACHER (Гантмахер)

PRINTED IN THE UNITED STATES OF AMERICA

PREFACE

THE MATRIX CALCULUS is widely applied nowadays in various branches of mathematics, mechanics, theoretical physics, theoretical electrical engineering, etc. However, neither in the Soviet nor the foreign literature is there a book that gives a sufficiently complete account of the problems of matrix theory and of its diverse applications. The present book is an attempt to fill this gap in the mathematical literature.

The book is based on lecture courses on the theory of matrices and its applications that the author has given several times in the course of the last seventeen years at the Universities of Moscow and Tiflis and at the Moscow Institute of Physical Technology.

The book is meant not only for mathematicians (undergraduates and research students) but also for specialists in allied fields (physics, engineering) who are interested in mathematics and its applications. Therefore the author has endeavoured to make his account of the material as accessible as possible, assuming only that the reader is acquainted with the theory of determinants and with the usual course of higher mathematics within the programme of higher technical education. Only a few isolated sections in the last chapters of the book require additional mathematical knowledge on the part of the reader. Moreover, the author has tried to keep the individual chapters as far as possible independent of each other. For example, Chapter V, *Functions of Matrices,* does not depend on the material contained in Chapters II and III. At those places of Chapter V where fundamental concepts introduced in Chapter IV are being used for the first time, the corresponding references are given. Thus, a reader who is acquainted with the rudiments of the theory of matrices can immediately begin with reading the chapters that interest him.

The book consists of two parts, containing fifteen chapters.

In Chapters I and III, information about matrices and linear operators is developed *ab initio* and the connection between operators and matrices is introduced.

Chapter II expounds the theoretical basis of Gauss's elimination method and certain associated effective methods of solving a system of n linear equations, for large n. In this chapter the reader also becomes acquainted with the technique of operating with matrices that are divided into rectangular 'blocks.'

In Chapter IV we introduce the extremely important 'characteristic' and 'minimal' polynomials of a square matrix, and the 'adjoint' and 'reduced adjoint' matrices.

In Chapter V, which is devoted to functions of matrices, we give the general definition of $f(A)$ as well as concrete methods of computing it—where $f(\lambda)$ is a function of a scalar argument λ and A is a square matrix. The concept of a function of a matrix is used in §§ 5 and 6 of this chapter for a complete investigation of the solutions of a system of linear differential equations of the first order with constant coefficients. Both the concept of a function of a matrix and this latter investigation of differential equations are based entirely on the concept of the minimal polynomial of a matrix and—in contrast to the usual exposition—do not use the so-called theory of elementary divisors, which is treated in Chapters VI and VII.

These five chapters constitute a first course on matrices and their applications. Very important problems in the theory of matrices arise in connection with the reduction of matrices to a normal form. This reduction is carried out on the basis of Weierstrass' theory of elementary divisors. In view of the importance of this theory we give two expositions in this book: an analytic one in Chapter VI and a geometric one in Chapter VII. We draw the reader's attention to §§ 7 and 8 of Chapter VI, where we study effective methods of finding a matrix that transforms a given matrix to normal form. In § 8 of Chapter VII we investigate in detail the method of A. N. Krylov for the practical computation of the coefficients of the characteristic polynomial.

In Chapter VIII certain types of matrix equations are solved. We also consider here the problem of determining all the matrices that are permutable with a given matrix and we study in detail the many-valued functions of matrices $\sqrt[m]{A}$ and $\ln A$.

Chapters IX and X deal with the theory of linear operators in a unitary space and the theory of quadratic and hermitian forms. These chapters do not depend on Weierstrass' theory of elementary divisors and use, of the preceding material, only the basic information on matrices and linear operators contained in the first three chapters of the book. In § 9 of Chapter X we apply the theory of forms to the study of the principal oscillations of a system with n degrees of freedom. In § 11 of this chapter we give an account of Frobenius' deep results on the theory of Hankel forms. These results are used later, in Chapter XV, to study special cases of the Routh-Hurwitz problem.

The last five chapters form the second part of the book [the second volume, in the present English translation]. In Chapter XI we determine normal forms for complex symmetric, skew-symmetric, and orthogonal matrices and establish interesting connections of these matrices with real matrices of the same classes and with unitary matrices.

In Chapter XII we expound the general theory of pencils of matrices of the form $A + \lambda B$, where A and B are arbitrary rectangular matrices of the same dimensions. Just as the study of regular pencils of matrices $A + \lambda B$ is based on Weierstrass' theory of elementary divisors, so the study of singular pencils is built upon Kronecker's theory of minimal indices, which is, as it were, a further development of Weierstrass's theory. By means of Kronecker's theory—the author believes that he has succeeded in simplifying the exposition of this theory—we establish in Chapter XII canonical forms of the pencil of matrices $A + \lambda B$ in the most general case. The results obtained there are applied to the study of systems of linear differential equations with constant coefficients.

In Chapter XIII we explain the remarkable spectral properties of matrices with non-negative elements and consider two important applications of matrices of this class: 1) homogeneous Markov chains in the theory of probability and 2) oscillatory properties of elastic vibrations in mechanics. The matrix method of studying homogeneous Markov chains was developed in the book [46] by V. I. Romanovskii and is based on the fact that the matrix of transition probabilities in a homogeneous Markov chain with a finite number of states is a matrix with non-negative elements of a special type (a 'stochastic' matrix).

The oscillatory properties of elastic vibrations are connected with another important class of non-negative matrices—the 'oscillation matrices.' These matrices and their applications were studied by M. G. Krein jointly with the author of this book. In Chapter XIII, only certain basic results in this domain are presented. The reader can find a detailed account of the whole material in the monograph [17].

In Chapter XIV we compile the applications of the theory of matrices to systems of differential equations with variable coefficients. The central place (§§ 5-9) in this chapter belongs to the theory of the multiplicative integral (Produktintegral) and its connection with Volterra's infinitesimal calculus. These problems are almost entirely unknown in Soviet mathematical literature. In the first sections and in § 11, we study reducible systems (in the sense of Lyapunov) in connection with the problem of stability of motion; we also give certain results of N. P. Erugin. Sections 9-11 refer to the analytic theory of systems of differential equations. Here we clarify an inaccuracy in Birkhoff's fundamental theorem, which is usually applied to the investigation of the solution of a system of differential equations in the neighborhood of a singular point, and we establish a canonical form of the solution in the case of a regular singular point.

In § 12 of Chapter XIV we give a brief survey of some results of the fundamental investigations of I. A. Lappo-Danilevskiĭ on analytic functions of several matrices and their applications to differential systems.

The last chapter, Chapter XV, deals with the applications of the theory of quadratic forms (in particular, of Hankel forms) to the Routh-Hurwitz problem of determining the number of roots of a polynomial in the right half-plane ($\mathrm{Re}\, z > 0$). The first sections of the chapter contain the classical treatment of the problem. In § 5 we give the theorem of A. M. Lyapunov in which a stability criterion is set up which is equivalent to the Routh-Hurwitz criterion. Together with the stability criterion of Routh-Hurwitz we give, in § 11 of this chapter, the comparatively little known criterion of Liénard and Chipart in which the number of determinant inequalities is only about half of that in the Routh-Hurwitz criterion.

At the end of Chapter XV we exhibit the close connection between stability problems and two remarkable theorems of A. A. Markov and P. L. Chebyshev, which were obtained by these celebrated authors on the basis of the expansion of certain continued fractions of special types in series of decreasing powers of the argument. Here we give a matrix proof of these theorems.

This, then, is a brief summary of the contents of this book.

F. R. Gantmacher

PUBLISHERS' PREFACE

THE PUBLISHERS WISH TO thank Professor Gantmacher for his kindness in communicating to the translator new versions of several paragraphs of the original Russian-language book.

The Publishers also take pleasure in thanking the VEB Deutscher Verlag der Wissenschaften, whose many published translations of Russian scientific books into the German language include a counterpart of the present work, for their kind spirit of cooperation in agreeing to the use of their formulas in the preparation of the present work.

No material changes have been made in the text in translating the present work from the Russian except for the replacement of several paragraphs by the new versions supplied by Professor Gantmacher. Some changes in the references and in the Bibliography have been made for the benefit of the English-language reader.

CONTENTS

CHAPTER XI

COMPLEX SYMMETRIC, SKEW-SYMMETRIC, AND ORTHOGONAL MATRICES

In Volume I, Chapter IX, in connection with the study of linear operators in a euclidean space, we investigated real symmetric, skew-symmetric, and orthogonal matrices, i.e., real square matrices characterized by the relations†.

$$S^{\mathsf{T}} = S,\ K^{\mathsf{T}} = -K, \text{ and } Q^{\mathsf{T}} = Q^{-1},$$

respectively (here Q^{T} denotes the transpose of the matrix Q). We have shown that in the field of complex numbers all these matrices have linear elementary divisors and we have set up normal forms for them, i.e., 'simplest' real symmetric, skew-symmetric, and orthogonal matrices to which arbitrary matrices of the types under consideration are real-similar and orthogonally similar.

The present chapter deals with the investigation of complex symmetric, skew-symmetric, and orthogonal matrices. We shall clarify the question of what elementary divisors these matrices can have and shall set up normal forms for them. These forms have a considerably more complicated structure than the corresponding normal forms in the real case. As a preliminary, we shall establish in the first section interesting connections between complex orthogonal and unitary matrices on the one hand, and real symmetric, skew-symmetric, and orthogonal matrices on the other hand.

§ 1. Some Formulas for Complex Orthogonal and Unitary Matrices

1. We begin with a lemma:

LEMMA 1:[1] 1. *If a matrix G is both hermitian and orthogonal ($G^{\mathsf{T}} = \bar{G} = G^{-1}$), then it can be represented in the form*

$$G = Ie^{iK}, \tag{1}$$

where I is a real symmetric involutory matrix and K a real skew-symmetric matrix permutable with it:

[1] See [169], pp. 223-225.

† In this and in the following chapters, a matrix denoted by the letter Q is not necessarily orthogonal.

$$I=\bar{I}=I^{\mathsf{T}}, \; I^2=E, \; K=\bar{K}=-K^{\mathsf{T}}. \tag{2}$$

2. *If, in addition, G is a positive-definite hermitian matrix,*[2] *then in* (1) $I=E$ *and*

$$G=e^{iK}. \tag{3}$$

Proof. Let

$$G=S+iT, \tag{4}$$

where S and T are real matrices. Then

$$\bar{G}=S-iT \text{ and } G^{\mathsf{T}}=S^{\mathsf{T}}+iT^{\mathsf{T}}. \tag{5}$$

Therefore the equation $\bar{G}=G^{\mathsf{T}}$ implies that $S=S^{\mathsf{T}}$ and $T=-T^{\mathsf{T}}$, i.e., S is symmetric and T skew-symmetric.

Moreover, when the expressions for G and $\bar{G}$ from (4) and (5) are substituted in the complex equation $G\bar{G}=E$, it breaks up into two real equations:

$$S^2+T^2=E \text{ and } ST=TS. \tag{6}$$

The second of these equations shows that S and T commute.

By Theorem 12′ of Chapter IX (Vol. I, p. 292), the commuting normal matrices S and T can be carried simultaneously into quasi-diagonal form by a real orthogonal transformation. Therefore[3]

$$\begin{gathered} S=Q\{s_1, s_1, s_2, s_2, \dots, s_q, s_q, s_{2q+1}, \dots, s_n\}Q^{-1}, \\ T=Q\left\{\left\|\begin{matrix} 0 & t_1 \\ -t_1 & 0 \end{matrix}\right\|, \left\|\begin{matrix} 0 & t_2 \\ -t_2 & 0 \end{matrix}\right\|, \dots, \left\|\begin{matrix} 0 & t_q \\ -t_q & 0 \end{matrix}\right\|, 0, \dots, 0\right\}Q^{-1} \end{gathered} \qquad (Q=\bar{Q}=Q^{\mathsf{T}-1}) \tag{7}$$

(the numbers s_i and t_i are real). Hence

$$G=S+iT=Q\left\{\left\|\begin{matrix} s_1 & it_1 \\ -it_1 & s_1 \end{matrix}\right\|, \left\|\begin{matrix} s_2 & it_2 \\ -it_2 & s_2 \end{matrix}\right\|, \dots, \left\|\begin{matrix} s_q & it_q \\ -it_q & s_q \end{matrix}\right\|, s_{2q+1}, \dots, s_n\right\}Q^{-1}. \tag{8}$$

On the other hand, when we compare the expressions (7) for S and T with the first of the equations (6), we find:

$$s_1^2-t_1^2=1, \quad s_2^2-t_2^2=1, \dots, \quad s_q^2-t_q^2=1, \quad s_{2q+1}=\pm 1, \dots, \quad s_n=\pm 1. \tag{9}$$

---

[2] I.e., G is the coefficient matrix of a positive-definite hermitian form (see Vol. I, Chapter X, § 9).

[3] See also the Note following Theorem 12′ of Vol. I, Chapter IX (p. 293).

Now it is easy to verify that a matrix of the type $\begin{Vmatrix} s & it \\ -it & s \end{Vmatrix}$ with $s^2 - t^2 = 1$ can always be represented in the form

$$\begin{Vmatrix} s & it \\ -it & s \end{Vmatrix} = \varepsilon e^{i\begin{Vmatrix} 0 & \varphi \\ -\varphi & 0 \end{Vmatrix}},$$

where

$$|s| = \cosh \varphi, \quad \varepsilon t = \sinh \varphi, \quad \varepsilon = \operatorname{sign} s.$$

Therefore we have from (8) and (9):

$$G = Q\{\pm e^{i\begin{Vmatrix} 0 & \varphi_1 \\ -\varphi_1 & 0 \end{Vmatrix}}, \pm e^{i\begin{Vmatrix} 0 & \varphi_2 \\ -\varphi_2 & 0 \end{Vmatrix}}, \ldots, \pm e^{i\begin{Vmatrix} 0 & \varphi_q \\ -\varphi_q & 0 \end{Vmatrix}}, \pm 1, \ldots, \pm 1\}Q^{-1}, \quad (10)$$

i.e.,

$$G = Ie^{iK},$$

where

$$\left.\begin{aligned} I &= Q\{\pm 1, \pm 1, \ldots, \pm 1\}Q^{-1}, \\ K &= Q\left\{\begin{Vmatrix} 0 & \varphi_1 \\ -\varphi_1 & 0 \end{Vmatrix}, \ldots, \begin{Vmatrix} 0 & \varphi_q \\ -\varphi_q & 0 \end{Vmatrix}, 0, \ldots, 0\right\}Q^{-1} \end{aligned}\right\} \quad (11)$$

and

$$IK = KI.$$

From (11) there follows the equation (2).

2. If, in addition, it is known that G is a positive-definite hermitian matrix, then we can state that all the characteristic values of G are positive (see Volume I, Chapter IX, p. 270). But by (10) these characteristic values are

$$\pm e^{\varphi_1}, \pm e^{-\varphi_1}, \pm e^{\varphi_2}, \pm e^{-\varphi_2}, \ldots, \pm e^{\varphi_q}, \pm e^{-\varphi_q}, \pm 1, \ldots, \pm 1$$

(here the signs correspond to the signs in (10)).

Therefore in the formula (10) and the first formula of (11), wherever the sign $\pm$ occurs, the $+$ sign must hold. Hence

$$I = Q\{1, 1, \ldots, 1\}Q^{-1} = E,$$

and this is what we had to prove.

This completes the proof of the lemma.

By means of the lemma we shall now prove the following theorem:

THEOREM 1: *Every complex orthogonal matrix Q can be represented in the form*

$$Q = Re^{iK}, \tag{12}$$

where R is a real orthogonal matrix and K a real skew-symmetric matrix

$$R = \overline{R} = R^{\mathsf{T}-1}, \quad K = \overline{K} = -K^{\mathsf{T}}. \tag{13}$$

Proof. Suppose that (12) holds. Then

$$Q^* = \overline{Q}^{\mathsf{T}} = e^{iK}R^{\mathsf{T}}$$

and

$$Q^*Q = e^{iK}R^{\mathsf{T}}Re^{iK} = e^{2iK}.$$

By the preceding lemma the required real skew-symmetric matrix K can be determined from the equation

$$Q^*Q = e^{2iK} \tag{14}$$

because the matrix Q^*Q is positive-definite hermitian and orthogonal. After K has been determined from (14) we can find R from (12):

$$R = Qe^{-iK}. \tag{15}$$

Then

$$R^*R = e^{-iK}Q^*Qe^{-iK} = E;$$

i.e., R is unitary. On the other hand, it follows from (15) that R, as the product of two orthogonal matrices, is itself orthogonal: $R^{\mathsf{T}}R = E$. Thus R is at the same time unitary and orthogonal, and hence real. The formula (15) can be written in the form (12).

This proves the theorem.[4]

Now we establish the following lemma:

LEMMA 2: *If a matrix D is both symmetric and unitary* ($D = D^{\mathsf{T}} = \overline{D}^{-1}$), *then it can be represented in the form*

$$D = e^{iS}, \tag{16}$$

where S is a real symmetric matrix ($S = \overline{S} = S^{\mathsf{T}}$).

[4] The formula (12), like the polar decomposition of a complex matrix (in connection with the formulas (87), (88) on p. 278 of Vol. I) has a close connection with the important Theorem of Cartan which establishes a certain representation for the automorphisms of the complex Lie groups; see [169], pp. 232-233.

Proof. We set

$$D = U + iV \quad (U = \overline{U}, V = \overline{V}). \tag{17}$$

Then

$$\overline{D} = U - iV, \quad D^{\mathsf{T}} = U^{\mathsf{T}} + iV^{\mathsf{T}}.$$

The complex equation $D = D^{\mathsf{T}}$ splits into the two real equations

$$U = U^{\mathsf{T}}, \quad V = V^{\mathsf{T}}.$$

Thus, U and V are real symmetric matrices.

The equation $D\overline{D} = E$ implies:

$$U^2 + V^2 = E, \quad UV = VU. \tag{18}$$

By the second of these equations, U and V commute. When we apply Theorem 12′ (together with the Note) of Chapter IX (Vol. I, pp. 292-3) to them, we obtain:

$$U = Q\{s_1, s_2, \ldots, s_n\}Q^{-1}, \quad V = Q\{t_1, t_2, \ldots, t_n\}Q^{-1}. \tag{19}$$

Here s_k and t_k $(k = 1, 2, \ldots, n)$ are real numbers. Now the first of the equations (18) yields:

$$s_k^2 + t_k^2 = 1 \quad (k = 1, 2, \ldots, n).$$

Therefore there exist real numbers φ_k $(k = 1, 2, \ldots, n)$ such that

$$s_k = \cos\varphi_k, \quad t_k = \sin\varphi_k \quad (k = 1, 2, \ldots, n).$$

Substituting these expressions for s_k and t_k in (19) and using (17), we find:

$$D = Q\{e^{i\varphi_1}, e^{i\varphi_2}, \ldots, e^{i\varphi_n}\}Q^{-1} = e^{iS},$$

where

$$S = Q\{\varphi_1, \varphi_2, \ldots, \varphi_n\}Q^{-1}. \tag{20}$$

From (20) it follows that $S = \overline{S} = S^{\mathsf{T}}$.

This proves the lemma.

Using the lemma we shall now prove the following theorem:

THEOREM 2: *Every unitary matrix U can be represented in the form*

$$U = Re^{iS}, \tag{21}$$

where R is a real orthogonal matrix and S a real symmetric matrix

$$R = \overline{R} = R^{\mathsf{T}-1}, \; S = \overline{S} = S^{\mathsf{T}}. \tag{22}$$

Proof. From (21) it follows that

$$U^{\mathsf{T}} = e^{iS} R^{\mathsf{T}}. \tag{23}$$

Multiplying (21) and (23), we obtain from (22):

$$U^{\mathsf{T}} U = e^{iS} R^{\mathsf{T}} R e^{iS} = e^{2iS}.$$

By Lemma 2, the real symmetric matrix S can be determined from the equation

$$U^{\mathsf{T}} U = e^{2iS} \tag{24}$$

because $U^{\mathsf{T}} U$ is symmetric and unitary. After S has been determined, we determine R by the equation

$$R = U e^{-iS}. \tag{25}$$

Then

$$R^{\mathsf{T}} = e^{-iS} U^{\mathsf{T}}, \tag{26}$$

and so from (24), (25), and (26) it follows that

$$R^{\mathsf{T}} R = e^{-iS} U^{\mathsf{T}} U e^{-iS} = E,$$

i.e., R is orthogonal.

On the other hand, by (25) R is the product of two unitary matrices and is therefore itself unitary. Since R is both orthogonal and unitary, it is real. Formula (25) can be written in the form (21).

This proves the theorem.

§ 2. Polar Decomposition of a Complex Matrix

We shall prove the following theorem:

THEOREM 3: *If $A = \| a_{ik} \|_1^n$ is a non-singular matrix with complex elements, then*

$$A = SQ \tag{27}$$

and

$$A = Q_1 S_1, \tag{28}$$

where S and S_1 are complex symmetric matrices, Q and Q_1 complex orthogonal matrices. Moreover,

$$S = \sqrt{AA^{\mathsf{T}}} = f(AA^{\mathsf{T}}), \quad S_1 = \sqrt{A^{\mathsf{T}} A} = f_1(A^{\mathsf{T}} A),$$

where $f(\lambda)$, $f_1(\lambda)$ are polynomials in λ.

The factors S and Q in (27) (Q_1 and S_1 in (28)) are permutable if and only if A and A^{T} are permutable.

Proof. It is sufficient to establish (27), for when we apply this decomposition to the matrix A^{T} and determine A from the formula thus obtained, we arrive at (28).

If (27) holds, then

$$A = SQ, \quad A^{\mathsf{T}} = Q^{-1}S$$

and therefore

$$AA^{\mathsf{T}} = S^2. \tag{29}$$

Conversely, since AA^{T} is non-singular ($|AA^{\mathsf{T}}| = |A|^2 \neq 0$), the function $\sqrt{\lambda}$ is defined on the spectrum of this matrix[5] and therefore an interpolation polynomial $f(\lambda)$ exists such that

$$\sqrt{AA^{\mathsf{T}}} = f(AA^{\mathsf{T}}). \tag{30}$$

We denote the symmetric matrix (30) by

$$S = \sqrt{AA^{\mathsf{T}}}.$$

Then (29) holds, and so $|S| \neq 0$. Determining Q from (27)

$$Q = S^{-1}A,$$

we verify easily that it is an orthogonal matrix. Thus (27) is established.

If the factors S and Q in (27) are permutable, then the matrices

$$A = SQ \quad \text{and} \quad A^{\mathsf{T}} = Q^{-1}S$$

are permutable, since

$$AA^{\mathsf{T}} = S^2, \quad A^{\mathsf{T}}A = Q^{-1}S^2Q.$$

Conversely, if $AA^{\mathsf{T}} = A^{\mathsf{T}}A$, then

$$S^2 = Q^{-1}S^2Q,$$

i.e., Q is permutable with $S^2 = AA^{\mathsf{T}}$. But then Q is also permutable with the matrix $S = f(AA^{\mathsf{T}})$.

Thus the theorem is proved completely.

2. Using the polar decomposition we shall now prove the following theorem:

---

[5] See Vol. I, Chapter V, § 1. We choose a single-valued branch of the function $\sqrt{\lambda}$ in a simply connected domain containing all the characteristic values of AA^{T}, but not the number 0.

THEOREM 4: *If two complex symmetric or skew-symmetric or orthogonal matrices are similar:*

$$B = T^{-1}AT, \tag{31}$$

then they are orthogonally similar; i.e., there exists an orthogonal matrix Q such that

$$B = Q^{-1}AQ. \tag{32}$$

Proof. From the conditions of the theorem there follows the existence of a polynomial $q(\lambda)$ such that

$$A^{\mathsf{T}} = q(A), \quad B^{\mathsf{T}} = q(B). \tag{33}$$

In the case of symmetric matrices this polynomial $q(\lambda)$ is identically equal to λ and, in the case of skew-symmetric matrices, to $-\lambda$. If A and B are orthogonal matrices, then $q(\lambda)$ is the interpolation polynomial for $1/\lambda$ on the common spectrum of A and B.

Using (33), we conduct the proof of our theorem exactly as we did the proof of the corresponding Theorem 10 of Chapter IX in the real case (Vol. I, p. 289). From (31) we deduce

$$q(B) = T^{-1}q(A)T$$

or by (33)

$$B^{\mathsf{T}} = T^{-1}A^{\mathsf{T}}T.$$

Hence

$$B = T^{\mathsf{T}}AT^{\mathsf{T}-1}.$$

Comparing this equation with (31), we easily find:

$$TT^{\mathsf{T}}A = ATT^{\mathsf{T}}. \tag{34}$$

Let us apply the polar decomposition to the non-singular matrix T

$$T = SQ \quad (S = S^{\mathsf{T}} = f(TT^{\mathsf{T}}), \quad Q^{\mathsf{T}} = Q^{-1}).$$

Since by (34) the matrix TT^{T} is permutable with A, the matrix $S = f(TT^{\mathsf{T}})$ is also permutable with A. Therefore, when we substitute the product SQ for T in (31), we have

$$B = Q^{-1}S^{-1}ASQ = Q^{-1}AQ.$$

This completes the proof of the theorem.

§ 3. The Normal Form of a Complex Symmetric Matrix

1. We shall prove the following theorem:

THEOREM 5: *There exists a complex symmetric matrix with arbitrary preassigned elementary divisors.*[6]

Proof. We consider the matrix H of order n in which the elements of the first superdiagonal are 1 and all the remaining elements are zero. We shall show that there exists a symmetric matrix S similar to H:

$$S = THT^{-1}. \tag{35}$$

We shall look for the transforming matrix T starting from the conditions:

$$S = THT^{-1} = S^{\mathsf{T}} = T^{\mathsf{T}-1}H^{\mathsf{T}}T^{\mathsf{T}}.$$

This equation can be rewritten as

$$VH = H^{\mathsf{T}}V, \tag{36}$$

where V is the symmetric matrix connected with T by the equation[7]

$$T^{\mathsf{T}}T = -2iV. \tag{37}$$

Recalling properties of the matrices H and $F = H^{\mathsf{T}}$ (Vol. I, pp. 13-14) we find that every solution V of the matrix equation (36) has the following form:

$$V = \left\| \begin{matrix} 0 & \cdots & 0 & a_0 \\ \vdots & & a_0 & a_1 \\ & \begin{matrix}\cdot\end{matrix}^{\cdot^{\cdot}} & & \vdots \\ 0 & a_0 & \cdots & \vdots \\ a_0 & a_1 & \cdots & a_{n-1} \end{matrix} \right\|, \tag{38}$$

where $a_0, a_1, \ldots, a_{n-1}$ are arbitrary complex numbers.

Since it is sufficient for us to find a single transforming matrix T, we set $a_0 = 1$, $a_1 = \ldots = a_{n-1} = 0$ in this formula and define V by the equation[8]

$$V = \left\| \begin{matrix} 0 & \ldots & 0 & 1 \\ 0 & \ldots & 1 & 0 \\ \cdot & \cdot & \cdot & \cdot \\ 1 & \ldots & 0 & 0 \end{matrix} \right\|. \tag{39}$$

---

[6] In connection with the contents of the present section as well as the two sections that follow, §§ 4 and 5, see [378].

[7] To simplify the following formulas it is convenient to introduce the factor $-2i$.

[8] The matrix V is both symmetric and orthogonal.

Furthermore, we shall require the transforming matrix T to be symmetric:

$$T = T^{\mathsf{T}}. \tag{40}$$

Then the equation (37) for T can be written as:

$$T^2 = -2iV. \tag{41}$$

We shall now look for the required matrix T in the form of a polynomial in V. Since $V^2 = E$, this can be taken as a polynomial of the first degree:

$$T = \alpha E + \beta V.$$

From (41), taking into account that $V^2 = E$, we find:

$$\alpha^2 + \beta^2 = 0, \quad 2\alpha\beta = -2i.$$

We can satisfy these relations by setting $\alpha = 1$, $\beta = -i$. Then

$$T = E - iV. \tag{42}$$

T is a non-singular symmetrix matrix.[9] At the same time, from (41):

$$T^{-1} = \frac{1}{2} iV^{-1}T = \frac{1}{2} iVT.$$

i.e.,

$$T^{-1} = \frac{1}{2}(E + iV). \tag{43}$$

Thus, a symmetric form S of H is determined by

$$S = THT^{-1} = \frac{1}{2}(E - iV)H(E + iV), \quad V = \begin{Vmatrix} 0 & \ldots & 0 & 1 \\ 0 & \ldots & 1 & 0 \\ \cdot & \cdot & \cdot & \cdot \\ 1 & \ldots & 0 & 0 \end{Vmatrix}. \tag{44}$$

Since S satisfies the equation (36) and $V^2 = E$, the equation (44) can be rewritten as follows:

$$2S = (H + H^{\mathsf{T}}) + i(HV - VH)$$

$$= \begin{Vmatrix} 0 & 1 & \ldots & 0 \\ 1 & \ddots & \ddots & \vdots \\ \vdots & \ddots & \ddots & 1 \\ 0 & \ldots & 1 & 0 \end{Vmatrix} + i \begin{Vmatrix} 0 & \ldots & 1 & 0 \\ \vdots & ⋰ & ⋰ & -1 \\ 1 & ⋰ & ⋰ & \vdots \\ 0 & -1 & \ldots & 0 \end{Vmatrix}. \tag{45}$$

---

[9] The fact that T is non-singular follows, in particular, from (41), because V is non-singular.

The formula (45) determines a symmetric form S of the matrix H.

In what follows, if n is the order of H, $H = H^{(n)}$, then we shall denote the corresponding matrices T, V, and S by $T^{(n)}$, $V^{(n)}$ and $S^{(n)}$.

Suppose that arbitrary elementary divisors are given:

$$(\lambda-\lambda_1)^{p_1}, \quad (\lambda-\lambda_2)^{p_2}, \quad \ldots, \quad (\lambda-\lambda_u)^{p_u}. \tag{46}$$

We form the corresponding Jordan matrix

$$J = \{ \lambda_1 E^{(p_1)} + H^{(p_1)}, \ \lambda_2 E^{(p_2)} + H^{(p_2)}, \ \ldots, \ \lambda_u E^{(p_u)} + H^{(p_u)} \}.$$

For every matrix $H^{(p_j)}$ we introduce the corresponding symmetric form $S^{(p_j)}$. From

$$S^{(p_j)} = T^{(p_j)} H^{(p_j)} [T^{(p_j)}]^{-1} \qquad (j = 1, 2, \ldots, u)$$

it follows that

$$\lambda_j E^{(p_j)} + S^{(p_j)} = T^{(p_j)} [\lambda_j E^{(p_j)} + H^{(p_j)}] [T^{(p_j)}]^{-1}.$$

Therefore setting

$$\tilde{S} = \{\lambda_1 E^{(p_1)} + S^{(p_1)}, \ \lambda_2 E^{(p_2)} + S^{(p_2)}, \ \ldots, \ \lambda_u E^{(p_u)} + S^{(p_u)} \}, \tag{47}$$

$$T = \{ T^{(p_1)}, \ T^{(p_2)}, \ \ldots, \ T^{(p_u)} \}, \tag{48}$$

we have:

$$\tilde{S} = TJT^{-1}.$$

$\tilde{S}$ is a symmetric form of J. $\tilde{S}$ is similar to J and has the same elementary divisors (46) as J. This proves the theorem.

COROLLARY 1. *Every square complex matrix $A = \| a_{ik} \|_1^n$ is similar to a symmetric matrix.*

Applying Theorem 4, we obtain:

COROLLARY 2. *Every complex symmetric matrix $S = \| a_{ik} \|_1^n$ is orthogonally similar to a symmetric matrix with the normal form $\tilde{S}$, i.e., there exists an orthogonal matrix Q such that*

$$\tilde{S} = QSQ^{-1}. \tag{49}$$

The normal form of a complex symmetric matrix has the quasi-diagonal form

$$\tilde{S} = \{ \lambda_1 E^{(p_1)} + S^{(p_1)}, \ \lambda_2 E^{(p_2)} + S^{(p_2)}, \ \ldots, \ \lambda_u E^{(p_u)} + S^{(p_u)} \}, \tag{50}$$

where the blocks $S^{(p)}$ are defined as follows (see (44), (45)):

$$S^{(p)} = \frac{1}{2}\,[E^{(p)} - iV^{(p)}]\; H^{(p)}\,[E^{(p)} + iV^{(p)}]$$

$$= \frac{1}{2}\,[H^{(p)} + H^{(p)\mathsf{T}} + i\,(H^{(p)}V^{(p)} - V^{(p)}H^{(p)})]$$

$$= \frac{1}{2}\left\{ \left\| \begin{array}{ccccc} 0 & 1 & \cdots & & 0 \\ 1 & \ddots & \ddots & & \vdots \\ \vdots & \ddots & \ddots & \ddots & \vdots \\ \vdots & & \ddots & \ddots & 1 \\ 0 & \cdots & & 1 & 0 \end{array} \right\| + i \left\| \begin{array}{ccccc} 0 & \cdots & & 1 & 0 \\ \vdots & & \cdot^{\cdot^{\cdot}} & \cdot^{\cdot^{\cdot}} & -1 \\ \vdots & \cdot^{\cdot^{\cdot}} & \cdot^{\cdot^{\cdot}} & \cdot^{\cdot^{\cdot}} & \vdots \\ 1 & \cdot^{\cdot^{\cdot}} & \cdot^{\cdot^{\cdot}} & & \vdots \\ 0 & -1 & \cdots & & 0 \end{array} \right\| \right\}. \qquad (51)$$

§ 4. The Normal Form of a Complex Skew-symmetric Matrix

1. We shall examine what restrictions the skew symmetry of a matrix imposes on its elementary divisors. In this task we shall make use of the following theorem:

Theorem 6: *A skew-symmetric matrix always has even rank.*

Proof. Let r be the rank of the skew-symmetric matrix K. Then K has r linearly independent rows, say those numbered $i_1, i_2, \ldots, i_r$; all the remaining rows are linear combinations of these r rows. Since the columns of K are obtained from the corresponding rows by multiplying the elements by -1, every column of K is a linear combination of the columns numbered $i_1, i_2, \ldots, i_r$. Therefore every minor of order r of K can be represented in the form

$$\alpha K\begin{pmatrix} i_1 & i_2 & \ldots & i_r \\ i_1 & i_2 & \ldots & i_r \end{pmatrix},$$

where α is a constant.

Hence it follows that

$$K\begin{pmatrix} i_1 & i_2 & \ldots & i_r \\ i_1 & i_2 & \ldots & i_r \end{pmatrix} \neq 0.$$

But a skew-symmetric determinant of odd order is always zero. Therefore r is even, and the theorem is proved.

Theorem 7: *If λ_0 is a characteristic value of the skew-symmetric matrix K with the corresponding elementary divisors*

$$(\lambda - \lambda_0)^{f_1}, \quad (\lambda - \lambda_0)^{f_2}, \; \ldots, \; (\lambda - \lambda_0)^{f_t},$$

then $-\lambda_0$ is also a characteristic value of K with the same number and the same powers of the corresponding elementary divisors of K

$$(\lambda+\lambda_0)^{f_1}, \quad (\lambda+\lambda_0)^{f_2}, \ \ldots, \ (\lambda+\lambda_0)^{f_t}.$$

2. *If zero is a characteristic value of the skew-symmetric matrix K,*[10] *then in the system of elementary divisors of K all those of even degree corresponding to the characteristic value zero are repeated an even number of times.*

Proof. 1. The transposed matrix K^{T} has the same elementary divisors as K. But $K^{\mathsf{T}} = -K$, and the elementary divisors of $-K$ are obtained from those of K by replacing the characteristic values $\lambda_1, \lambda_2, \ldots$ by $-\lambda_1, -\lambda_2, \ldots$. Hence the first part of our theorem follows.

2. Suppose that to the characteristic value zero of K there correspond δ_1 elementary divisors of the form λ, δ_2 of the form λ^2, etc. In general, we denote by δ_p the number of elementary divisors of the form λ^p $(p = 1, 2, \ldots)$. We shall show that $\delta_2, \delta_4, \ldots$ are even numbers.

The defect d of K is equal to the number of linearly independent characteristic vectors corresponding to the characteristic value zero or, what is the same, to the number of elementary divisors of the form $\lambda, \lambda^2, \lambda^3, \ldots$. Therefore

$$d = \delta_1 + \delta_2 + \delta_3 + \cdots. \tag{52}$$

Since, by Theorem 6, the rank of K is even and $d = n - r$, d has the same parity as n. The same statement can be made about the defects $d_3, d_5, \ldots$ of the matrices $K^3, K^5, \ldots$, because odd powers of a skew-symmetric matrix are themselves skew-symmetric. Therefore all the numbers $d_1 = d, d_3, d_5, \ldots$ have the same parity.

On the other hand, when K is raised to the m-th power, every elementary divisor λ^p for $p < m$ splits into p elementary divisors (of the first degree) and for $p \geqq m$ into m elementary divisors.[11] Therefore the number of elementary divisors of the matrices $K, K^3, \ldots$ that are powers of λ are determined by the formulas[12]

$$\begin{aligned} d_3 &= \delta_1 + 2\delta_2 + 3(\delta_3 + \delta_4 + \cdots), \\ d_5 &= \delta_1 + 2\delta_2 + 3\delta_3 + 4\delta_4 + 5(\delta_5 + \delta_6 + \cdots), \\ &\cdots\cdots\cdots\cdots\cdots\cdots \end{aligned} \tag{53}$$

Comparing (52) with (53) and bearing in mind that all the numbers $d_1 = d, d_3, d_5, \ldots$ are of the same parity, we conclude easily that $\delta_2, \delta_4, \ldots$ are even numbers.

This completes the proof of the theorem.

---

[10] I.e., if $|K| = 0$. For odd n we always have $|K| = 0$.

[11] See Vol. I, Chapter VI, Theorem 9, p. 158.

[12] These formulas were introduced (without reference to Theorem 9) in Vol. I, Chapter VI (see formulas (49) on p. 155).

2. THEOREM 8: *There exists a skew-symmetric matrix with arbitrary pre-assigned elementary divisors subject to the restrictions* 1., 2. *of the preceding theorem.*

Proof. To begin with, we shall find a skew-symmetric form for the quasi-diagonal matrix of order $2p$:

$$J_{\lambda_0}^{(pp)} = \{ \lambda_0 E + H, -\lambda_0 E - H \} \tag{54}$$

having two elementary divisors $(\lambda - \lambda_0)^p$ and $(\lambda + \lambda_0)^p$; here $E = E^{(p)}$, $H = H^{(p)}$.

We shall look for a transforming matrix T such that

$$T J_{\lambda^0}^{(pp)} T^{-1}$$

is skew-symmetric, i.e., such that the following equation holds:

$$T J_{\lambda_0}^{(pp)} T^{-1} + T^{\mathsf{T}-1} [J_{\lambda_0}^{(pp)}]^{\mathsf{T}} T^{\mathsf{T}} = O$$

or

$$W J_{\lambda_0}^{(pp)} + [J_{\lambda_0}^{(pp)}]^{\mathsf{T}} W = O, \tag{55}$$

where W is the symmetric matrix connected with T by the equation[13]

$$T^{\mathsf{T}} T = -2iW. \tag{56}$$

We dissect W into four square blocks each of order p:

$$W = \begin{pmatrix} W_{11} & W_{12} \\ W_{21} & W_{22} \end{pmatrix}.$$

Then (55) can be written as follows:

$$\begin{pmatrix} W_{11} & W_{12} \\ W_{21} & W_{22} \end{pmatrix} \begin{pmatrix} \lambda_0 E + H & O \\ O & -\lambda_0 E - H \end{pmatrix} + \begin{pmatrix} \lambda_0 E + H^{\mathsf{T}} & O \\ O & -\lambda_0 E - H^{\mathsf{T}} \end{pmatrix} \begin{pmatrix} W_{11} & W_{12} \\ W_{21} & W_{22} \end{pmatrix} = O. \tag{57}$$

When we perform the indicated operations on the partitioned matrices on the left-hand side of (57), we replace this equation by four matrix equations:

$$\begin{aligned} &1.\ H^{\mathsf{T}} W_{11} + W_{11} (2\lambda_0 E + H) = O, \\ &2.\ H^{\mathsf{T}} W_{12} - W_{12} H = O, \\ &3.\ H^{\mathsf{T}} W_{21} - W_{21} H = O, \\ &4.\ H^{\mathsf{T}} W_{22} + W_{22} (2\lambda_0 E + H) = O. \end{aligned} \tag{58}$$

---

[13] See footnote 7 on p. 9.

The equation $AX - XB = O$, where A and B are square matrices without common characteristic values, has only the trivial solution $X = O$.[14] Therefore the first and fourth of the equations (58) yield: $W_{11} = W_{22} = O$.[15] As regards the second of these equations, it can be satisfied, as we have seen in the proof of Theorem 5, by setting

$$W_{12} = V = \begin{Vmatrix} 0 & \cdot & \cdot & \cdot & 0 & 1 \\ 0 & \cdot & \cdot & \cdot & 1 & 0 \\ \cdot & \cdot & \cdot & \cdot & \cdot & \cdot \\ 1 & \cdot & \cdot & \cdot & 0 & 0 \end{Vmatrix}, \tag{59}$$

since (cf. (36))

$$VH - H^{\mathsf{T}} V = O.$$

From the symmetry of W and V it follows that

$$W_{21} = W_{12}^{\mathsf{T}} = V.$$

The third equation is then automatically satisfied.

Thus,

$$W = \begin{pmatrix} O & V \\ V & O \end{pmatrix} = V^{(2p)}. \tag{60}$$

But then, as has become apparent on page 10, the equation (56) will be satisfied if we set

$$T = E^{(2p)} - iV^{(2p)}. \tag{61}$$

Then

$$T^{-1} = \frac{1}{2}(E^{(2p)} + iV^{(2p)}). \tag{62}$$

Therefore, the required skew-symmetric matrix can be found by the formula[16]

$$\begin{aligned} K_{\lambda_0}^{(pp)} &= \frac{1}{2}[E^{(2p)} - iV^{(2p)}]\, J_{\lambda_0}^{(pp)}\, [E^{(2p)} + iV^{(2p)}] \\ &= \frac{1}{2}[J_{\lambda_0}^{(pp)} - J_{\lambda_0}^{(pp)\mathsf{T}} + i(J_{\lambda_0}^{(pp)} V^{(2p)} - V^{(2p)} J_{\lambda_0}^{(pp)})]. \end{aligned} \tag{63}$$

When we substitute for $J_{\lambda_0}^{(pp)}$ and $V^{(2p)}$ the corresponding partitioned matrices from (54) and (60), we find:

---

[14] See Vol. I, Chapter VIII, § 1.

[15] For $\lambda_0 \neq 0$ the equations 1. and 4. have no solutions other than zero. For $\lambda_0 = 0$ there exist other solutions, but we choose the zero solution.

[16] Here we use equations (55) and (60). From these it follows that

$$V^{(2p)} J_{\lambda_0}^{(pp)} V^{(2p)} = -J_{\lambda_0}^{(pp)\mathsf{T}}.$$

$$K_{\lambda_0}^{(pp)} = \frac{1}{2}\left[\begin{pmatrix} H-H^{\mathsf T} & O \\ O & H^{\mathsf T}-H\end{pmatrix} + i\begin{pmatrix}\lambda_0E+H & O\\ O & -\lambda_0E-H\end{pmatrix}\begin{pmatrix}O&V\\V&O\end{pmatrix} \right.$$

$$\left. - i\begin{pmatrix}O&V\\V&O\end{pmatrix}\begin{pmatrix}\lambda_0E+H & O\\ O & -\lambda_0E-H\end{pmatrix}\right]$$

$$= \frac{1}{2}\begin{pmatrix} H-H^{\mathsf T} & i(2\lambda_0V+HV+VH)\\ -i(2\lambda_0 V+HV+VH) & H^{\mathsf T}-H\end{pmatrix}, \tag{64}$$

i.e.,

$$K_{\lambda_0}^{(pp)} = \frac{1}{2}\left\Vert\begin{array}{cccccc:cccccc}
0 & 1 & \cdots & \cdots & \cdots & 0 & 0 & \cdots & \cdots & \cdots & i & 2\lambda_0 \\
-1 & 0 & \ddots & & & \vdots & \vdots & & & \iddots & 2\lambda_0 & i \\
\vdots & \ddots & \ddots & \ddots & & \vdots & \vdots & & \iddots & \iddots & \iddots & \vdots \\
\vdots & & \ddots & \ddots & \ddots & \vdots & \vdots & \iddots & \iddots & \iddots & & \vdots \\
\vdots & & & \ddots & \ddots & 1 & i & \iddots & \iddots & & & \vdots \\
0 & \cdots & \cdots & \cdots & -1 & 0 & 2\lambda_0 & \cdot & i & \cdots & \cdots & 0 \\
\hdashline
0 & \cdots & \cdots & \cdots & -i & -2\lambda_0 & 0 & -1 & \cdots & \cdots & \cdots & 0 \\
\vdots & & & \iddots & -2\lambda_0 & -i & 1 & 0 & \ddots & & & \vdots \\
\vdots & & \iddots & \iddots & \iddots & \vdots & \vdots & \ddots & \ddots & \ddots & & \vdots \\
\vdots & \iddots & \iddots & \iddots & & \vdots & \vdots & & \ddots & \ddots & \ddots & \vdots \\
-i & \iddots & \iddots & & & \vdots & \vdots & & & \ddots & \ddots & -1 \\
-2\lambda_0 & -i & \cdots & \cdots & \cdots & 0 & 0 & \cdots & \cdots & \cdots & 1 & 0
\end{array}\right\Vert . \tag{65}$$

We shall now construct a skew-symmetric matrix $K^{(q)}$ of order q having one elementary divisor λ^q, where q is odd. Obviously, the required skew-symmetric matrix will be similar to the matrix

$$J^{(q)} = \left\Vert\begin{array}{ccccccc}
0 & 1 & 0 & \cdots & \cdots & \cdots & 0 \\
0 & 0 & 1 & & & & \\
 & & \ddots & \ddots & & & \\
 & & & \ddots & \ddots & & \\
 & & & & \ddots & -1 & 0 \\
 & & & & & 0 & -1 \\
0 & & \cdots & & & 0 & 0
\end{array}\right\Vert . \tag{66}$$

In this matrix all the elements outside the first superdiagonal are equal to zero, and along the first superdiagonal there are at first $(q-1)/2$ elements 1 and then $(q-1)/2$ elements -1. Setting

$$K^{(q)} = TJ^{(q)} T^{-1}, \tag{67}$$

we find from the condition of skew-symmetry:

$$W_1 J^{(q)} + J^{(q)\mathsf{T}} W_1 = 0, \tag{68}$$

where

$$T^{\mathsf{T}} T = -2iW_1. \tag{69}$$

By direct verification we can convince ourselves that the matrix

$$W_1 = V^{(q)} = \begin{Vmatrix} 0 & \dots & 0 & 1 \\ 0 & \dots & 1 & 0 \\ \cdot & \cdot & \cdot & \cdot \\ 1 & \dots & 0 & 0 \end{Vmatrix}$$

satisfies the condition (68). Taking this value for W_1 we find from (69), as before:

$$T = E^{(q)} - iV^{(q)},\ T^{-1} = \frac{1}{2}[E^{(q)} + iV^{(q)}], \tag{70}$$

$$\begin{aligned} K^{(q)} &= \frac{1}{2}[E^{(q)} - iV^{(q)}]\, J^{(q)}\, [E^{(q)} + iV^{(q)}] \\ &= \frac{1}{2}[J^{(q)} - J^{(q)\mathsf{T}} + i\,(J^{(q)} V^{(q)} - V^{(q)} J^{(q)})]. \end{aligned} \tag{71}$$

When we perform the corresponding computation, we find:

$$2K^{(q)} = \begin{Vmatrix} 0 & 1 & \cdots & & 0 \\ -1 & 0 & \ddots & & \vdots \\ & \ddots & \ddots & \ddots & \vdots \\ & & \ddots & \ddots & -1 \\ 0 & & & 1 & 0 \end{Vmatrix} + i \begin{Vmatrix} 0 & \cdots & & 1 & 0 \\ & & \iddots & \iddots & 1 \\ & \iddots & \iddots & \iddots & \vdots \\ -1 & \iddots & \iddots & & \vdots \\ 0 & -1 & \cdots & & 0 \end{Vmatrix}. \tag{72}$$

Suppose that arbitrary elementary divisors are given, subject to the conditions of Theorem 7:

$$\left.\begin{array}{c}(\lambda-\lambda_j)^{p_j},\ (\lambda+\lambda_j)^{p_j}\ \ (j=1,2,\ldots,u),\\ \lambda^{q_k}\ (k=1,2,\ldots,v;\ q_1,q_2,\ldots,q_v \text{ are odd numbers}).^{17}\end{array}\right\} \tag{73}$$

Then the quasi-diagonal skew-symmetric matrix

$$\tilde{K}=\{K_{\lambda_1}^{(p_1p_1)},\ \ldots,\ K_{\lambda_u}^{(p_up_u)};\ K^{(q_1)},\ \ldots,\ K^{(q_v)}\} \tag{74}$$

has the elementary divisors (73).

This concludes the proof of the theorem.

Corollary: *Every complex skew-symmetric matrix K is orthogonally similar to a skew-symmetric matrix having the normal form $\tilde{K}$ determined by* (74), (65), *and* (72); *i.e., there exists a (complex) orthogonal matrix Q such that*

$$K=Q\tilde{K}Q^{-1}. \tag{75}$$

Note. If K is a real skew-symmetric matrix, then it has linear elementary divisors (see Vol. I, Chapter IX, § 13).

$$\lambda-i\varphi_1,\ \lambda+i\varphi_1,\ \ldots,\ \lambda-i\varphi_u,\ \lambda+i\varphi_u,\ \underbrace{\lambda,\ \ldots,\ \lambda}_{v \text{ times}}\quad (\varphi_j \text{ are real numbers}).$$

In this case, setting all the $p_j=1$ and all the $q_k=1$ in (74), we obtain as the normal form of a real skew-symmetric matrix

$$\tilde{K}=\left\{\left\|\begin{matrix}0 & \varphi_1\\ -\varphi_1 & 0\end{matrix}\right\|,\ \ldots,\ \left\|\begin{matrix}0 & \varphi_u\\ -\varphi_u & 0\end{matrix}\right\|,\ 0,\ \ldots,\ 0\right\}.$$

§ 5. The Normal Form of a Complex Orthogonal Matrix

1. Let us begin by examining what restrictions the orthogonality of a matrix imposes on its elementary divisors.

Theorem 9: 1. *If λ_0 ($\lambda_0^2\neq 1$) is a characteristic value of an orthogonal matrix Q and if the elementary divisors*

$$(\lambda-\lambda_0)^{f_1},\ (\lambda-\lambda_0)^{f_2},\ \ldots,\ (\lambda-\lambda_0)^{f_t}$$

[17] Some of the numbers $\lambda_1, \lambda_2, \ldots, \lambda_u$ may be zero. Moreover, one of the numbers u and v may be zero; i.e., in some cases there may be elementary divisors of only one type.

correspond to this characteristic value, then $1/\lambda_0$ *is also a characteristic value of* Q *and it has the same corresponding elementary divisors*:

$$(\lambda-\lambda_0^{-1})^{f_1},\ (\lambda-\lambda_0^{-1})^{f_2},\ \ldots,\ (\lambda-\lambda_0^{-1})^{f_t}.$$

2. *If* $\lambda_0=\pm 1$ *is a characteristic value of the orthogonal matrix* Q, *then the elementary divisors of even degree corresponding to* λ_0 *are repeated an even number of times.*

Proof. 1. For every non-singular matrix Q on passing from Q to Q^{-1} each elementary divisor $(\lambda-\lambda_0)^f$ is replaced by the elementary divisor $(\lambda-\lambda_0^{-1})^f$.[18] On the other hand, the matrices Q and Q^{T} always have the same elementary divisors. Therefore the first part of our theorem follows at once from the orthogonality condition $Q^{\mathsf{T}}=Q^{-1}$.

2. Let us assume that the number 1 is a characteristic value of Q, while -1 is not ($|E-Q|=0,\ |E+Q|\neq 0$). Then we apply Cayley's formulas (see Vol. I, Chapter IX, § 14), which remain valid for complex matrices. We define a matrix K by the equation

$$K=(E-Q)\,(E+Q)^{-1}. \tag{76}$$

Direct verification shows that $K^{\mathsf{T}}=-K$, so that K is skew-symmetric. When we solve the equation (76) for Q, we find:[19]

$$Q=(E-K)\,(E+K)^{-1}.$$

Setting $f(\lambda)=\frac{1-\lambda}{1+\lambda}$, we have $f'(\lambda)=-\frac{2}{(1+\lambda)^2}\neq 0$. Therefore in the transition from K to $Q=f(K)$ the elementary divisors do not split.[20] Hence in the system of elementary divisors of Q those of the form $(\lambda-1)^{2p}$ are repeated an even number of times, because this holds for the elementary divisors of the form λ^{2p} of K (see Theorem 7).

The case where Q has the characteristic value -1, but not $+1$, is reduced to the preceding case by considering the orthogonal matrix $-Q$.

We now proceed to the most complicated case, where Q has both the characteristic value $+1$ and -1. We denote by $\psi(\lambda)$ the minimal polynomial of Q. Using the first part of the theorem, which has already been proved, we can write $\psi(\lambda)$ in the form

---

[18] See Vol. I, Chapter VI, § 7. Setting $f(\lambda)=1/\lambda$, we have $f'(\lambda)=-1/\lambda^2\neq 0$. Hence it follows that in the transition from Q to Q^{-1} the elementary divisors do not split (see Vol. I, p. 158).

[19] Note that (76) implies that $E+K=2(E+Q)^{-1}$ and therefore

$$|E+K|=2^n\,|E+Q|^{-1}\neq 0.$$

[20] See Vol. I, p. 158.

$$\psi(\lambda)=(\lambda-1)^{m_1}(\lambda+1)^{m_2}\prod_{j=1}^{u}(\lambda-\lambda_j)^{p_j}(\lambda-\lambda_j^{-1})^{p_j}\quad(\lambda_j^2\neq 1;\ j=1,2,\ldots,u).$$

We consider the polynomial $g(\lambda)$ of degree less than m (m is the degree of $\psi(\lambda)$) for which $g(1)=1$ and all the remaining $m-1$ values on the spectrum of Q are zero; and we set:[21]

$$P=g(Q). \tag{77}$$

Note that the functions $(g(\lambda))^2$ and $g(1/\lambda)$ assume on the spectrum of Q the same values as $g(\lambda)$. Therefore

$$P^2=P,\qquad P^{\mathsf{T}}=g(Q^{\mathsf{T}})=g(Q^{-1})=P, \tag{78}$$

i.e., P is a symmetric projective matrix.[22]

We define a polynomial $h(\lambda)$ and a matrix N by the equations

$$h(\lambda)=(\lambda-1)\,g(\lambda), \tag{79}$$

$$N=h(Q)=(Q-E)\,P. \tag{80}$$

Since $(h(\lambda))^{m_1}$ vanishes on the spectrum of Q, it is divisible by $\psi(\lambda)$ without remainder. Hence:

$$N^{m_1}=O,$$

i.e., N is a nilpotent matrix with m_1 as index of nilpotency.

From (80) we find:[23]

$$N^{\mathsf{T}}=(Q^{\mathsf{T}}-E)\,P. \tag{81}$$

---

[21] From the fundamental formula (see Vol. I, p. 104)

$$g(A)=\sum_{k=1}^{s}[g(\lambda_k)Z_{k1}+g'(\lambda_k)Z_{k2}+\cdots]$$

it follows that

$$p=Z_{11}.$$

[22] A hermitian operator $\boldsymbol{P}$ is called *projective* if $\boldsymbol{P}^2=\boldsymbol{P}$. In accordance with this, a hermitian matrix P for which $P^2=P$ is called *projective*. An example of a projective operator $\boldsymbol{P}$ in a unitary space $\boldsymbol{R}$ is the operator of the orthogonal projection of a vector $\boldsymbol{x}\in\boldsymbol{R}$ into a subspace $\boldsymbol{S}=\boldsymbol{PR}$, i.e., $\boldsymbol{Px}=\boldsymbol{x}_{\boldsymbol{S}}$, where $\boldsymbol{x}_{\boldsymbol{S}}\in\boldsymbol{S}$ and $(\boldsymbol{x}-\boldsymbol{x}_{\boldsymbol{S}})\perp\boldsymbol{S}$ (see Vol. I, p. 248).

[23] All the matrices that occur here, P, N, N^{T}, $Q^{\mathsf{T}}=Q^{-1}$, are permutable among each other and with Q, since they are all functions of Q.

Let us consider the matrix

$$R = N(N^{\mathsf{T}} + 2E). \tag{82}$$

From (78), (80), and (81) it follows that

$$R = NN^{\mathsf{T}} + 2N = (Q - Q^{\mathsf{T}})P.$$

From this representation of R it is clear that R is *skew-symmetric*.

On the other hand, from (82)

$$R^k = N^k (N^{\mathsf{T}} + 2E)^k \quad (k = 1, 2, \ldots). \tag{83}$$

But N^{T}, like N, is nilpotent, and therefore

$$|N^{\mathsf{T}} + 2E| \neq 0.$$

Hence it follows from (83) that the matrices R^k and N^k have the same rank for every k.

Now for odd k the matrix R^k is skew-symmetric and therefore (see p. 12) has even rank. Therefore each of the matrices

$$N, N^3, N^5, \ldots$$

has odd rank.

By repeating verbatim for N the arguments that were used on p. 13 for K we may therefore state that among the elementary divisors of N those of the form λ^{2p} are repeated an even number of times. But to each elementary divisor λ^{2p} of N there corresponds an elementary divisor $(\lambda - 1)^{2p}$ of Q, and vice versa.[24] Hence it follows that among the elementary divisors of Q those of the form $(\lambda - 1)^{2p}$ are repeated an even number of times.

We obtain a similar statement for the elementary divisors of the form $(\lambda + 1)^{2p}$ by applying what has just been proved to the matrix $-Q$.

Thus, the proof of the theorem is complete.

2. We shall now prove the converse theorem.

[24] Since $h(1) = 0$, $h'(1) \neq 0$, in passing from Q to $N = h(Q)$ the elementary divisors of the form $(\lambda - 1)^{2p}$ of Q do not split and are therefore replaced by elementary divisors λ^{2p} (see Vol. I, Chapter VI, § 7).

THEOREM 10: *Every system of powers of the form*

$$\left.\begin{array}{l} (\lambda-\lambda_j)^{p_j},\ (\lambda-\lambda_j^{-1})^{p_j}\ (\lambda_j\neq 0;\ j=1,2,\ldots,u), \\ (\lambda-1)^{q_1},\ (\lambda-1)^{q_2},\ \ldots,\ (\lambda-1)^{q_v}, \\ (\lambda+1)^{t_1},\ (\lambda+1)^{t_2},\ \ldots,\ (\lambda+1)^{t_w} \\ (q_1,\ldots,q_v,\ t_1,\ldots,t_w \text{ are odd numbers}) \end{array}\right\} \tag{84}$$

is the system of elementary divisors of some complex orthogonal matrix Q.[25]

Proof. We denote by μ_j the numbers connected with the numbers λ_j $(j=1,2,\ldots,n)$ by the equations

$$\lambda_j = e^{\mu_j}\quad (j=1,2,\ldots,u)$$

We now introduce the 'canonical' skew-symmetric matrices (see the preceding section)

$$K_{\mu_j}^{(p_jp_j)}\ (j=1,2,\ldots,u);\ K^{(q_1)},\ldots,K^{(q_v)};\ K^{(t_1)},\ldots,K^{(t_w)},$$

with the elementary divisors

$$(\lambda-\mu_j)^{p_j},\ (\lambda+\mu_j)^{p_j}\ (j=1,2,\ldots,u)\ \lambda^{q_1},\ldots,\lambda^{q_v};\ \lambda^{t_1},\ldots,\lambda^{t_w}.$$

If K is a skew-symmetric matri*x*, then

$$Q = e^{K}$$

is orthogonal ($Q^{\mathsf T} = e^{K^{\mathsf T}} = e^{-K} = Q^{-1}$). Moreover, to each elementary divisor $(\lambda-\mu)^p$ of K there corresponds an elementary divisor $(\lambda - e^{\mu})^p$ of Q.[26]

Therefore the quasi-diagonal matrix

$$\tilde{Q} = \left\{ e^{K_{\mu_1}^{(p_1p_1)}},\ldots,e^{K_{\mu_u}^{(p_up_u)}};\ e^{K^{(q_1)}},\ldots,e^{K^{(q_v)}};\ -e^{K^{(t_1)}},\ldots,-e^{K^{(t_w)}} \right\} \tag{85}$$

is orthogonal and has the elementary divisors (84).

This proves the theorem.

From Theorems 4, 9, and 10 we obtain:

---

[25] Some (or even all) of the numbers λ_j may be ± 1. One or two of the numbers u, v, w may be zero. Then the elementary divisors of the corresponding type are absent in Q.

[26] This follows from the fact that for $f(\lambda) = e^{\lambda}$ we have $f'(\lambda) = e^{\lambda} \neq 0$ for every λ.

Corollary: *Every (complex) orthogonal matrix Q is orthogonally similar to an orthogonal matrix having the normal form $\tilde{Q}$;* i.e., *there exists an orthogonal matrix Q_1 such that*

$$Q = Q_1 \tilde{Q} Q_1^{-1}. \tag{86}$$

Note. Just as we have given a concrete form to the diagonal blocks in the skew-symmetric matrix $\tilde{K}$, so we could for the normal form $\tilde{Q}$.[27]

[27] See [378].

CHAPTER XII

SINGULAR PENCILS OF MATRICES

§ 1. Introduction

1. The present chapter deals with the following problem:

Given four matrices A, B, A_1, B_1 all of dimension $m \times n$ with elements from a number field F, *it is required to find under what conditions there exist two square non-singular matrices P and Q of orders m and n, respectively, such that*[1]

$$PAQ = A_1, \quad PBQ = B_1 \tag{1}$$

By introduction of the pencils of matrices $A + \lambda B$ and $A_1 + \lambda B_1$ the two matrix equations (1) can be replaced by the single equation

$$P(A + \lambda B)Q = A_1 + \lambda B_1 \tag{2}$$

Definition 1: *Two pencils of rectangular matrices $A + \lambda B$ and $A_1 + \lambda B_1$ of the same dimensions $m \times n$ connected by the equation (2) in which P and Q are constant square non-singular matrices (i.e., matrices independent of λ) of orders m and n, respectively, will be called strictly equivalent.*[2]

According to the general definition of equivalence of λ-matrices (see Vol. I, Chapter VI, p. 132), the pencils $A + \lambda B$ and $A_1 + \lambda B_1$ are equivalent if an equation of the form (2) holds in which P and Q are two square λ-matrices with constant non-vanishing determinants. For strict equivalence it is required in addition that P and Q do not depend on λ.[3]

A criterion for equivalence of the pencils $A + \lambda B$ and $A_1 + \lambda B_1$ follows from the general criterion for equivalence of λ-matrices and consists in the equality of the invariant polynomials or, what is the same, of the elementary divisors of the pencils $A + \lambda B$ and $A_1 + \lambda B_1$ (see Vol. I, Chapter VI, p. 141).

---

[1] If such matrices P and Q exist, then their elements can be taken from the field F. This follows from the fact that the equations (1) can be written in the form $PA = A_1Q^{-1}$, $PB = B_1Q^{-1}$ and are therefore equivalent to a certain system of linear homogeneous equations for the elements of P and Q^{-1} with coefficients in F.

[2] See Vol. I, Chapter VI, p. 145.

[3] We have replaced the term 'equivalent pencils' that occurs in the literature by 'strictly equivalent pencils,' in order to draw a sharp distinction between Definition 1 and the definition of equivalence in Vol. I, Chapter VI.

In this chapter, we shall establish a criterion for strict equivalence of two pencils of matrices and we shall determine for each pencil a strictly equivalent canonical form.

2. The task we have set ourselves has a natural geometrical interpretation. We consider a pencil of linear operators $\boldsymbol{A} + \lambda \boldsymbol{B}$ mapping $\boldsymbol{R}_n$ into $\boldsymbol{R}_m$. For a definite choice of bases in these spaces the pencil of operators $\boldsymbol{A} + \lambda \boldsymbol{B}$ corresponds to a pencil of rectangular matrices $A + \lambda B$ (of dimension $m \times n$); under a change of bases in $\boldsymbol{R}_n$ and $\boldsymbol{R}_m$ the pencil $A + \lambda B$ is replaced by a strictly equivalent pencil $P(A + \lambda B)Q$, where P and Q are square non-singular matrices of order m and n (see Vol. I, Chapter III, §§ 2 and 4). Thus, a criterion for strict equivalence gives a characterization of that class of matrix pencils $A + \lambda B$ (of dimension $m \times n$) which describe one and the same pencil of operators $\boldsymbol{A} + \lambda \boldsymbol{B}$ mapping $\boldsymbol{R}_n$ into $\boldsymbol{R}_m$ for various choices of bases in these spaces.

In order to obtain a canonical form for a pencil it is necessary to find bases for $\boldsymbol{R}_n$ and $\boldsymbol{R}_m$ in which the pencil of operators $\boldsymbol{A} + \lambda \boldsymbol{B}$ is described by matrices of the simplest possible form.

Since a pencil of operators is given by two operators $\boldsymbol{A}$ and $\boldsymbol{B}$, we can also say that: *The present chapter deals with the simultaneous investigation of two operators $\boldsymbol{A}$ and $\boldsymbol{B}$ mapping $\boldsymbol{R}_n$ into $\boldsymbol{R}_m$.*

3. All the pencils of matrices $A + \lambda B$ of dimension $m \times n$ fall into two basic types: *regular* and *singular* pencils.

DEFINITION 2: *A pencil of matrices $A + \lambda B$ is called regular if*

1) *A and B are square matrices of the same order n; and*

2) *The determinant $| A + \lambda B |$ does not vanish identically.*

In all other cases ($m \neq n$, or $m = n$ but $| A + \lambda B | \equiv 0$), *the pencil is called singular.*

A criterion for strict equivalence of regular pencils of matrices and also a canonical form for such pencils were established by Weierstrass in 1867 [377] on the basis of his theory of elementary divisors, which we have expounded in Chapters VI and VII. The analogous problems for singular pencils were solved later, in 1890, by the investigations of Kronecker [249].[4] Kronecker's results form the primary content of this chapter.

§ 2. Regular Pencils of Matrices

1. We consider the special case where the pencils $A + \lambda B$ and $A_1 + \lambda B_1$ consist of square matrices ($m = n$) $| B | \neq 0$, $| B_1 | \neq 0$. In this case, as we have shown in Chapter VI (Vol. I, pp. 145-146), the two concepts of 'equiv-

[4] Of more recent papers dealing with singular pencils of matrices we mention [234], [369], and [255].

alence' and 'strict equivalence' of pencils coincide. Therefore, by applying to the pencils the general criterion for equivalence of λ-matrices (Vol. I, p. 141) we are led to the following theorem:

THEOREM 1: *Two pencils of square matrices of the same order $A+\lambda B$ and $A_1+\lambda B_1$ for which $|B|\neq 0$ and $|B_1|\neq 0$ are strictly equivalent if and only if the pencils have the same elementary divisors in* F.

A pencil of square matrices $A+\lambda B$ with $|B|\neq 0$ was called regular in Chapter VI, because it represents a special case of a regular matrix polynomial in λ (see Vol. I, Chapter IV, p. 76). In the preceding section of this chapter we have given a wider definition of regularity. According to this definition it is quite possible in a regular pencil to have $|B|=0$ (and even $|A|=|B|=0$).

In order to find out whether Theorem 1 remains valid for regular pencils (with the extended Definition 1), we consider the following example:

$$A+\lambda B=\left\|\begin{matrix}2&1&3\\3&2&5\\3&2&6\end{matrix}\right\|+\lambda\left\|\begin{matrix}1&1&2\\1&1&2\\1&1&3\end{matrix}\right\|,\ A_1+\lambda B_1=\left\|\begin{matrix}2&1&1\\1&2&1\\1&1&1\end{matrix}\right\|+\lambda\left\|\begin{matrix}1&1&1\\1&1&1\\1&1&1\end{matrix}\right\|. \tag{3}$$

It is easy to see that here each of the pencils $A+\lambda B$ and $A_1+\lambda B_1$ has only one elementary divisor, $\lambda+1$. However, the pencils are not strictly equivalent, since the matrices B and B_1 are of ranks 2 and 1, respectively; whereas if an equation (2) were to hold, it would follow from it that the ranks of B and B_1 are equal. Nevertheless, the pencils (3) are regular according to Definition 1, since

$$|A+\lambda B|\equiv|A_1+\lambda B_1|\equiv\lambda+1.$$

This example shows that Theorem 1 is not true with the extended definition of regularity of a pencil.

2. In order to preserve Theorem 1, we have to introduce the concept of 'infinite' elementary divisors of a pencil. We shall give the pencil $A+\lambda B$ in terms of 'homogeneous' parameters λ, μ: $\mu A+\lambda B$. Then the determinant $\Delta(\lambda,\mu)\equiv|\mu A+\lambda B|$ is a homogeneous function of λ, μ. By determining the greatest common divisor $D_k(\lambda,\mu)$ of all the minors of order k of the matrix $\mu A+\lambda B$ $(k=1,2,\ldots,n)$, we obtain the invariant polynomials by the well known formulas

$$i_1(\lambda,\mu)=\frac{D_n(\lambda,\mu)}{D_{n-1}(\lambda,\mu)},\quad i_2(\lambda,\mu)=\frac{D_{n-1}(\lambda,\mu)}{D_{n-2}(\lambda,\mu)},\ \ldots;$$

here all the $D_k(\lambda,\mu)$ and $i_j(\lambda,\mu)$ are homogeneous polynomials in λ and μ.

Splitting the invariant polynomials into powers of homogeneous polynomials irreducible over F, we obtain the elementary divisors $e_\alpha(\lambda, \mu)$ $(\alpha = 1, 2, \ldots)$ of the pencil $\mu A + \lambda B$ in F.

It is quite obvious that if we set $\mu = 1$ in $e_\alpha(\lambda, \mu)$ we are back to the elementary divisors $e_\alpha(\lambda)$ of the pencil $A + \lambda B$. Conversely, from each elementary divisor $e_\alpha(\lambda)$ of degree q we obtain the correspondingly elementary divisor $e_\alpha(\lambda, \mu)$ by the formula $e_\alpha(\lambda, \mu) = \mu^q e_\alpha\left(\frac{\lambda}{\mu}\right)$. We can obtain in this way all the elementary divisors of the pencil $\mu A + \lambda B$ apart from those of the form μ^q.

Elementary divisors of the form μ^q exist if and only if $|B| = 0$ and are called 'infinite' elementary divisors of the pencil $A + \lambda B$.

Since strict equivalence of the pencils $A + \lambda B$ and $A_1 + \lambda B_1$ implies strict equivalence of the pencils $\mu A + \lambda B$ and $\mu A_1 + \lambda B_1$, we see that for strictly equivalent pencils $A + \lambda B$ and $A_1 + \lambda B_1$ not only their 'finite,' but also their 'infinite' elementary divisors must coincide.

Suppose now that $A + \lambda B$ and $A_1 + \lambda B_1$ are two regular pencils for which all the elementary divisors coincide (including the infinite ones). We introduce homogeneous parameters: $\mu A + \lambda B$, $\mu A_1 + \lambda B_1$. Let us now transform the parameters

$$\lambda = \alpha_1 \tilde{\lambda} + \alpha_2 \tilde{\mu}, \quad \mu = \beta_1 \tilde{\lambda} + \beta_2 \tilde{\mu} \quad (\alpha_1 \beta_2 - \alpha_2 \beta_1 \neq 0).$$

In the new parameters the pencils are written as follows:

$$\tilde{\mu} \tilde{A} + \tilde{\lambda} \tilde{B}, \tilde{\mu} \tilde{A}_1 + \tilde{\lambda} \tilde{B}_1, \text{ where } \tilde{B} = \beta_1 A + \alpha_1 B, \tilde{B}_1 = \beta_1 A_1 + \alpha_1 B_1.$$

From the regularity of the pencils $\mu A + \lambda B$ and $\mu A_1 + \lambda B_1$ it follows that we can choose the numbers α_1 and β_1 such that $|\tilde{B}| \neq 0$ and $|\tilde{B}_1| \neq 0$.

Therefore by Theorem 1 the pencils $\tilde{\mu} \tilde{A} + \tilde{\lambda} \tilde{B}$ and $\tilde{\mu} \tilde{A}_1 + \tilde{\lambda} \tilde{B}_1$ and consequently the original pencils $\mu A + \lambda B$ and $\mu A_1 + \lambda B_1$ (or, what is the same, $A + \lambda B$ and $A_1 + \lambda B_1$) are strictly equivalent. Thus, we have arrived at the following generalization of Theorem 1:

THEOREM 2: *Two regular pencils $A + \lambda B$ and $A_1 + \lambda B_1$ are strictly equivalent if and only if they have the same ('finite' and 'infinite') elementary divisors.*

In our example above the pencils (3) had the same 'finite' elementary divisor $\lambda + 1$, but different 'infinite' elementary divisors (the first pencil has one 'infinite' elementary divisor μ^2; the second has two: μ, μ). Therefore these pencils turn out to be not strictly equivalent.

3. Suppose now that $A + \lambda B$ is an arbitrary regular pencil. Then there exists a number c such that $| A + cB | \neq 0$. We represent the given pencil in the form $A_1 + (\lambda - c)B$, where $A_1 = A + cB$, so that $| A_1 | \neq 0$. We multiply the pencil on the left by A_1^{-1}: $E + (\lambda - c)A_1^{-1}B$. By a similarity transformation we put the pencil in the form[5]

$$E + (\lambda - c)\{J_0, J_1\} = \{E - cJ_0 + \lambda J_0,\ E - cJ_1 + \lambda J_1\}, \tag{4}$$

where $\{J_0, J_1\}$ is the quasi-diagonal normal form of $A_1^{-1}B$, J_0 is a nilpotent Jordan matrix,[6] and $| J_1 | \neq 0$.

We multiply the first diagonal block on the right-hand side of (4) by $(E - cJ_0)^{-1}$ and obtain: $E + \lambda(E - cJ_0)^{-1}J_0$. Here the coefficient of λ is a nilpotent matrix.[7] Therefore by a similarity transformation we can put this pencil into the form[8]

$$E + \lambda\widehat{J}_0 = \{N^{(u_1)}, N^{(u_2)}, \ldots, N^{(u_s)}\}\ (N^{(u)} = E^{(u)} + \lambda H^{(u)}). \tag{5}$$

We multiply the second diagonal block on the right-hand side of (4) by J_1^{-1}; it can then be put into the form $J + \lambda E$ by a similarity transformation, where J is a matrix of normal form[9] and E the unit matrix. We have thus arrived at the following theorem:

THEOREM 3: *Every regular pencil $A + \lambda B$ can be reduced to a (strictly equivalent) canonical quasi-diagonal form*

$$\{N^{(u_1)}, N^{(u_2)}, \ldots, N^{(u_s)}, J + \lambda E\}\ (N^{(u)} = E^{(u)} + \lambda H^{(u)}), \tag{6}$$

where the first s diagonal blocks correspond to infinite elementary divisors $\mu^{u_1}, \mu^{u_2}, \ldots, \mu^{u_s}$ of the pencil $A + \lambda B$ and where the normal form of the last diagonal block $J + \lambda E$ is uniquely determined by the finite elementary divisors of the given pencil.

---

[5] The unit matrices E in the diagonal blocks on the right-hand side of (4) have the same order as J_0 and J_1.

[6] I.e., $J_0^l = O$ for some integer $l > 0$.

[7] From $J_0^l = O$ it follows that $[(E - cJ_0)^{-1}J_0]^l = O$.

[8] Here $E^{(u)}$ is a unit matrix of order u and $H^{(u)}$ is a matrix of order u whose elements in the first superdiagonal are 1, while the remaining elements are zero.

[9] Since the matrix J can be replaced here by an arbitrary similar matrix, we may assume that J has one of the normal forms (for example, the natural form of the first or second kind or the Jordan form (see Vol. I, Chapter VI, § 6)).

§ 3. Singular Pencils. The Reduction Theorem

1. We now proceed to consider a singular pencil of matrices $A + \lambda B$ of dimension $m \times n$. We denote by r the *rank of the pencil*, i.e., the largest of the orders of minors that do not vanish identically. From the singularity of the pencil it follows that at least one of the inequalities $r < n$ and $r < m$ holds, say $r < n$. Then the columns of the λ-matrix $A + \lambda B$ are linearly dependent, i.e., the equation

$$(A + \lambda B)\, x = o, \tag{7}$$

where x is an unknown column matrix, has a non-zero solution. Every non-zero solution of this equation determines some dependence among the columns of $A + \lambda B$. We restrict ourselves to only such solutions $x(\lambda)$ of (7) as are polynomials in λ,[10] and among these solutions we choose one of least possible degree ε:

$$x(\lambda) = x_0 - \lambda x_1 + \lambda^2 x_2 - \cdots + (-1)^{\varepsilon} \lambda^{\varepsilon} x_{\varepsilon} \qquad (x_{\varepsilon} \neq 0). \tag{8}$$

Substituting this solution in (7) and equating to zero the coefficients of the powers of λ, we obtain:

$$Ax_0 = o,\ Bx_0 - Ax_1 = o,\ Bx_1 - Ax_2 = o,\ \ldots,\ Bx_{\varepsilon-1} - Ax_{\varepsilon} = o,\ Bx_{\varepsilon} = o. \tag{9}$$

Considering this as a system of linear homogeneous equations for the elements of the columns $x_0, -x_1, +x_2 \ldots, (-1)^{\varepsilon} x_{\varepsilon}$, we deduce that the coefficient matrix of the system

$$M_{\varepsilon} = M_{\varepsilon}[A + \lambda B] = \overbrace{\begin{pmatrix} A & O & \ldots & O \\ B & A & & \vdots \\ O & B & \ddots & \vdots \\ \vdots & & \ddots & A \\ O & O & \ldots & B \end{pmatrix}}^{\varepsilon+1} \tag{10}$$

is of rank $\varrho_{\varepsilon} < (\varepsilon + 1)n$. At the same time, by the minimal property of ε, the ranks $\varrho_0, \varrho_1, \ldots, \varrho_{\varepsilon-1}$ of the matrices

---

[10] For the actual determination of the elements of the column x satisfying (7) it is convenient to solve a system of linear homogeneous equations in which the coefficients of the unknown depend linearly on λ. The fundamental linearly independent solutions x can always be chosen such that their elements are polynomials in λ.

$$M_0=\begin{pmatrix}A\\B\end{pmatrix},\quad M_1=\begin{pmatrix}A&O\\B&A\\O&B\end{pmatrix},\ \ldots,\ M_{\varepsilon-1}=\begin{pmatrix}\overbrace{\begin{matrix}A&O&\ldots&O\end{matrix}}^{\varepsilon}\\\begin{matrix}B&A&&\cdot\end{matrix}\\\vdots\quad\ddots\quad\vdots\\\begin{matrix}\cdot&&\cdot&A\end{matrix}\\\begin{matrix}O&&\ldots&B\end{matrix}\end{pmatrix} \tag{10'}$$

satisfy the equations $\varrho_0 = n$, $\varrho_1 = 2n$, ..., $\varrho_{\varepsilon-1} = \varepsilon n$.

Thus: *The number ε is the least value of the index k for which the sign $<$ holds in the relation $\varrho_k \leqq (k+1)n$.*

Now we can formulate and prove the following fundamental theorem:

2. THEOREM 4: *If the equation* (7) *has a solution of minimal degree ε and $\varepsilon > 0$, then the given pencil $A + \lambda B$ is strictly equivalent to a pencil of the form*

$$\begin{pmatrix}L_\varepsilon & O\\ O & \hat{A}+\lambda\hat{B}\end{pmatrix}, \tag{11}$$

where

$$L_\varepsilon = \left.\begin{Vmatrix}\overbrace{\begin{matrix}\lambda & 1 & 0 & \ldots & 0 & 0\end{matrix}}^{\varepsilon+1}\\ \begin{matrix}0 & \lambda & 1 & & \cdot & \cdot\end{matrix}\\ \begin{matrix}\cdot & \cdot & \ddots & \ddots & \cdot & \cdot\end{matrix}\\ \begin{matrix}0 & 0 & \ldots & & \lambda & 1\end{matrix}\end{Vmatrix}\right\}\varepsilon, \tag{12}$$

and $\hat{A} + \lambda\hat{B}$ is a pencil of matrices for which the equation analogous to (7) *has no solution of degree less than ε.*

Proof. We shall conduct the proof of the theorem in three stages. First, we shall show that the given pencil $A + \lambda B$ is strictly equivalent to a pencil of the form

$$\begin{pmatrix}L_\varepsilon & D+\lambda F\\ O & \hat{A}+\lambda\hat{B}\end{pmatrix}, \tag{13}$$

where D, F, $\hat{A}$, $\hat{B}$ are constant rectangular matrices of the appropriate dimensions. Then we shall establish that the equation $(\hat{A}+\lambda\hat{B})\hat{x} = O$ has no solution $x(\lambda)$ of degree less than ε. Finally, we shall prove that by further transformations the pencil (13) can be brought into the quasi-diagonal form (11).

1. The first part of the proof will be couched in geometrical terms. Instead of the pencil of matrices $A + \lambda B$ we consider a pencil of operators $\boldsymbol{A} + \lambda \boldsymbol{B}$ mapping $\boldsymbol{R}_n$ into $\boldsymbol{R}_m$ and show that with a suitable choice of bases in the spaces the matrix corresponding to the operator $A + \lambda B$ assumes the form (13).

Instead of (7) we take the vector equation

$$(\boldsymbol{A} + \lambda \boldsymbol{B})\, \boldsymbol{x} = \boldsymbol{o} \tag{14}$$

with the vector solution

$$\boldsymbol{x}(\lambda) = \boldsymbol{x}_0 - \lambda \boldsymbol{x}_1 + \lambda^2 \boldsymbol{x}_2 - \cdots + (-1)^{\varepsilon} \lambda^{\varepsilon} \boldsymbol{x}_{\varepsilon}; \tag{15}$$

the equations (9) are replaced by the vector equations

$$\boldsymbol{A}\boldsymbol{x}_0 = \boldsymbol{o}, \quad \boldsymbol{A}\boldsymbol{x}_1 = \boldsymbol{B}\boldsymbol{x}_0, \quad \boldsymbol{A}\boldsymbol{x}_2 = \boldsymbol{B}\boldsymbol{x}_1, \ldots, \boldsymbol{A}\boldsymbol{x}_{\varepsilon} = \boldsymbol{B}\boldsymbol{x}_{\varepsilon-1}, \quad \boldsymbol{B}\boldsymbol{x}_{\varepsilon} = \boldsymbol{o} \tag{16}$$

Below we shall show that the vectors

$$\boldsymbol{A}\boldsymbol{x}_1, \boldsymbol{A}\boldsymbol{x}_2, \ldots, \boldsymbol{A}\boldsymbol{x}_{\varepsilon} \tag{17}$$

are linearly independent. Hence it will be easy to deduce the linear independence of the vectors

$$\boldsymbol{x}_0, \boldsymbol{x}_1, \ldots, \boldsymbol{x}_{\varepsilon}. \tag{18}$$

For since $\boldsymbol{A}\boldsymbol{x}_0 = \boldsymbol{o}$ we have from $\alpha_0 \boldsymbol{x}_0 + \alpha_1 \boldsymbol{x}_1 + \cdots + \alpha_s \boldsymbol{x}_s = \boldsymbol{o}$ that $\alpha_1 \boldsymbol{A}\, \boldsymbol{x}_1 + \cdots + \alpha_{\varepsilon} \boldsymbol{A}\, \boldsymbol{x}_{\varepsilon} = \boldsymbol{o}$, so that by the linear independence of the vectors (17) $\alpha_1 = \alpha_2 = \ldots = \alpha_{\varepsilon} = 0$. But $\boldsymbol{x}_0 \neq 0$, since otherwise $\frac{1}{\lambda}\boldsymbol{x}(\lambda)$ would be a solution of (14) of degree $\varepsilon - 1$, which is impossible. Therefore $\alpha_0 = 0$ also.

Now if we take the vectors (17) and (18) as the first $\varepsilon + 1$ vectors for new bases in $\boldsymbol{R}_m$ and $\boldsymbol{R}_n$, respectively, then in these new bases the operators $\boldsymbol{A}$ and $\boldsymbol{B}$, by (16), will correspond to the matrices

$$\tilde{A} = \begin{Vmatrix} \overbrace{0 & 1 & & \ldots & 0}^{\varepsilon+1} & * & \ldots & * \\ 0 & 0 & 1 & \ldots & 0 & * & \ldots & * \\ \cdot & \cdot & & \cdot & \cdot & & & \\ \cdot & \cdot & & \cdot & \cdot & \cdot & \cdot & \cdot \\ 0 & 0 & & \ldots & 1 & * & \ldots & * \\ 0 & 0 & & \ldots & 0 & * & \ldots & * \\ \cdot & \cdot & \cdot & \cdot & \cdot & \cdot & \cdot & \cdot \\ 0 & 0 & & \ldots & 0 & * & \ldots & * \end{Vmatrix}, \quad \tilde{B} = \begin{Vmatrix} \overbrace{1 & 0 & \ldots & 0 & 0}^{\varepsilon+1} & * & \ldots & * \\ 0 & 1 & \ldots & 0 & 0 & * & \ldots & * \\ \cdot & \cdot & & & \cdot & & & \\ \cdot & \cdot & & & \cdot & \cdot & \cdot & \cdot \\ 0 & 0 & \ldots & 1 & 0 & * & \ldots & * \\ 0 & 0 & \ldots & 0 & 0 & * & \ldots & * \\ \cdot & \cdot & \ldots & 0 & 0 & * & \ldots & * \\ 0 & 0 & \ldots & 0 & 0 & * & \ldots & * \end{Vmatrix};$$

hence the λ-matrix $\tilde{A}+\lambda\tilde{B}$ is of the form (13). All the preceding arguments will be justified if we can show that the vectors (17) are linearly independent. Assume the contrary and let $\boldsymbol{Ax}_h$ ($h \geqq 1$) be the first vector in (17) that is linearly dependent on the preceding ones:

$$\boldsymbol{Ax}_h = \alpha_1 \boldsymbol{Ax}_{h-1} + \alpha_2 \boldsymbol{Ax}_{h-2} + \cdots + \alpha_{h-1}\boldsymbol{Ax}_1 .$$

By (16) this equation can be rewritten as follows:

$$\boldsymbol{Bx}_{h-1} = \alpha_1 \boldsymbol{Bx}_{h-2} + \alpha_2 \boldsymbol{Bx}_{h-3} + \cdots + \alpha_{h-1}\boldsymbol{Bx}_0 ,$$

i.e.,

$$\boldsymbol{Bx}^*_{h-1} = \boldsymbol{o},$$

where

$$\boldsymbol{x}^*_{h-1} = \boldsymbol{x}_{h-1} - \alpha_1 \boldsymbol{x}_{h-2} - \alpha_2 \boldsymbol{x}_{h-3} - \cdots - \alpha_{h-1}\boldsymbol{x}_0 .$$

Furthermore, again by (16),

$$\boldsymbol{Ax}^*_{h-1} = \boldsymbol{B}(\boldsymbol{x}_{h-2} - \alpha_1 \boldsymbol{x}_{h-3} - \cdots - \alpha_{h-2}\boldsymbol{x}_0) = \boldsymbol{Bx}^*_{h-2} ,$$

where

$$\boldsymbol{x}^*_{h-2} = \boldsymbol{x}_{h-2} - \alpha_1 \boldsymbol{x}_{h-3} - \cdots - \alpha_{h-2}\boldsymbol{x}_0 .$$

Continuing the process and introducing the vectors

$$\boldsymbol{x}^*_{h-3} = \boldsymbol{x}_{h-3} - \alpha_1 \boldsymbol{x}_{h-4} - \cdots - \alpha_{h-3}\boldsymbol{x}_0, \ \ldots, \ \boldsymbol{x}^*_1 = \boldsymbol{x}_1 - \alpha_1 \boldsymbol{x}_0, \ \boldsymbol{x}^*_0 = \boldsymbol{x}_0 ,$$

we obtain a chain of equations

$$\boldsymbol{Bx}^*_{h-1} = \boldsymbol{o}, \quad \boldsymbol{Ax}^*_{h-1} = \boldsymbol{Bx}^*_{h-2}, \ \ldots, \ \boldsymbol{Ax}^*_1 = \boldsymbol{Bx}^*_0, \quad \boldsymbol{Ax}^*_0 = \boldsymbol{o}. \tag{19}$$

From (19) it follows that

$$\boldsymbol{x}^*(\lambda) = \boldsymbol{x}^*_0 - \lambda \boldsymbol{x}^*_1 + \cdots + (-1)^{h-1}\boldsymbol{x}^*_{h-1} \quad (\boldsymbol{x}^*_0 = \boldsymbol{x}_0 \neq \boldsymbol{o})$$

is a non-zero solution of (14) of degree $\leqq h-1 < \varepsilon$, which is impossible. Thus, the vectors (17) are linearly independent.

2 We shall now show that the equation $(\hat{A} + \lambda\hat{B})\,\hat{x} = o$ has no solutions of degree less than ε. To begin with, we observe that the equation $L_\varepsilon y = o$, like (7), has a non-zero solution of least degree ε. We can see this immediately, if we replace the matrix equation $L_\varepsilon y = o$ by the system of ordinary equations

$$\lambda y_1 + y_2 = 0, \quad \lambda y_2 + y_3 = 0, \ \ldots, \ \lambda y_\varepsilon + y_{\varepsilon+1} = 0 \quad (y = (y_1, y_2, \ldots, y_{\varepsilon+1}));$$

$$y_k = (-1)^{k-1} y_1 \lambda^{k-1} \quad (k = 1, 2, \ldots, \varepsilon + 1).$$

On the other hand, if the pencil has the 'triangular' form (13) then the corresponding matrix pencil M_k ($k = 0, 1, \ldots, \varepsilon$) (see (10) and (10′) on pp. 29 and 30) can also be brought into triangular form, after a suitable permutation of rows and columns:

$$\begin{pmatrix} M_k[L_\varepsilon] & M_k[D+\lambda F] \\ O & M_k[\hat{A}+\lambda\hat{B}] \end{pmatrix}. \tag{20}$$

For $k = \varepsilon - 1$ all the columns of this matrix, like those of $M_{\varepsilon-1}[L_\varepsilon]$, are linearly independent.[11] But $M_{\varepsilon-1}[L_\varepsilon]$ is a square matrix of order $\varepsilon(\varepsilon+1)$. Therefore in $M_{\varepsilon-1}[\hat{A}+\lambda\hat{B}]$ also, all the columns are linearly independent and, as we have explained at the beginning of the section, this means that the equation $(\hat{A}+\lambda\hat{B})\,\hat{x} = o$ has no solution of degree less than or equal to $\varepsilon - 1$, which is what we had to prove.

3. Let us replace the pencil (13) by the strictly equivalent pencil

$$\begin{pmatrix} E_1 & Y \\ O & E_2 \end{pmatrix}\begin{pmatrix} L_\varepsilon & D+\lambda F \\ O & \hat{A}+\lambda\hat{B} \end{pmatrix}\begin{pmatrix} E_3 & -X \\ O & E_4 \end{pmatrix} = \begin{pmatrix} L_\varepsilon & D+\lambda F+Y(\hat{A}+\lambda\hat{B})-L_\varepsilon X \\ O & \hat{A}+\lambda\hat{B} \end{pmatrix}, \tag{21}$$

where E_1, E_2, E_3, and E_4 are square unit matrices of orders ε, $m-\varepsilon$, $\varepsilon+1$, and $n-\varepsilon-1$, respectively, and X, Y are arbitrary constant rectangular matrices of the appropriate dimensions. Our theorem will be completely proved if we can show that the matrices X and Y can be chosen such that the matrix equation

$$L_\varepsilon X = D + \lambda F + Y(\hat{A}+\lambda\hat{B}) \tag{22}$$

holds.

We introduce a notation for the elements of D, F, X and also for the rows of Y and the columns of $\hat{A}$ and $\hat{B}$:

$$D = \| d_{ik} \|, \quad F = \| f_{ik} \|, \quad X = \| x_{jk} \|$$

$$(i = 1, 2, \ldots, \varepsilon; \quad k = 1, 2, \ldots, n-\varepsilon-1; \quad j = 1, 2, \ldots, \varepsilon+1),$$

$$Y = \begin{pmatrix} y_1 \\ y_2 \\ \cdot \\ \cdot \\ \cdot \\ y_\varepsilon \end{pmatrix}, \quad \hat{A} = (a_1, a_2, \ldots, a_{n-\varepsilon-1}), \quad \hat{B} = (b_1, b_2, \ldots, b_{n-\varepsilon-1}).$$

Then the matrix equation (22) can be replaced by a system of scalar equations that expresses the equality of the elements of the k-th column on the right-hand and left-hand sides of (22) ($k = 1, 2, \ldots, n-\varepsilon-1$):

---

[11] This follows from the fact that the rank of the matrix (20) for $k = \varepsilon - 1$ is equal to εn; a similar equation holds for the rank of the matrix $M_{\varepsilon-1}[L_\varepsilon]$.

$$
\begin{aligned}
x_{2k} + \lambda x_{1k} &= d_{1k} + \lambda f_{1k} + y_1 a_k + \lambda y_1 b_k, \\
x_{3k} + \lambda x_{2k} &= d_{2k} + \lambda f_{2k} + y_2 a_k + \lambda y_2 b_k, \\
x_{4k} + \lambda x_{3k} &= d_{3k} + \lambda f_{3k} + y_3 a_k + \lambda y_3 b_k, \\
&\cdots\cdots\cdots\cdots\cdots \\
x_{\varepsilon+1,k} + \lambda x_{\varepsilon k} &= d_{\varepsilon k} + \lambda f_{\varepsilon k} + y_\varepsilon a_k + \lambda y_\varepsilon b_k \\
&(k = 1, 2, \ldots, n-\varepsilon-1).
\end{aligned} \tag{23}
$$

The left-hand sides of these equations are linear binomials in λ. The free term of each of the first $\varepsilon - 1$ of these binomials is equal to the coefficient of λ in the next binomial. But then the right-hand sides must also satisfy this condition. Therefore

$$
\begin{aligned}
y_1 a_k - y_2 b_k &= f_{2k} - d_{1k}, \\
y_2 a_k - y_3 b_k &= f_{3k} - d_{2k}, \\
&\cdots\cdots\cdots \\
y_{\varepsilon-1} a_k - y_\varepsilon b_k &= f_{\varepsilon k} - d_{\varepsilon-1,k} \\
(k = 1, 2, &\ldots, n-\varepsilon-1).
\end{aligned} \tag{24}
$$

If (24) holds, then the required elements of X can obviously be determined from (23).

It now remains to show that the system of equations (24) for the elements of Y always has a solution for arbitrary d_{ik} and f_{ik} ($i = 1, 2, \ldots, \varepsilon$; $k = 1, 2, \ldots, n-\varepsilon-1$). Indeed, the matrix formed from the coefficients of the unknown elements of the rows $y_1, -y_2, y_3, -y_4, \ldots$, can be written, after transposition, in the form

$$
\left(\begin{array}{cccc}
\overbrace{\hat{A} \quad O \quad \ldots \quad O}^{\varepsilon-1} \\
\hat{B} & \hat{A} & & \vdots \\
O & \hat{B} & \ddots & \vdots \\
\vdots & & \ddots & \hat{A} \\
O & O & \ldots & \hat{B}
\end{array}\right).
$$

But this is the matrix $M_{\varepsilon-2}$ for the pencil of rectangular matrices $\hat{A} + \lambda \hat{B}$ (see (10′) on p. 30). The rank of the matrix is $(\varepsilon - 1)(n - \varepsilon - 1)$, because the equation $(\hat{A} + \lambda \hat{B})\,\hat{x} = o$, by what we have shown, has no solutions of degree less than ε. Thus, the rank of the system of equations (24) is equal to the number of equations and such a system is consistent (noncontradictory) for arbitrary free terms.

This completes the proof of the theorem.

§ 4. The Canonical Form of a Singular Pencil of Matrices

1. Let $A + \lambda B$ be an arbitrary singular pencil of matrices of dimension $m \times n$. To begin with, we shall assume that neither among the columns nor among the rows of the pencil is there a linear dependence with constant coefficients.

Let $r < n$, where r is the rank of the pencil, so that the columns of $A + \lambda B$ are linearly dependent. In this case the equation $(A + \lambda B)x = o$ has a non-zero solution of minimal degree ε_1. From the restriction made at the beginning of this section it follows that $\varepsilon_1 > 0$. Therefore by Theorem 4 the given pencil can be transformed into the form

$$\begin{pmatrix} L_{\varepsilon_1} & O \\ O & A_1 + \lambda B_1 \end{pmatrix},$$

where the equation $(A_1 + \lambda B_1)\, x^{(1)} = o$ has no solution $x^{(1)}$ of degree less than ε_1.

If this equation has a non-zero solution of minimal degree ε_2 (where, necessarily, $\varepsilon_2 \geqq \varepsilon_1$), then by applying Theorem 4 to the pencil $A_1 + \lambda B_1$ we can transform the given pencil into the form

$$\begin{pmatrix} L_{\varepsilon_1} & O & O \\ O & L_{\varepsilon_2} & O \\ O & O & A_2 + \lambda B_2 \end{pmatrix}.$$

Continuing this process, we can put the given pencil into the quasi-diagonal form

$$\begin{pmatrix} L_{\varepsilon_1} & & & & & O \\ & L_{\varepsilon_2} & & & & \\ & & \cdot & & & \\ & & & \cdot & & \\ & & & & L_{\varepsilon_p} & \\ O & & & & & A_p + \lambda B_p \end{pmatrix}, \tag{25}$$

where $0 < \varepsilon_1 \leqq \varepsilon_2 \leqq \ldots \leqq \varepsilon_p$ and the equation $(A_p + \lambda B_p)\, x^{(p)} = o$ has no non-zero solution, so that the columns of $A_p + \lambda B_p$ are linearly independent.[12]

If the rows of $A_p + \lambda B_p$ are linearly dependent, then the transposed pencil $A_p^{\mathsf{T}} + \lambda B_p^{\mathsf{T}}$ can be put into the form (25), where instead of $\varepsilon_1, \varepsilon_2, \ldots, \varepsilon_p$ there occur the numbers $(0 <)\, \eta_1 \leqq \eta_2 \leqq \cdots \leqq \eta_q$.[13] But then the given pencil $A + \lambda B$ turns out to be transformable into the quasi-diagonal form

---

[12] In the special case where $\varepsilon_1 + \varepsilon_2 + \ldots + \varepsilon_p = m$ the block $A_p + \lambda B_p$ is absent.

[13] Since no linear dependence with constant coefficients exists among the rows of the pencil $A + \lambda B$ and consequently of $A_p + \lambda B_p$, we have $\eta_1 > 0$.

$$\begin{pmatrix} L_{\varepsilon_1} & & & & & & & & O \\ & L_{\varepsilon_2} & & & & & & & \\ & & \ddots & & & & & & \\ & & & L_{\varepsilon_p} & & & & & \\ & & & & L^{\mathsf{T}}_{\eta_1} & & & & \\ & & & & & L^{\mathsf{T}}_{\eta_2} & & & \\ & & & & & & \ddots & & \\ & & & & & & & L^{\mathsf{T}}_{\eta_q} & \\ O & & & & & & & & A_0 + \lambda B_0 \end{pmatrix}, \tag{26}$$

$$(0 < \varepsilon_1 \leqq \varepsilon_2 \leqq \cdots \leqq \varepsilon_p, \quad 0 < \eta_1 \leqq \eta_2 \leqq \cdots \leqq \eta_q)$$

where both the columns and the rows of $A_0 + \lambda B_0$ are linearly independent, i.e., $A_0 + \lambda B_0$ is a regular pencil.[16]

2. We now consider the general case where the rows and the columns of the given pencil may be connected by linear relations with constant coefficients. We denote the maximal number of constant independent solutions of the equations

$$(A + \lambda B)x = o \quad \text{and} \quad (A^{\mathsf{T}} + \lambda B^{\mathsf{T}}) = o$$

by g and h, respectively. Instead of the first of these equations we consider, just as in the proof of Theorem 4, the corresponding vector equation $(\boldsymbol{A} + \lambda \boldsymbol{B})\boldsymbol{x} = \boldsymbol{o}$ ($\boldsymbol{A}$ and $\boldsymbol{B}$ are operators mapping $\boldsymbol{R}_n$ into $\boldsymbol{R}_m$). We denote linearly independent constant solutions of this equation by $\boldsymbol{e}_1, \boldsymbol{e}_2, \ldots, \boldsymbol{e}_g$ and take them as the first g basis vectors in $\boldsymbol{R}_n$. Then the first g columns of the corresponding matrix $\tilde{A} + \lambda B$ consist of zeros

$$\tilde{A} + \lambda \tilde{B} = (\overbrace{O}^{g}, \tilde{A}_1 + \lambda \tilde{B}_1). \tag{27}$$

Similarly, the first h rows of the pencil $\tilde{A}_1 + \lambda \tilde{B}_1$ can be made into zeros. The given pencil then assumes the form

$$\begin{pmatrix} {}^{h}[\overbrace{O}^{g} & O \\ O & A^0 + \lambda B^0 \end{pmatrix}, \tag{28}$$

[16] If in the given pencil $r = n$, i.e., if the columns of the pencil are linearly independent, then the first p diagonal blocks in (26) of the form L_ε are absent ($p = 0$). In the same way, if $r = m$, i.e., if the rows of $A + \lambda B$ are linearly independent, then in (26) the diagonal blocks of the form L^{T}_η are absent ($q = 0$).

where there is no longer any linear dependence with constant coefficients among the rows or the columns of the pencil $A^0 + \lambda B^0$. The pencil $A^0 + \lambda B^0$ can now be represented in the form (26). Thus, in the general case, the pencil $A + \lambda B$ can always be put into the canonical quasi-diagonal form

$$\{{}_h[\overbrace{O}^{g},\ L_{\varepsilon_{g+1}}, \ldots, L_{\varepsilon_p},\ L^{\mathsf{T}}_{\eta_{h+1}}, \ldots, L^{\mathsf{T}}_{\eta_q},\ A_0 + \lambda B_0\}. \tag{29}$$

The choice of indices for ε and η is due to the fact that it is convenient here to take $\varepsilon_1 = \varepsilon_2 = \cdots = \varepsilon_g = 0$ and $\eta_1 = \eta_2 = \cdots = \eta_h = 0$.

When we replace the regular pencil $A_0 + \lambda B_0$ in (29) by its canonical form (6) (see § 2, p. 28), we finally obtain the following quasi-diagonal matrix

$$\{{}_h[\overbrace{O}^{g};\ L_{\varepsilon_{g+1}}, \ldots, L_{\varepsilon_p};\ L^{\mathsf{T}}_{\eta_{h+1}}, \ldots, L^{\mathsf{T}}_{\eta_q};\ N^{(u_1)}, \ldots, N^{(u_s)};\ J + \lambda E\}, \tag{30}$$

where the matrix J is of Jordan normal form or of natural normal form and $N^{(u)} = E^{(u)} + \lambda H^{(u)}$.

The matrix (30) *is the canonical form of the pencil* $A + \lambda B$ *in the most general case.*

In order to determine the canonical form (30) of a given pencil immediately, without carrying out the successive reduction processes, we shall, following Kronecker, introduce in the next section the concept of minimal indices of a pencil.

§ 5. The Minimal Indices of a Pencil. Criterion for Strong Equivalence of Pencils

1. Let $A + \lambda B$ be an arbitrary singular pencil of rectangular matrices. Then the k polynomial columns $x_1(\lambda), x_2(\lambda), \ldots, x_k(\lambda)$ that are solutions of the equation

$$(A + \lambda B)x = o \tag{31}$$

are linearly dependent if the rank of the polynomial matrix formed from these columns $X = [x_1(\lambda), x_2(\lambda), \ldots, x_k(\lambda)]$ is less than k. In that case there exist k polynomials $p_1(\lambda), p_2(\lambda), \ldots, p_k(\lambda)$, not all identically zero, such that

$$p_1(\lambda)\, x_1(\lambda) + p_2(\lambda)\, x_2(\lambda) + \cdots + p_k(\lambda)\, x_k(\lambda) \equiv o\,.$$

But if the rank of X is k, then such a dependence does not exist and the solutions $x_1(\lambda), x_2(\lambda), \ldots, x_k(\lambda)$ are linearly independent.

Among all the solutions of (31) we choose a non-zero solution $x_1(\lambda)$ of least degree ε_1. Among all the solutions of the same equation that are linearly independent of $x_1(\lambda)$ we take a solution $x_2(\lambda)$ of least degree ε_2. Obviously, $\varepsilon_1 \leqq \varepsilon_2$. We continue the process, choosing among the solutions that are linearly independent of $x_1(\lambda)$ and $x_2(\lambda)$ a solution $x_3(\lambda)$ of minimal degree ε_3, etc. Since the number of linearly independent solutions of (31) is always at most n, the process must come to an end. We obtain a *fundamental series of solutions* of (31)

$$x_1(\lambda), x_2(\lambda), \ldots, x_p(\lambda) \tag{32}$$

having the degrees

$$\varepsilon_1 \leqq \varepsilon_2 \leqq \cdots \leqq \varepsilon_p. \tag{33}$$

In general, a fundamental series of solutions is not uniquely determined (to within scalar factors) by the pencil $A + \lambda B$. However, *two distinct fundamental series of solutions always have one and the same series of degrees* $\varepsilon_1, \varepsilon_2, \ldots, \varepsilon_p$. For let us consider in addition to (32) another fundamental series of solutions $\tilde{x}_1(\lambda), \tilde{x}_2(\lambda), \ldots$ with the degrees $\tilde{\varepsilon}_1, \tilde{\varepsilon}_2, \ldots$. Suppose that in (33)

$$\varepsilon_1 = \cdots = \varepsilon_{n_1} < \varepsilon_{n_1+1} = \cdots = \varepsilon_{n_2} < \cdots$$

and similarly, in the series $\tilde{\varepsilon}_1, \tilde{\varepsilon}_2, \ldots$,

$$\tilde{\varepsilon}_1 = \cdots = \tilde{\varepsilon}_{\tilde{n}_1} < \tilde{\varepsilon}_{\tilde{n}_1+1} = \cdots = \tilde{\varepsilon}_{\tilde{n}_2} < \cdots.$$

Obviously, $\varepsilon_1 = \tilde{\varepsilon}_1$. Every column $\tilde{x}_i(\lambda)$ $(i = 1, 2, \ldots, \tilde{n}_1)$ is a linear combination of the columns $x_1(\lambda), x_2(\lambda), \ldots, x_{n_1}(\lambda)$, since otherwise the solution $x_{n_1+1}(\lambda)$ in (32) could be replaced by $x_i(\lambda)$, which is of smaller degree. It is obvious that, conversely, every column $x_i(\lambda)$ $(i = 1, 2, \ldots, n_1)$ is a linear combination of the columns $\tilde{x}_1(\lambda), \tilde{x}_2(\lambda), \ldots, \tilde{x}_{n_1+1}(\lambda)$. Therefore $n_1 = \tilde{n}_1$ and $\varepsilon_{n_1+1} = \tilde{\varepsilon}_{\tilde{n}_1+1}$. Now by a similar argument we obtain that $n_2 = \tilde{n}_2$ and $\varepsilon_{n_2+1} = \tilde{\varepsilon}_{\tilde{n}_2+1}$, etc.

2. Every solution $x_k(\lambda)$ of the fundamental series (32) yields a linear dependence of degree ε_k among the columns of $A + \lambda B$ $(k = 1, 2, \ldots, p)$. Therefore the numbers $\varepsilon_1, \varepsilon_2, \ldots, \varepsilon_p$ are called the *minimal indices for the columns* of the pencil $A + \lambda B$.

The *minimal indices* $\eta_1, \eta_2, \ldots, \eta_q$ *for the rows* of the pencil $A + \lambda B$ are introduced similarly. Here the equation $(A + \lambda B)x = o$ is replaced by $(A^{\mathsf{T}} + \lambda B^{\mathsf{T}})y = o$, and $\eta_1, \eta_2, \ldots, \eta_q$ are defined as minimal indices for the columns of the transposed pencil $A^{\mathsf{T}} + \lambda B^{\mathsf{T}}$.

Strictly equivalent pencils have the same minimal indices. For let $A + \lambda B$ and $P(A + \lambda B)Q$ be two such pencils (P and Q are non-singular square matrices). Then the equation (31) for the first pencil can be written, after multiplication on the left by P, as follows:

$$P(A + \lambda B)Q \cdot Q^{-1}x = o.$$

Hence it is clear that all the solutions of (31), after multiplication on the left by Q^{-1}, give rise to a complete system of solutions of the equation

$$P(A + \lambda B)Qz = o.$$

Therefore the pencils $A + \lambda B$ and $P(A + \lambda B)Q$ have the same minimal indices for the columns. That the minimal indices for the rows also coincide can be established by going over to the transposed pencils.

Let us compute the minimal indices for the canonical quasi-diagonal matrix

$$\left\{ {}_h[\overset{g}{\overline{O}}, \quad L_{\varepsilon_{g+1}}, \ldots, L_{\varepsilon_p}; \quad L^{\mathsf{T}}_{\eta_{h+1}}, \ldots, L^{\mathsf{T}}_{\eta_q}, \quad A_0 + \lambda B_0 \right\} \tag{34}$$

($A_0 + \lambda B_0$ is a regular pencil having the normal form (6)).

We note first of all that: *The complete system of indices for the columns (rows) of a quasi-diagonal matrix is obtained as the union of the corresponding systems of minimal indices of the individual diagonal blocks.* The matrix L_ε has only one index ε for the columns, and its rows are linearly independent. Similarly, the matrix L^{T}_η has only one index η for the rows, and its columns are linearly independent. Therefore the matrix (34) has as its minimal indices for the columns

$$\varepsilon_1 = \cdots = \varepsilon_g = 0, \quad \varepsilon_{g+1}, \ldots, \varepsilon_p$$

and for the rows

$$\eta_1 = \cdots = \eta_h = 0, \quad \eta_{h+1}, \ldots, \eta_q.$$

We note further that L_ε has no elementary divisors, since among its minors of maximal order ε there is one equal to 1 and one equal to λ^ε. The same statement is, of course, true for the transposed matrix $L^{\mathsf{T}}_\varepsilon$. Since the elementary divisors of a quasi-diagonal matrix are obtained by combining those of the individual diagonal blocks (see Volume I, Chapter VI, p. 141), *the elementary divisors of the λ-matrix* (34) *coincide with those of its regular 'kernel'* $A_0 + \lambda B_0$.

The canonical form of the pencil (34) *is completely determined by the minimal indices* $\varepsilon_1, \ldots, \varepsilon_p, \eta_1, \ldots, \eta_q$ *and the elementary divisors of the pencil or, what is the same, of the strictly equivalent pencil* $A + \lambda B$. Since two pencils having one and the same canonical form are strictly equivalent, we have proved the following theorem:

THEOREM 5 (Kronecker): *Two arbitrary pencils $A + \lambda B$ and $A_1 + \lambda B_1$ of rectangular matrices of the same dimension $m \times n$ are strictly equivalent if and only if they have the same minimal indices and the same (finite and infinite) elementary divisors.*

In conclusion, we write down, for purposes of illustration, the canonical form of a pencil $A + \lambda B$ with the minimal indices $\varepsilon_1 = 0$, $\varepsilon_2 = 1$, $\varepsilon_3 = 2$, $\eta_1 = 0$, $\eta_2 = 0$, $\eta_3 = 2$ and the elementary divisors λ^2, $(\lambda + 2)^2$, μ^3:[15]

$$\left\| \begin{array}{ccccccccccccccc}
0 & & & & & & & & & & & & & & \\
0 & & & & & & & & & & & & & & \\
 & \lambda & 1 & & & & & & & & & & & & \\
 & & & \lambda & 1 & 0 & & & & & & & & & \\
 & & & 0 & \lambda & 1 & & & & & & & & & \\
 & & & & & & \lambda & 0 & & & & & & & \\
 & & & & & & 1 & \lambda & & & & & & & \\
 & & & & & & 0 & 1 & & & & & & & \\
 & & & & & & & & 1 & \lambda & 0 & & & & \\
 & & & & & & & & 0 & 1 & \lambda & & & & \\
 & & & & & & & & 0 & 0 & 1 & & & & \\
 & & & & & & & & & & & \lambda & 1 & & \\
 & & & & & & & & & & & 0 & \lambda & & \\
 & & & & & & & & & & & & & \lambda+2 & 1 \\
 & & & & & & & & & & & & & 0 & \lambda+2
\end{array} \right\| . \tag{35}$$

§ 6. Singular Pencils of Quadratic Forms

1. Suppose given two complex quadratic forms:

$$A(x, x) = \sum_{i,k=1}^{n} a_{ik} x_i x_k, \qquad B(x, x) = \sum_{i,k=1}^{n} b_{ik} x_i x_k; \tag{36}$$

they generate a pencil of quadratic forms $A(x, x) + \lambda B(x, x)$. This pencil of forms corresponds to a pencil of symmetric matrices $A + \lambda B$ ($A^\mathsf{T} = A$, $B^\mathsf{T} = B$). If we subject the variables in the pencil of forms $A(x,x) + \lambda B(x,x)$ to a non-singular linear transformation $x = Tz$ ($|T| \neq 0$), then the transformed pencil of forms $\tilde{A}(z, z) + \lambda \tilde{B}(z, z)$ corresponds to the pencil of matrices

[15] All the elements of the matrix that are not mentioned expressly are zero.

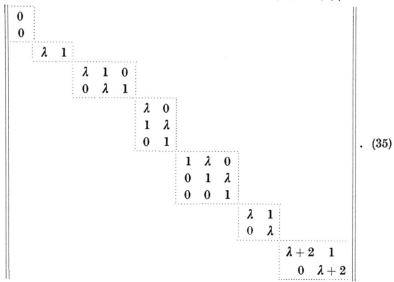

$$\tilde{A} + \lambda \tilde{B} = T^{\mathsf{T}} (A + \lambda B)\, T\ ; \tag{37}$$

here T is a constant (i.e., independent of λ) non-singular square matrix of order n.

Two pencils of matrices $A + \lambda B$ and $\tilde{A} + \lambda \tilde{B}$ that are connected by a relation (36) are called *congruent* (see Definition 1 of Chapter X; Vol. I, p. 296).

Obviously, congruence is a special case of equivalence of pencils of matrices. However, if congruence of two pencils of symmetric (or skew-symmetric) matrices is under consideration, then the concept of congruence coincides with that of equivalence. This is the content of the following theorem.

THEOREM 6: *Two strictly equivalent pencils of complex symmetric (or skew-symmetric) matrices are always congruent.*

Proof. Let $\Lambda \equiv A + \lambda B$ and $\tilde{\Lambda} \equiv \tilde{A} + \lambda \tilde{B}$ be two strictly equivalent pencils of symmetric (skew-symmetric) matrices:

$$\tilde{\Lambda} = P \Lambda Q \quad (\Lambda^{\mathsf{T}} = \pm \Lambda,\ \tilde{\Lambda}^{\mathsf{T}} = \pm \tilde{\Lambda};\ \ |P| \neq 0,\ \ |Q| \neq 0). \tag{38}$$

By going over to the transposed matrices we obtain:

$$\tilde{\Lambda} = Q^{\mathsf{T}} \Lambda P^{\mathsf{T}}. \tag{39}$$

From (38) and (39), we have

$$\Lambda Q P^{\mathsf{T}-1} = P^{-1} Q^{\mathsf{T}} \Lambda. \tag{40}$$

Setting

$$U = Q P^{\mathsf{T}-1}, \tag{41}$$

we rewrite (40) as follows:

$$\Lambda U = U^{\mathsf{T}} \Lambda. \tag{42}$$

From (42) it follows easily that

$$\Lambda U^k = U^{\mathsf{T}k} \Lambda \quad (k = 0, 1, 2, \ldots)$$

and, in general,

$$\Lambda S = S^{\mathsf{T}} \Lambda, \tag{43}$$

where

$$S = f(U), \tag{44}$$

and $f(\lambda)$ is an arbitrary polynomial in λ. Let us assume that this polynomial is chosen such that $|S| \neq 0$. Then we have from (43):

$$\Lambda = S^{\mathsf{T}} \Lambda S^{-1}. \tag{45}$$

Substituting this expression for Λ in (38), we have:

$$\tilde{\Lambda} = PS^{\mathsf{T}}\Lambda S^{-1}Q. \tag{46}$$

If this relation is to be a congruence transformation, the following equation must be satisfied:

$$(PS^{\mathsf{T}})^{\mathsf{T}} = S^{-1}Q\,,$$

which can be rewritten as

$$S^2 = QP^{\mathsf{T}-1} = U.$$

Now the matrix $S = f(U)$ satisfies this equation if we take as $f(\lambda)$ the interpolation polynomial for $\sqrt{\lambda}$ on the spectrum of U. This can be done, because the many-valued function $\sqrt{\lambda}$ has a single-valued branch determined on the spectrum of U, since $|U| \neq 0$.

The equation (46) now becomes the condition for congruence

$$\tilde{\Lambda} = T^{\mathsf{T}}\Lambda T \quad \left(T = SQ = \sqrt{QP^{\mathsf{T}-1}}\,Q\right). \tag{47}$$

From this theorem and Theorem 5 we deduce:

COROLLARY: *Two pencils of quadratic forms*

$$A(x, x) + \lambda B(x, x) \quad \textit{and} \quad \tilde{A}(z, z) + \lambda \tilde{B}(z, z)$$

can be carried into one another by a transformation $x = Tz$ $(|T| \neq 0)$ *if and only if the pencils of symmetric matrices* $A + \lambda B$ *and* $\tilde{A} + \lambda\tilde{B}$ *have the same elementary divisors (finite and infinite) and the same minimal indices.*

Note. For pencils of symmetric matrices the rows and columns have the same minimal indices:

$$p = q; \quad \varepsilon_1 = \eta_1, \ldots, \varepsilon_p = \eta_p. \tag{48}$$

2. Let us raise the following question: *Given two arbitrary complex quadratic forms*

$$A(x, x) = \sum_{i,k=1}^{n} a_{ik}x_i x_k, \qquad B(x, x) = \sum_{i,k=1}^{n} b_{ik}x_i x_k.$$

Under what conditions can the two forms be reduced simultaneously to sums of squares

$$\sum_{i=1}^{n} a_i z_i^2 \quad \textit{and} \quad \sum_{i=1}^{n} b_i z_i^2 \tag{49}$$

by a non-singular transformation of the variables $x = Tz$ $(|T| \neq 0)$?

Let us assume that the quadratic forms $A(x, x)$ and $B(x, x)$ have this property. Then the pencil of matrices $A + \lambda B$ is congruent to the pencil of diagonal matrices

$$\{a_1 + \lambda b_1, a_2 + \lambda b_2, \ldots, a_n + \lambda b_n\}. \tag{50}$$

Suppose that among the diagonal binomials $a_i + \lambda b_i$ there are precisely r $(r \leqq n)$ that are not identically zero. Without loss of generality we can assume that

$$a_1 = b_1 = 0, \ldots, a_{n-r} = b_{n-r} = 0, \ a_i + \lambda b_i \not\equiv 0 \quad (i = n - r + 1, \ldots, n). \tag{51}$$

Setting

$$A_0 + \lambda B_0 = \{a_{n-r+1} + \lambda b_{n-r+1}, \ldots, a_n + \lambda b_n\}, \tag{52}$$

we represent the matrix (51) in the form

$$\{\overbrace{O}^{n-r}, \quad A_0 + \lambda B_0\}. \tag{53}$$

Comparing (52) with (34) (p.39), we see that in this case all the minimal indices are zero. Moreover, all the elementary divisors are linear. Thus we have obtained the following theorem:

THEOREM 7: *Two quadratic forms $A(x, x)$ and $B(x, x)$ can be reduced simultaneously to sums of squares* (49) *by a transformation of the variables if and only if in the pencil of matrices $A + \lambda B$ all the elementary divisors (finite and infinite) are linear and all the minimal indices are zero.*

In order to reduce two quadratic forms $A(x, x)$ and $B(x, x)$ *simultaneously* to some canonical form in the general case, we have to replace the pencil of matrices $A + \lambda B$ by a strictly equivalent 'canonical' pencil of symmetric matrices.

Suppose the pencil of symmetric matrices $A + \lambda B$ has the minimal indices $\varepsilon_1 = \ldots = \varepsilon_g = 0$, $\varepsilon_{g+1} \neq 0, \ldots, \varepsilon_p \neq 0$, the infinite elementary divisors $\mu^{u_1}, \mu^{u_2}, \ldots, \mu^{u_s}$ and the finite ones $(\lambda + \lambda_1)^{c_1}$, $(\lambda + \lambda_2)^{c_2}, \ldots, (\lambda + \lambda_t)^{c_t}$ Then, in the canonical form (30), $g = h$, $p = q$ and $\varepsilon_{g+1} = \eta_{g+1}, \ldots, \varepsilon_p = \eta_p$. We replace in (30) every two diagonal blocks of the form L_ε and $L_\varepsilon^{\mathsf{T}}$ by a single diagonal block $\begin{pmatrix} 0 & L_\varepsilon^{\mathsf{T}} \\ L_\varepsilon & 0 \end{pmatrix}$ and each block of the form $N^{(u)} = E^{(u)} + \lambda H^{(u)}$ by the strictly equivalent symmetric block

$$\tilde{N}^{(u)} = V^{(u)}N^{(u)} = \begin{Vmatrix} 0 & 0 & \dots & 0 & 1 \\ 0 & 0 & \dots & 1 & \lambda \\ \cdot & \cdot & \cdot & \cdot & \cdot \\ 1 & \lambda & \dots & 0 & 0 \end{Vmatrix} \quad \text{with} \quad V^{(u)} = \begin{Vmatrix} 0 & 0 & \dots & 0 & 1 \\ 0 & 0 & \dots & 1 & 0 \\ \cdot & & & \cdot & \cdot \\ \cdot & & \cdot & & \cdot \\ \cdot & 1 & & & \cdot \\ 1 & 0 & \dots & 0 & 0 \end{Vmatrix}. \tag{54}$$

Moreover, instead of the regular diagonal block $J + \lambda E$ in (30) (J is a Jordan matrix)

$$J + \lambda E = \{(\lambda + \lambda_1) E^{(c_1)} + H^{(c_1)}, \dots, (\lambda + \lambda_t) E^{(c_t)} + H^{(c_t)}\},$$

we take the strictly equivalent block

$$\{Z_{\lambda_1}^{(c_1)}, \dots, Z_{\lambda_t}^{(c_t)}\}, \tag{55}$$

where

$$Z_{\lambda_i}^{(c_i)} = V^{(c_i)} [(\lambda + \lambda_i) E^{(c_i)} + H^{(c_i)}]$$

$$= \begin{Vmatrix} 0 & \dots & 0 & \lambda + \lambda_i \\ 0 & \dots & \lambda + \lambda_i & 1 \\ \cdot & \cdot & \cdot & \cdot \\ \cdot & \cdot & \cdot & \cdot \\ \cdot & \cdot & \cdot & \cdot \\ \lambda + \lambda_i & 1 & \dots & 0 \end{Vmatrix} \quad (i = 1, 2, \dots, t). \tag{56}$$

The pencil $A + \lambda B$ is strictly equivalent to the symmetric pencil

$$\tilde{A} + \lambda \tilde{B}$$
$$= \left\{ O, \begin{pmatrix} O & L_{\varepsilon_{g+1}}^{\mathsf{T}} \\ L_{\varepsilon_{g+1}} & O \end{pmatrix}, \dots, \begin{pmatrix} O & L_{\varepsilon_p}^{\mathsf{T}} \\ L_{\varepsilon_p} & O \end{pmatrix}; \tilde{N}^{(u_1)}, \dots, \tilde{N}^{(u_s)}; Z_{\lambda_1}^{(c_1)}, \dots, Z_{\lambda_t}^{(c_t)} \right\}. \tag{57}$$

Two quadratic forms with complex coefficients $A(x, x)$ and $B(x, x)$ can be simultaneously reduced to the canonical forms $\tilde{A}(z, z)$ and $\tilde{B}(z, z)$ defined in (57) by a transformation of the variables $x = Tz$ ($|T| \neq 0$).[17]

---

[17] In the Russian edition the author stated that propositions analogous to Theorems 6 and 7 hold for hermitian forms. A. I. Mal'cev has pointed out to the author that this is not the case. As regards singular pencils of hermitian forms, see [197 II].

§ 7. Application to Differential Equations

1. The results obtained will now be applied to a system of m linear differential equations of the first order in n unknown functions with constant coefficients:[18]

$$\sum_{k=1}^{n} a_{ik}x_k + \sum_{k=1}^{n} b_{ik}\frac{dx_k}{dt} = f_i(t) \quad (i = 1, 2, \ldots, m), \tag{58}$$

or in matrix notation:

$$Ax + B\frac{dx}{dt} = f(t); \tag{59}$$

here[19]

$$A = \|a_{ik}\|, \quad B = \|b_{ik}\| \quad (i = 1, 2, \ldots, m;\ k = 1, 2, \ldots, n),$$

$$x = (x_1, x_2, \ldots, x_n), \quad f = (f_1, f_2, \ldots, f_m).$$

We introduce new unknown functions $z_1, z_2, \ldots, z_n$ that are connected with the old $x_1, x_2, \ldots, x_n$ by a linear non-singular transformation with constant coefficients:

$$x = Qz \quad (z = (z_1, z_2, \ldots, z_n);\ |Q| \neq 0). \tag{60}$$

Moreover, instead of the equations (58) we can take m arbitrary independent combinations of these equations, which is equivalent to multiplying the matrices A, B, f on the left by a square non-singular matrix P of order m. Substituting Qz for x in (59) and multiplying (59) on the left by P, we obtain:

$$\tilde{A}z + \tilde{B}\frac{dz}{dt} = \tilde{f}(t), \tag{61}$$

where

$$\tilde{A} = PAQ, \quad \tilde{B} = PBQ, \quad \tilde{f} = Pf = (\tilde{f}_1, \tilde{f}_2, \ldots, \tilde{f}_n). \tag{62}$$

The matrix pencils $A + \lambda B$ and $\tilde{A} + \lambda\tilde{B}$ are strictly equivalent:

$$\tilde{A} + \lambda\tilde{B} = P(A + \lambda B)Q. \tag{63}$$

We choose the matrices P and Q such that the pencil $\tilde{A} + \lambda\tilde{B}$ has the canonical quasi-diagonal form

---

[18] The particular case where $m = n$ and the system (58) is solved with respect to the derivatives has been treated in detail in Vol. I, Chapter V, § 5.

It is well known that a system of linear differential equations with constant coefficients of arbitrary order s can be reduced to the form (58) if all the derivatives of the unknown functions up to and including the order $s - 1$ are included as additional unknown functions.

[19] We recall that parentheses denote column matrices. Thus, $x = (x_1, x_2, \ldots, x_n)$ is the column with the elements $x_1, x_2, \ldots, x_n$.

$$\tilde{A} + \lambda \tilde{B} = \{O,\ L_{\varepsilon_{g+1}}, \ldots, L_{\varepsilon_p},\ L^{\mathsf{T}}_{\eta_{h+1}}, \ldots, L^{\mathsf{T}}_{\eta_q},\ N^{(u_1)}, \ldots, N^{(u_s)},\ J + \lambda E\}. \tag{64}$$

In accordance with the diagonal blocks in (64) the system of differential equations splits into $\nu = p - g + q - h + s + 2$ separate systems of the form

$$O \cdot \overset{1}{z} = \overset{1}{\tilde{f}}, \tag{65}$$

$$L_{\varepsilon_{g+i}}\left(\frac{d}{dt}\right) \overset{1+i}{z} = \overset{1+i}{\tilde{f}} \qquad (i = 1, 2, \ldots, p-g), \tag{66}$$

$$L^{\mathsf{T}}_{\eta_{h+j}}\left(\frac{d}{dt}\right) \overset{p-g+1+j}{z} = \overset{p-g+1+j}{\tilde{f}} \qquad (j = 1, 2, \ldots, q-h), \tag{67}$$

$$N^{(u_k)}\left(\frac{d}{dt}\right) \overset{p-g+q-h+1+k}{z} = \overset{p-g+q-h+1+k}{\tilde{f}} \qquad (k = 1, 2, \ldots, s), \tag{68}$$

$$\left(J + \frac{d}{dt}\right) \overset{\nu}{z} = \overset{\nu}{\tilde{f}}, \tag{69}$$

where

$$z = \begin{pmatrix} \overset{1}{z} \\ \overset{2}{z} \\ \cdot \\ \cdot \\ \cdot \\ \overset{\nu}{z} \end{pmatrix}, \quad \tilde{f} = \begin{pmatrix} \overset{1}{\tilde{f}} \\ \overset{2}{\tilde{f}} \\ \cdot \\ \cdot \\ \cdot \\ \overset{\nu}{\tilde{f}} \end{pmatrix}, \tag{70}$$

$$\overset{1}{z} = (z_1, \ldots, z_g),\ \overset{1}{\tilde{f}} = (\tilde{f}_1, \ldots, \tilde{f}_h),\ \overset{2}{z} = (z_{g+1}, \ldots),\ \overset{2}{\tilde{f}} = (\tilde{f}_{h+1}, \ldots) \quad \text{etc.}, \tag{71}$$

$$A\left(\frac{d}{dt}\right) = A + B\frac{d}{dt}, \qquad \text{if} \quad A(\lambda) \equiv A + \lambda B. \tag{72}$$

Thus, the integration of the system (59) in the most general case is reduced to the integration of the special systems (65)-(69) of the same type. In these systems the matrix pencil $A + \lambda B$ has the form O, L_ε, L^{T}_η, $N^{(u)}$, and $J + \lambda E$, respectively.

1) The system (65) is not inconsistent if and only if

$$\overset{1}{\tilde{f}} \equiv o,$$

i.e.,

$$\tilde{f}_1 \equiv 0, \ldots, \tilde{f}_h \equiv 0. \tag{73}$$

In that case we can take arbitrary functions of t as the unknown functions $z_1, z_2, \ldots, z_g$ that form the columns $\overset{1}{z}$.

2) The system (66) is of the form

$$L_\varepsilon\left(\frac{d}{dt}\right)z=\tilde{f} \tag{74}$$

or, more explicitly,[20]

$$\frac{dz_1}{dt}+z_2=\tilde{f}_1(t),\ \frac{dz_2}{dt}+z_3=\tilde{f}_2(t),\ \ldots,\ \frac{dz_\varepsilon}{dt}+z_{\varepsilon+1}=\tilde{f}_\varepsilon(t). \tag{75}$$

Such a system is always consistent. If we take for $z_{\varepsilon+1}(t)$ an arbitrary function of t, then all the remaining unknown functions $z_\varepsilon, z_{\varepsilon-1}, \ldots, z_1$ can be determined from (75) by successive quadratures.

3) The system (67) is of the form

$$L_\eta^{\mathsf{T}}\left(\frac{d}{dt}\right)z=\tilde{f} \tag{76}$$

or, more explicitly,[21]

$$\frac{dz_1}{dt}=\tilde{f}_1(t),\ \frac{dz_2}{dt}+z_1=\tilde{f}_2(t),\ \ldots,\ \frac{dz_\eta}{dt}+z_{\eta-1}=\tilde{f}_\eta(t),\ \ z_\eta=\tilde{f}_{\eta+1}(t). \tag{77}$$

From all the equations (77) except the first we determine $z_\eta, z_{\eta-1}, \ldots, z_1$ uniquely:

$$\begin{aligned} z_\eta &= \tilde{f}_{\eta+1}, \\ z_{\eta-1} &= \tilde{f}_\eta - \frac{d\tilde{f}_{\eta+1}}{dt}, \\ &\cdots\cdots\cdots\cdots \\ z_1 &= \tilde{f}_2 - \frac{d\tilde{f}_3}{dt} + \cdots + (-1)^{\eta-1}\frac{d^{\eta-1}\tilde{f}_{\eta+1}}{dt^{\eta-1}}. \end{aligned} \tag{78}$$

Substituting this expression for z_1 into the first equation, we obtain the condition for consistency:

$$\tilde{f}_1-\frac{d\tilde{f}_2}{dt}+\frac{d^2\tilde{f}_3}{dt^2}-\cdots+(-1)^\eta\frac{d^\eta\tilde{f}_{\eta+1}}{dt^\eta}=0. \tag{79}$$

---

[20] We have changed the indices of z and $\tilde{f}$ to simplify the notation. In order to return from (75) to (66) we have to replace ε by ε_i and add to each index of z the number $g+\varepsilon_{g+1}+\cdots+\varepsilon_{g+i-1}+i-1$, to each index of $\tilde{f}$ the number $h+\varepsilon_{g+1}+\cdots+\varepsilon_{g+i-1}$.

[21] Here, as in the preceding case, we have changed the notation. See the preceding footnote.

4) The system (68) is of the form

$$N^{(u)}\left(\frac{d}{dt}\right)z=\tilde{f} \tag{80}$$

or, more explicitly,

$$\frac{dz_2}{dt}+z_1=\tilde{f}_1,\ \frac{dz_3}{dt}+z_2=\tilde{f}_2,\ \ldots,\ \frac{dz_u}{dt}+z_{u-1}=\tilde{f}_{u-1},\ z_u=\tilde{f}_u. \tag{81}$$

Hence we determine successively the unique solutions

$$\begin{aligned} z_u &=\tilde{f}_u, \\ z_{u-1} &=\tilde{f}_{u-1}-\frac{d\tilde{f}_u}{dt}, \\ &\cdots\cdots\cdots\cdots\cdots \\ z_1 &=\tilde{f}_1-\frac{d\tilde{f}_2}{dt}+\frac{d^2\tilde{f}_3}{dt_2}-\cdots+(-1)^{u-1}\frac{d^{u-1}\tilde{f}_u}{dt^{u-1}}. \end{aligned} \tag{82}$$

5) The system (69) is of the form

$$Jz+\frac{dz}{dt}=\tilde{f}. \tag{83}$$

As we have proved in Vol. I, Chapter V, § 5, the general solution of such a system has the form

$$z=e^{-Jt}z_0+\int_0^t e^{-J(t-\tau)}f(\tau)\,d\tau; \tag{84}$$

here z_0 is a column matrix with arbitrary elements (the initial values of the unknown functions for $t=0$).

The inverse transition from the system (61) to (59) is effected by the formulas (60) and (62), according to which each of the functions $x_1, \ldots, x_n$ is a linear combination of the functions $z_1, \ldots, z_n$ and each of the functions $\tilde{f}_1(t), \ldots, \tilde{f}_m(t)$ is expressed linearly (with constant coefficients) in terms of the functions $f_1(t), \ldots, f_m(t)$.

2. The preceding analysis shows that: *In general, for the consistency of the system* (58) *certain well-defined linear dependence relations (with constant coefficients) must hold among the right-hand sides of the equations and the derivatives of these right-hand sides.*

If these relations are satisfied, then the general solution of the system contains both arbitrary constants and arbitrary functions linearly.

The character of the consistency conditions and the character of the solutions (in particular, the number of arbitrary constants and arbitrary functions) are determined by the minimal indices and the elementary divisors of the pencil $A + \lambda B$*, because the canonical form* (65)-(69) *of the system of differential equations depends on these minimal indices and elementary divisors.*

CHAPTER XIII

MATRICES WITH NON-NEGATIVE ELEMENTS

In this chapter we shall study properties of real matrices with non-negative elements. Such matrices have important applications in the theory of probability, where they are used for the investigation of Markov chains ('stochastic matrices,' see [46]), and in the theory of small oscillations of elastic systems ('oscillation matrices,' see [17]).

§ 1. General Properties

1. We begin with some definitions.

DEFINITION 1: *A rectangular matrix A with real elements*

$$A = \| a_{ik} \| \qquad (i = 1, 2, \dots, m;\ k = 1, 2, \dots, n)$$

is called non-negative (notation: $A \geqq O$) *or positive* (notation: $A > O$) *if all the elements of A are non-negative* ($a_{ik} \geqq 0$) *or positive* ($a_{ik} > 0$).

DEFINITION 2: *A square matrix* $A = \| a_{ik} \|_1^n$ *is called reducible if the index set* $1, 2, \dots, n$ *can be split into two complementary sets (without common indices)* $i_1, i_2, \dots, i_\mu;\ k_1, k_2, \dots, k_\nu$ $(\mu + \nu = n)$ *such that*

$$a_{i_\alpha k_\beta} = 0 \quad (\alpha = 1, 2, \dots, \mu;\ \beta = 1, 2, \dots, \nu).$$

Otherwise the matrix is called *irreducible.*

By a *permutation* of a square matrix $A = \| a_{ik} \|_1^n$ we mean a permutation of the rows of A combined with the *same* permutation of the columns.

The definition of a reducible matrix and an irreducible matrix can also be formulated as follows:

DEFINITION 2′: *A matrix* $A = \| a_{ik} \|_1^n$ *is called reducible if there is a permutation that puts it into the form*

$$\tilde{A} = \begin{pmatrix} B & O \\ C & D \end{pmatrix},$$

where B and D are square matrices. Otherwise A is called irreducible.

Suppose that $A = \| a_{ik} \|_1^n$ corresponds to a linear operator $\boldsymbol{A}$ in an n-dimensional vector space $\boldsymbol{R}$ with the basis $\boldsymbol{e}_1, \boldsymbol{e}_2, \ldots, \boldsymbol{e}_n$. To a permutation of A there corresponds a renumbering of the basis vectors, i.e., a transition from the basis $\boldsymbol{e}_1, \boldsymbol{e}_2, \ldots, \boldsymbol{e}_n$ to a new basis $\boldsymbol{e}'_1 = \boldsymbol{e}_{j_1}, \boldsymbol{e}'_2 = \boldsymbol{e}_{j_2}, \ldots, \boldsymbol{e}'_n = \boldsymbol{e}_{j_n}$, where $(j_1, j_2, \ldots, j_n)$ is a permutation of the indices $1, 2, \ldots, n$. The matrix A then goes over into a similar matrix $\tilde{A} = T^{-1}AT$. (Each row and each column of the transforming matrix T contains a single element 1, and the remaining elements are zero.)

2. By a ν-dimensional *coordinate subspace* of $\boldsymbol{R}$ we mean a subspace of $\boldsymbol{R}$ with a basis $\boldsymbol{e}_{k_1}, \boldsymbol{e}_{k_2}, \ldots, \boldsymbol{e}_{k_\nu}$ $(1 \leqq k_1 < k_2 < \ldots < k_\nu \leqq n)$. There are $\binom{n}{\nu}$ ν-dimensional coordinate subspaces of $\boldsymbol{R}$ connected with a given basis $\boldsymbol{e}_1, \boldsymbol{e}_2, \ldots, \boldsymbol{e}_n$. The definition of a reducible matrix can also be given in the following form:

Definition 2″: *A matrix $A = \| a_{ik} \|_1^n$ is called reducible if and only if the corresponding operator $\boldsymbol{A}$ has a ν-dimensional invariant coordinate subspace with $\nu < n$.*

We shall now prove the following lemma:

Lemma 1: *If $A \geqq O$ is an irreducible matrix of order n, then*

$$(E + A)^{n-1} > O. \tag{1}$$

Proof. For the proof of the lemma it is sufficient to show that for every vector[1] (i.e., column) $y \geqq o$ $(y \neq o)$ the inequality

$$(E + A)^{n-1} y > o$$

holds.

This inequality will be established if we can only show that *under the conditions $y \geqq o$ and $y \neq o$ the vector $z = (E + A)y$ always has fewer zero coordinates than y does.* Let us assume the contrary. Then y and z have the same zero coordinates.[2] Without loss of generality we may assume that the columns y and z have the form[3]

---

[1] Here and throughout this chapter we mean by a vector a column of n numbers. In this way we identify, as it were, a vector with the column of its coordinates in that basis in which the given matrix $A = \| a_{ik} \|_1^n$ determines a certain linear operator.

[2] Here we start from the fact that $z = y + Ay$ and $Ay \geqq o$; therefore to positive coordinates of y there correspond positive coordinates of z.

[3] The columns y and z can be brought into this form by means of a suitable renumbering of the coordinates (the same for y and z).

$$y=\begin{pmatrix} u \\ o \end{pmatrix},\quad z=\begin{pmatrix} v \\ o \end{pmatrix}\quad (u>o,\ v>o),$$

where the columns u and v are of the same dimension.

Setting

$$A=\begin{pmatrix} A_{11} & A_{12} \\ A_{21} & A_{22} \end{pmatrix},$$

we have

$$\begin{pmatrix} u \\ o \end{pmatrix}+\begin{pmatrix} A_{11} & A_{12} \\ A_{21} & A_{22} \end{pmatrix}\begin{pmatrix} u \\ o \end{pmatrix}=\begin{pmatrix} v \\ o \end{pmatrix};$$

and hence

$$A_{21}u=o.$$

Since $u > o$, it follows that

$$A_{21}=O.$$

This equation contradicts the irreducibility of A.

Thus the lemma is proved.

We introduce the powers of A:

$$A^q=\| a_{ik}^{(q)} \|_1^n \quad (q=1, 2, \ldots).$$

Then the lemma has the following corollary:

COROLLARY: *If $A \geqq O$ is an irreducible matrix, then for every index pair i, k ($1 \leqq i, k \leqq n$) there exists a positive integer q such that*

$$a_{ik}^{(q)}>0. \tag{2}$$

Moreover, q can always be chosen within the bounds

$$\left.\begin{array}{ll} q \leqq m-1 & \text{if } i \neq k, \\ q \leqq m & \text{if } i=k, \end{array}\right\} \tag{3}$$

where m is the degree of the minimal polynomial $\psi(\lambda)$ of A.

For let $r(\lambda)$ denote the remainder on dividing $(\lambda+1)^{n-1}$ by $\psi(\lambda)$. Then by (1) we have $r(A) > O$. Since the degree of $r(\lambda)$ is less than m, it follows from this inequality that for arbitrary i, k ($1 \leqq i, k \leqq n$) at least one of the non-negative numbers

$$\delta_{ik},\ a_{ik},\ a_{ik}^{(2)},\ \ldots,\ a_{ik}^{(m-1)}$$

is not zero. Since $\delta_{ik}=0$ for $i \neq k$, the first of the relations (3) follows.

The other relation (for $i=k$) is obtained similarly if the inequality $r(A) > O$ is replaced by $Ar(A) > O$.[4]

Note. This corollary of the lemma shows that in (1) the number $n-1$ can be replaced by $m-1$.

§ 2. Spectral Properties of Irreducible Non-negative Matrices

1. In 1907 Perron found a remarkable property of the spectra (i.e., the characteristic values and characteristic vectors) of positive matrices.[5]

THEOREM 1 (Perron): *A positive matrix $A=\| a_{ik} \|_1^n$ always has a real and positive characteristic value r which is a simple root of the characteristic equation and exceeds the moduli of all the other characteristic values. To this 'maximal' characteristic value r there corresponds a characteristic vector $z=(z_1, z_2, \ldots, z_n)$ of A with positive coordinates $z_i > 0$ $(i=1, 2, \ldots, n)$.*[6]

A positive matrix is a special case of an irreducible non-negative matrix. Frobenius[7] has generalized Perron's theorem by investigating the spectral properties of irreducible non-negative matrices.

THEOREM 2 (Frobenius): *An irreducible non-negative matrix $A=\| a_{ik} \|_1^n$ always has a positive characteristic value r that is a simple root of the characteristic equation. The moduli of all the other characteristic values do not exceed r. To the 'maximal' characteristic value r there corresponds a characteristic vector with positive coordinates.*

Moreover, if A has h characteristic values $\lambda_0=r, \lambda_1, \ldots, \lambda_{h-1}$ of modulus r, then these numbers are all distinct and are roots of the equation

$$\lambda^h - r^h = 0\,. \tag{4}$$

More generally: The whole spectrum $\lambda_0, \lambda_1, \ldots, \lambda_{n-1}$ of A, regarded as a system of points in the complex λ-plane, goes over into itself under a rotation

---

[4] The product of an irreducible non-negative matrix and a positive matrix is itself positive.

[5] See [316], [317], and [17], p. 100.

[6] Since r is a simple characteristic value, the characteristic vector z belonging to it is determined to within a scalar factor. By Perron's theorem all the coordinates of z are real, different from zero, and of like sign. By multiplying z by -1, if necessary, we can make all its coordinates positive. In the latter case the vector (column) $z=(z_1, z_2, z_3, \ldots, z_n)$ is called *positive* (as in Definition 1).

[7] See [165] and [166].

of the plane by the angle $2\pi/h$. If $h > 1$, then A can be put by means of a permutation into the following 'cyclic' form:

$$A = \begin{pmatrix} O & A_{12} & O & \dots & O \\ O & O & A_{23} & \dots & O \\ . & . & . & . & . \\ O & O & O & \dots & A_{h-1,h} \\ A_{h1} & O & O & \dots & O \end{pmatrix}, \tag{5}$$

where there are square blocks along the main diagonal.

Since Perron's theorem follows as a special case from Frobenius' theorem, we shall only prove the latter.[8] To begin with, we shall agree on some notation.

We write

$$C \leqq D \text{ or } D \geqq C,$$

where C and D are real rectangular matrices of the same dimensions $m \times n$

$$C = \| c_{ik} \|, \ D = \| d_{ik} \| \quad (i = 1, 2, \dots, m; \ k = 1, 2, \dots, n),$$

if and only if

$$c_{ik} \leqq d_{ik} \quad (i = 1, 2, \dots, m; \ k = 1, 2, \dots, n) \,. \tag{6}$$

If the equality sign can be omitted in *all* the inequalities (6), then we shall write

$$C < D \text{ or } D > C.$$

In particular, $C \geqq O$ ($> O$) means that all the elements of C are non-negative (positive).

Furthermore, we denote by C^+ the matrix mod C which arises from C when all the elements are replaced by their moduli.

2. *Proof of Frobenius' Theorem:*[9] Let $x = (x_1, x_2, \dots, x_n)$ $(x \neq o)$ be a fixed real vector. We set:

$$r_x = \min_{1 \leqq i \leqq n} \frac{(Ax)_i}{x_i} \quad \left((Ax)_i = \sum_{k=1}^{n} a_{ik} x_k; \ i = 1, 2, \dots, n\right).$$

In the definition of the minimum we exclude here the values of i for which $x_i = 0$. Obviously $r_x \geqq 0$, and r_x is the largest real number ϱ for which

$$\varrho x \leqq Ax.$$

---

[8] For a direct proof of Perron's theorem see [17], p. 100 ff.

[9] This proof is due to Wielandt [384].

We shall show that the function r_x assumes a maximum value r for some vector $z \geqq o$:

$$r = r_z = \max_{(x \geqq o)} r_x = \max_{(x \geqq o)} \min_{1 \leqq i \leqq n} \frac{(Ax)_i}{x_i}. \tag{7}$$

From the definition of r_x it follows that on multiplication of a vector $x \geqq o$ $(x \neq o)$ by a number $\lambda > 0$ the value of r_x does not change. Therefore, in the computation of the maximum of r_x we can restrict ourselves to the closed set M of vectors x for which

$$x \geqq o \quad \text{and} \quad (xx) \equiv \sum_{i=1}^{n} x_i^2 = 1.$$

If the function r_x were continuous on M, then the existence of a maximum would be guaranteed. However, though continuous at every 'point' $x > o$, r_x may have discontinuities at the boundary points of M at which one of the coordinates vanishes. Therefore, we introduce in place of M the set N of all the vectors y of the form

$$y = (E + A)^{n-1} x \quad (x \in M).$$

The set N, like M, is bounded and closed and by Lemma 1 consists of *positive* vectors only.

Moreover, when we multiply both sides of the inequality

$$r_x x \leqq Ax,$$

by $(E + A)^{n-1} > O$, we obtain:

$$r_x y \leqq Ay \qquad (y = (E + A)^{n-1} x).$$

Hence, from the definition of r_y we have

$$r_x \leqq r_y.$$

Therefore in the computation of the maximum of r_x we can replace M by the set N which consists of positive vectors only. On the bounded and closed set N the function r_x is continuous and therefore assumes a largest value for some vector $z > o$.

Every vector $z \geqq o$ for which

$$r_z = r \tag{8}$$

will be called *extremal*.

We shall now show that: 1) *The number* r *defined by* (7) *is positive and is a characteristic value of* A; 2) *Every extremal vector* z *is positive and is a characteristic vector of* A *for the characteristic value* r, *i.e.,*

$$r>0,\ z>o,\ Az=rz. \tag{9}$$

For if $u=\underbrace{(1, 1, \ldots, 1)}_{n}$, then $r_u=\min\limits_{1\leqq i\leqq n}\sum\limits_{k=1}^{n} a_{ik}$. But then $r_u>0$, because no row of an irreducible matrix can consist of zeros only. Therefore $r>0$, since $r\geqq r_u$. Now let

$$x=(E+A)^{n-1}z. \tag{10}$$

Then, by Lemma 1, $x>o$. Suppose that $Az-rz\neq o$. Then by (1), (8), and (10) we obtain successively:

$$Az-rz\geqq o,\ (E+A)^{n-1}(Az-rz)>o,\ Ax-rx>o.$$

The last inequality contradicts the definition of r, because it would imply that $Ax-(r+\varepsilon)x>o$ for sufficiently small $\varepsilon>0$, i.e., $r_x\geqq r+\varepsilon>r$. Therefore $Az=rz$. But then

$$o<x=(E+A)^{n-1}z=(1+r)^{n-1}z,$$

so that $z>o$.

We shall now show that *the moduli of all the characteristic values do not exceed* r. Let

$$Ay=\alpha y\quad (y\neq o). \tag{11}$$

Taking the moduli of both sides in (11), we obtain:[10]

$$|\alpha|\,y^+\leqq Ay^+. \tag{12}$$

Hence

$$|\alpha|\leqq r_{y^+}\leqq r.$$

Let y be some characteristic vector corresponding to r:

$$Ay=ry\quad (y\neq o).$$

Then setting $\alpha=r$ in (11) and (12) we conclude that y^+ is an extremal vector, so that $y^+>o$, i.e., $y=(y_1, y_2, \ldots, y_n)$, where $y_i\neq 0$ $(i=1, 2, \ldots, n)$. Hence it follows that *only one characteristic direction corresponds to the characteristic value* r; for if there were two linearly independent characteristic vectors z and z_1, we could chose numbers c and d such that the characteristic vector $y=cz+dz_1$ has a zero coordinate, and by what we have shown this is impossible.

---

[10] Regarding the notation y^+, see p. 54.

We now consider the adjoint matrix of the characteristic matrix $\lambda E - A$:

$$B(\lambda) = \| B_{ik}(\lambda) \|_1^n = \Delta(\lambda)(\lambda E - A)^{-1},$$

where $\Delta(\lambda)$ is the characteristic polynomial of A and $B_{ik}(\lambda)$ the algebraic complement of the element $\lambda\delta_{ki} - a_{ki}$ in the determinant $\Delta(\lambda)$. From the fact that only one characteristic vector $z = (z_1, z_2, \ldots, z_n)$ with $z_1 > 0$, $z_2 > 0, \ldots, z_n > 0$ corresponds to the characteristic value r (apart from a factor) it follows that $B(r) \neq O$ and that in every non-zero column of $B(r)$ all the elements are different from zero and are of the same sign. The same is true for the rows of $B(r)$, since in the preceding argument A can be replaced by the transposed matrix A^{T}. From these properties of the rows and columns of A it follows that *all* the $B_{ik}(r)$ $(i, k = 1, 2, \ldots, n)$ are different from zero and are of the same sign σ. Therefore

$$\sigma\Delta'(r) = \sigma \sum_{i=1}^{n} B_{ii}(r) > 0,$$

i.e., $\Delta'(r) \neq 0$ and *r is a simple root of the characteristic equation* $\Delta(\lambda) = 0$.

Since r is the maximal root of $\Delta(\lambda) = \lambda^n + \ldots$, $\Delta(\lambda)$ increases for $\lambda \geqq r$. Therefore $\Delta'(r) > 0$ and $\sigma = 1$, i.e.,

$$B_{ik}(r) > 0 \quad (i, k = 1, 2, \ldots, n). \tag{13}$$

3. Proceeding now to the proof of the second part of Frobenius' theorem, we shall make use of the following interesting lemma:[11]

LEMMA 2: *If $A = \| a_{ik} \|_1^r$ and $C = \| c_{ik} \|_1^n$ are two square matrices of the same order n, where A is irreducible and*[12]

$$C^+ \leqq A, \tag{14}$$

then for every characteristic value γ of C and the maximal characteristic value r of A we have the inequality

$$|\gamma| \leqq r. \tag{15}$$

In the relation (15) *the equality sign holds if and only if*

$$C = e^{i\varphi} DAD^{-1}, \tag{16}$$

where $e^{i\varphi} = \gamma/r$ and D is a diagonal matrix whose diagonal elements are of unit modulus $(D^+ = E)$.

---

[11] See [384].

[12] C is a complex matrix and $A \geqq O$.

Proof. We denote by y a characteristic vector of C corresponding to the characteristic value γ:

$$Cy = \gamma y \quad (\gamma \neq 0). \tag{17}$$

From (14) and (17) we find

$$|\gamma| \, y^+ \leqq C^+ y^+ \leqq Ay^+. \tag{18}$$

Therefore

$$|\gamma| \leqq r_{y^+} \leqq r.$$

Let us now examine the case $|\gamma| = r$ in detail. Here it follows from (18) that y^+ is an extremal vector for A, so that $y^+ > o$ and that y^+ is a characteristic vector of A for the characteristic value r. Therefore the relation (18) assumes the form

$$Ay^+ = C^+y^+ = ry^+, \quad y^+ > o. \tag{19}$$

Hence by (14)

$$C^+ = A. \tag{20}$$

Let $y = (y_1, y_2, \ldots, y_n)$, where

$$y_j = |y_j| \, e^{i\varphi_j} \qquad (j = 1, 2, \ldots, n).$$

We define a diagonal matrix D by the equation

$$D = \{ e^{i\varphi_1}, e^{i\varphi_2}, \ldots, e^{i\varphi_n} \}.$$

Then

$$y = Dy^+.$$

Substituting this expression for y in (17) and then setting $\gamma = re^{i\varphi}$, we find easily:

$$Fy^+ = ry^+, \tag{21}$$

where

$$F = e^{-i\varphi} D^{-1} CD. \tag{22}$$

Comparing (19) with (21), we obtain

$$Fy^+ = C^+y^+ = Ay^+. \tag{23}$$

But by (22) and (20)

$$F^+ = C^+ = A.$$

Therefore we find from (23)

$$Fy^+ = F^+ y^+.$$

Since $y^+ > o$, this equation can hold only if

$$F = F^+,$$

i.e.,

$$e^{-i\varphi} D^{-1} C D = A.$$

Hence

$$C = e^{i\varphi} D A D^{-1},$$

and the Lemma is proved.

4. We return to Frobenius' theorem and apply the lemma to an irreducible matrix $A \geqq O$ that has precisely h characteristic values of maximal modulus r:

$$\lambda_0 = re^{i\varphi_0},\ \lambda_1 = re^{i\varphi_1},\ \ldots,\ \lambda_{h-1} = re^{i\varphi_{h-1}}$$
$$(0 = \varphi_0 < \varphi_1 < \varphi_2 < \cdots < \varphi_{h-1} < 2\pi).$$

Then, setting $C = A$ and $\gamma = \lambda_k$ in the lemma, we have, for every $k = 0, 1, \ldots, h-1$,

$$A = e^{i\varphi_k} D_k A D_k^{-1}, \tag{24}$$

where D_k is a diagonal matrix with $D_k^+ = E$.

Again, let z be a positive characteristic vector of A corresponding to the maximal characteristic value r:

$$Az = rz \qquad (z > o). \tag{25}$$

Then setting

$$\overset{k}{y} = D_k z \qquad \left(\overset{k}{y}{}^+ = z > o\right), \tag{26}$$

we find from (25) and (26):

$$A\overset{k}{y} = \lambda_k \overset{k}{y} \qquad (\lambda_k = re^{i\varphi_k};\quad k = 0, 1, \ldots, h-1). \tag{27}$$

The last equation shows that the vectors $\overset{0}{y}, \overset{1}{y}, \ldots, \overset{h-1}{y}$ defined in (26) are characteristic vectors of A for the characteristic values $\lambda_0, \lambda_1, \ldots, \lambda_{h-1}$.

From (24) it follows not only that $\lambda_0 = r$, but also that each characteristic value $\lambda_1, \ldots, \lambda_{h-1}$ of A is simple. Therefore the characteristic vectors $\overset{k}{y}$ and hence the matrices D_k $(k = 0, 1, \ldots, h-1)$ are determined to within scalar factors. To define the matrices $D_0, D_1, \ldots, D_{h-1}$ uniquely we shall choose their first diagonal element to be 1. Then $D_0 = E$ and $\overset{0}{y} = z > o$.

Furthermore, from (24) it follows that

$$A = e^{i(\varphi_j \pm \varphi_k)} D_j D_k^{\pm 1} A D_k^{\pm 1} D_j^{-1} \quad (j, k = 0, 1, \ldots, h-1).$$

Hence we deduce similarly that the vector

$$D_j D_k^{\pm 1} z$$

is a characteristic vector of A corresponding to the characteristic value $re^{i(\varphi_j \pm \varphi_k)}$.

Therefore $e^{i(\varphi_j \pm \varphi_k)}$ coincides with one of the numbers $e^{i\varphi_l}$ and the matrix $D_j D_k^{\pm 1}$ with the corresponding matrix D_l; that is, we have, for some l_1, l_2 $(0 \leqq l_1, l_2 \leqq h-1)$

$$e^{i(\varphi_j + \varphi_k)} = e^{i\varphi_{l_1}}, \qquad e^{i(\varphi_j - \varphi_k)} = e^{i\varphi_{l_2}},$$
$$D_j D_k = D_{l_1}, \qquad D_j D_k^{-1} = D_{l_2}.$$

Thus: *The numbers $e^{i\varphi_0}, e^{i\varphi_1}, \ldots, e^{i\varphi_{h-1}}$ and the corresponding diagonal matrices $D_0, D_1, \ldots, D_{h-1}$ form two isomorphic multiplicative abelian groups.*

In every finite group consisting of h distinct elements the h-th power of every element is equal to the unit element of the group. Therefore $e^{i\varphi_0}, e^{i\varphi_1}, \ldots, e^{i\varphi_{h-1}}$ are h-th roots of unity. Since there are h such roots of unity and $\varphi_0 = 0 < \varphi_1 < \varphi_2 < \cdots < \varphi_{h-1} < 2\pi$,

$$\varphi_k = \frac{2k\pi}{h} \qquad (k = 0, 1, 2, \ldots, h-1)$$

and

$$e^{i\varphi_k} = \varepsilon^k \qquad (\varepsilon = e^{i\varphi_1} = e^{i\frac{2\pi}{h}}; \quad k = 0, 1, \ldots, h-1), \tag{28}$$

$$\lambda_k = r\varepsilon^k \qquad (k = 0, 1, \ldots, h-1). \tag{29}$$

The numbers $\lambda_0, \lambda_1, \ldots, \lambda_{h-1}$ form a complete system of roots of (4).

In accordance with (28), we have:[14]

$$D_k = D^k \qquad (D = D_1;\ k = 0, 1, \ldots, h-1). \tag{30}$$

The equation (24) now gives us (for $k = 1$):

$$A = e^{i\frac{2\pi}{h}} DAD^{-1}. \tag{31}$$

---

[14] Here we use the isomorphism of the multiplicative groups $e^{i\varphi_0}, e^{i\varphi_1}, \ldots, e^{i\varphi_{h-1}}$ and $D_0, D_1, \ldots, D_{h-1}$.

Hence it follows that the matrix A on multiplication by $e^{i\frac{2\pi}{h}}$ goes over into a similar matrix and, therefore, that *the whole system of n characteristic values of A on multiplication by $e^{i\frac{2\pi}{h}}$ goes over into itself.*[15]

Further,

$$D^h = E,$$

so that all the diagonal elements of D are h-th roots of unity. By a permutation of A (and similarly of D) we can arrange that D be of the following quasi-diagonal form:

$$D = \{ \eta_0 E_0, \eta_1 E_1, \ldots, \eta_{s-1} E_{s-1} \}, \tag{32}$$

where $E_0, E_1, \ldots, E_{s-1}$ are unit matrices and

$$\eta_p = e^{i\psi_p}, \ \psi_p = n_p \frac{2\pi}{h}$$

$$(n_p \text{ is an integer}; p = 0, 1, \ldots, s-1; 0 < n_0 < \ldots < n_{s-1} < h).$$

Obviously $s \leqq h$.

Writing A in block form (in accordance with (32))

$$A = \begin{pmatrix} A_{11} & A_{12} & \ldots & A_{1s} \\ A_{21} & A_{22} & \ldots & A_{2s} \\ \cdot & \cdot & \cdot & \cdot \\ A_{s1} & A_{s2} & \ldots & A_{ss} \end{pmatrix}, \tag{33}$$

we replace (31) by the system of equations

$$\varepsilon A_{pq} = \frac{\eta_{q-1}}{\eta_{p-1}} A_{pq} \qquad \left(p, q = 1, 2, \ldots, s; \ \varepsilon = e^{i\frac{2\pi}{h}}\right) \tag{34}$$

Hence for every p and q either $\frac{\eta_{q-1}}{\eta_{p-1}} = \varepsilon$ or $A_{pq} = O$.

Let us take $p = 1$. Since the matrices $A_{12}, A_{13}, \ldots, A_{1s}$ cannot vanish simultaneously, one of the numbers $\frac{\eta_1}{\eta_0}, \frac{\eta_2}{\eta_0}, \ldots, \frac{\eta_{s-1}}{\eta_0}$ $(\eta_0 = 1)$ must be equal to ε. This is only possible for $n_1 = 1$. Then $\frac{\eta_1}{\eta_0} = \varepsilon$ and $A_{11} = A_{13} = \ldots = A_{1s} = O$. Setting $p = 2$ in (34), we find similarly that $n_2 = 2$ and that $A_{21} = A_{22} = A_{24} = \ldots = A_{2s} = O$, etc. Finally, we obtain

[15] The number h is the largest integer having these properties, because A has precisely h characteristic values of maximal modulus r. Moreover, it follows from (31) that all the characteristic values of the matrix fall into systems (with h numbers in each) of the form $\mu_0, \mu_0\varepsilon, \ldots, \mu_0\varepsilon^{h-1}$ and that within each such system to any two characteristic values there correspond elementary divisors of equal degree. One such system is formed by the roots of the equation (4) $\lambda_0, \lambda_1, \ldots, \lambda_{h-1}$.

$$A=\begin{pmatrix} O & A_{12} & O & \dots & O \\ O & O & A_{23} & \dots & O \\ \cdot & \cdot & \cdot & \cdot & \cdot \\ O & O & O & \dots & A_{s-1,s} \\ A_{s1} & A_{s2} & A_{s3} & \dots & A_{ss} \end{pmatrix}.$$

Here $n_1=1$, $n_2=2$, $\dots$, $n_{s-1}=s-1$. But then for $p=s$ on the right-hand sides of (34) we have the factors

$$\frac{\eta_{q-1}}{\eta_{s-1}}=e^{(q-s)\frac{2\pi}{h}i} \qquad (q=1, 2, \dots, s).$$

One of these numbers must be equal to $\varepsilon=e^{i\frac{2\pi}{h}}$. This is only possible when $s=h$ and $q=1$; consequently, $A_{s2}=\dots=A_{ss}=O$.

Thus,

$$D=\{E_0, \varepsilon E_1, \varepsilon^2 E_2, \dots, \varepsilon^{h-1} E_{h-1}\},$$

and the matrix A has the form (5).

Frobenius' theorem is now completely proved.

5. We now make a few general comments on Frobenius' theorem.

Remark 1. In the proof of Frobenius' theorem we have established incidentally that for an irreducible matrix $A \geqq O$ with the maximal characteristic value r the adjoint matrix $B(\lambda)$ is positive for $\lambda=r$:

$$B(r)>O, \tag{35}$$

i.e.,

$$B_{ik}(r)>0 \qquad (i, k=1, 2, \dots, n), \tag{35'}$$

where $B_{ik}(r)$ is the algebraic complement of the element $r\delta_{ki}-a_{ki}$ in the determinant $|rE-A|$.

Let us now consider the reduced adjoint matrix (see Vol. I, Chapter IV, § 6)

$$C(\lambda)=\frac{B(\lambda)}{D_{n-1}(\lambda)},$$

where $D_{n-1}(\lambda)$ is the greatest common divisor of all the polynomials $B_{ik}(\lambda)$ $(i, k=1, 2, \dots, n)$. It follows from (35') that $D_{n-1}(r)\neq 0$. All the roots of $D_{n-1}(\lambda)$ are characteristic values[16] distinct from r. Therefore all the

[16] $D_{n-1}(\lambda)$ is a divisor of the characteristic polynomial $D_n(\lambda)\equiv|\lambda E-A|$.

roots of $D_{n-1}(\lambda)$ either are complex or are real and less than r. Hence $D_{n-1}(r) > 0$ and this, in conjunction with (35), yields:[17]

$$C(r) = \frac{B(r)}{D_{n-1}(r)} > O. \tag{36}$$

Remark 2. The inequality (35′) enables us to determine bounds for the maximal characteristic value r.

We introduce the notation

$$s_i = \sum_{k=1}^{n} a_{ik} \quad (i = 1, 2, \ldots, n), \qquad s = \min_{1 \leq i \leq n} s_i, \qquad S = \max_{1 \leq i \leq n} s_i.$$

Then: *For an irreducible matrix* $A \geqq O$

$$s \leqq r \leqq S, \tag{37}$$

and the equality sign on the left or the right of r *holds for* $s = S$ *only; i.e., holds only when all the 'row-sums'* $s_1, s_2, \ldots, s_n$ *are equal.*[18]

For if we add to the last column of the characteristic determinant

$$\Delta(r) = \begin{vmatrix} r - a_{11} & -a_{12} & \ldots & -a_{1n} \\ -a_{21} & r - a_{22} & \ldots & -a_{2n} \\ \cdot & \cdot & \cdot & \cdot \\ -a_{n1} & -a_{n2} & \ldots & r - a_{nn} \end{vmatrix}$$

all the preceding columns and then expand the determinant with respect to the elements of the last column, we obtain:

$$\sum_{k=1}^{n} (r - s_k) B_{nk}(r) = 0.$$

Hence (37) follows by (35′).

Remark 3. *An irreducible matrix* $A \geqq O$ *cannot have two linearly independent non-negative characteristic vectors.* For suppose that, apart from the positive characteristic vector $z > o$ corresponding to the maximal characteristic value r, the matrix A has another characteristic vector $y \geqq o$ (linearly independent of z) for the characteristic value α:

---

[17] In the following section it will be shown for an irreducible matrix $B(\lambda) > O$, that $C(\lambda) > O$ for every real $\lambda \geqq r$.

[18] Narrower bounds for r than (s, S) are established in the papers [256], [295] and [119, IV].

$$Ay = \alpha y \quad (y \neq o;\ y \geqq o).$$

Since r is a simple root of the characteristic equation $|\lambda E - A| = 0$,

$$\alpha \neq r.$$

We denote by u the positive characteristic vector of the transposed matrix A^{T} for $\lambda = r$:

$$A^{\mathsf{T}} u = ru \quad (u > o).$$

Then[19]

$$r(y, u) = (y, A^{\mathsf{T}} u) = (Ay, u) = \alpha (y, u);$$

hence, as $\alpha \neq r$,

$$(y, u) = 0,$$

and this is impossible for $u > o$, $y \geqq o$, $y \neq o$.

Remark 4. In the proof of Frobenius' Theorem we have established the following characterization of the maximal characteristic value r of an irreducible matrix $A \geqq O$:

$$r = \max_{(x \geqq o)} r_x,$$

where r_x is the largest number ϱ for which $\varrho x \leqq Ax$. In other words, since $r_x = \min_{1 \leqq i \leqq n} \frac{(Ax)_i}{x_i}$, we have

$$r = \max_{(x \geqq o)} \min_{1 \leqq i \leqq n} \frac{(Ax)_i}{x_i}.$$

Similarly, we can define for every vector $x \geqq o$ $(x \neq o)$ a number r^x as the least number σ for which

$$\sigma x \geqq Ax;$$

i.e., we set

$$r^x = \max_{1 \leqq i \leqq n} \frac{(Ax)_i}{x_i}.$$

If for some i we have here $x_i = 0$, $(Ax)_i \neq 0$, then we shall take $r^x = +\infty$.

As in the case of the function r_x, it turns out here that the function r^x assumes a least value $\widehat{r}$ for some vector $v > o$.

Let us show that the number $\widehat{r}$ defined by

$$\widehat{r} = \min_{(x \geqq o)} r^x = \min_{(x \geqq o)} \max_{1 \leqq i \leqq n} \frac{(Ax)_i}{x_i} \tag{38}$$

[19] If $y = (y_1, y_2, \ldots, y_n)$ and $u = (u_1, u_2, \ldots, u_n)$, then we mean by (y, u) the 'scalar product' $y^{\mathsf{T}} u = \sum_{i=1}^{n} y_i u_i$. Then $(y, A^{\mathsf{T}} u) = y^{\mathsf{T}} A^{\mathsf{T}} u$ and $(Ay, u) = (Ay)^{\mathsf{T}} u = y^{\mathsf{T}} A^{\mathsf{T}} u$.

coincides with r and that the vector $v \geqq o$ $(v \neq o)$ for which this minimum is assumed is a characteristic vector of A for $\lambda = r$.

For,

$$\hat{r}v - Av \geqq o \quad (v \geqq o, \quad v \neq o).$$

Suppose now that the sign $\geqq$ cannot be replaced by the equality sign. Then by Lemma 1

$$(E + A)^{n-1} (\hat{r}v - Av) > o, \quad (E + A)^{n-1} v > o. \tag{39}$$

Setting

$$u = (E + A)^{n-1} v > o,$$

we have

$$\hat{r}u > Au$$

and so for sufficiently small $\varepsilon > 0$

$$(\hat{r} - \varepsilon) u > Au \quad (u > o),$$

which contradicts the definition of $\hat{r}$. Thus

$$Av = \hat{r}v.$$

But then

$$u = (E + A)^{n-1} v = (1 + \hat{r})^{n-1} v.$$

Therefore $u > o$ implies that $v > o$.

Hence, by the Remark 3,

$$\hat{r} = r.$$

Thus we have for r the double characterization:

$$r = \max_{(x \geqq o)} \min_{1 \leqq i \leqq n} \frac{(Ax)_i}{x_i} = \min_{(x \geqq o)} \max_{1 \leqq i \leqq n} \frac{(Ax)_i}{x_i}. \tag{40}$$

Moreover we have shown that $\max\limits_{(x \geqq o)}$ or $\min\limits_{(x \geqq o)}$ is only assumed for a positive characteristic vector for $\lambda = r$.

From this characterization of r we obtain the inequality[20]

$$\min_{1 \leqq i \leqq n} \frac{(Ax)_i}{x_i} \leqq r \leqq \max_{1 \leqq i \leqq n} \frac{(Ax)_i}{x_i} \quad (x \geqq o, \ x \neq o). \tag{41}$$

Remark 5. Since in (40) $\max\limits_{(x \geqq o)}$ and $\min\limits_{(x \geqq o)}$ are only assumed for a positive characteristic vector of the irreducible matrix $A \geqq O$, the inequalities

[20] See [128] and also [17], p. 325 ff.

$$rz \leqq Az,\ z \geqq 0,\ z \neq o$$

or

$$rz \geqq Az,\ z \geqq o,\ z \neq o$$

always imply that

$$Az = rz,\ z > o.$$

§ 3. Reducible Matrices

1. The spectral properties of irreducible non-negative matrices that were established in the preceding section are not preserved when we go over to reducible matrices. However, since every non-negative matrix $A \geqq O$ can be represented as the limit of a sequence of irreducible positive matrices A_m

$$A = \lim_{m \to \infty} A_m \quad (A_m > O,\ m = 1, 2, \ldots), \tag{42}$$

some of the spectral properties of irreducible matrices hold in a weaker form for reducible matrices.

For an arbitrary non-negative matrix $A = \| a_{ik} \|_1^n$ we can prove the following theorem:

THEOREM 3: *A non-negative matrix $A = \| a_{ik} \|_1^n$ always has a non-negative characteristic value r such that the moduli of all the characteristic values of A do not exceed r. To this 'maximal' characteristic value r there corresponds a non-negative characteristic vector*

$$Ay = ry \quad (y \geqq o,\ y \neq o).$$

The adjoint matrix $B(\lambda) = \| B_{ik}(\lambda) \|_1^n = (\lambda E - A)^{-1} \Delta(\lambda)$ satisfies the inequalities

$$B(\lambda) \geqq O, \quad \frac{d}{d\lambda} B(\lambda) \geqq O \quad \textit{for} \quad \lambda \geqq r. \tag{43}$$

Proof. Let A be represented as in (42). We denote by $r^{(m)}$ and $y^{(m)}$ the maximal characteristic value of the positive matrix A_m and the corresponding normalized[21] positive characteristic vector:

$$A_m y^{(m)} = r^{(m)} y^{(m)} \quad ((y^{(m)}, y^{(m)}) = 1,\ y^{(m)} > o;\ m = 1, 2, \ldots). \tag{44}$$

Then it follows from (42) that the limit

$$\lim r^{(m)} = r$$

[21] By a normalized vector we mean a column $y = (y_1, y_2, \ldots, y_n)$ for which $(y, y) \equiv \sum_{i=1}^{n} y_i^2 = 1$.

exists, where r is a characteristic value of A. From the fact that $r^{(m)} > 0$ and $r^{(m)} > | \lambda_0^{(m)} |$, where $\lambda_0^{(m)}$ is an arbitrary characteristic value of A_m $(m = 1, 2, \ldots)$, we obtain by proceeding to the limit:

$$r \geqq 0,\ r \geqq | \lambda_0 |,$$

where λ_0 is an arbitrary characteristic value of A. This passage to the limit gives us in place of (35)

$$B(r) \geqq O. \tag{45}$$

Furthermore, from the sequence of normalized characteristic vectors $y^{(m)}$ $(m = 1, 2, \ldots)$ we can select a subsequence $y^{(m_p)}$ $(p = 1, 2, \ldots)$ that converges to some normalized (and therefore non-zero) vector y. When we go to the limit on both sides of (44) by giving to m the values m_p $(p = 1, 2, \ldots)$ successively, we obtain:

$$Ay = ry \quad (y \geqq o,\ y \neq o).$$

The inequalities (43) will be established by induction on the order n. For $n = 1$, they are obvious.[22] Let us establish them for a matrix $A = \| a_{ik} \|_1^n$ of order n on the assumption that they are true for matrices of order less than n.

Expanding the characteristic determinant $\Delta(\lambda) = | \lambda E - A |$ with respect to the elements of the last row and the last column, we obtain:

$$\Delta(\lambda) = (\lambda - a_{nn}) B_{nn}(\lambda) - \sum_{i,k=1}^{n-1} B_{ki}^{(n)}(\lambda) a_{in} a_{nk}. \tag{46}$$

Here $B_{nn}(\lambda) = | \lambda \delta_{ik} - a_{ik} |_1^{n-1}$ is the characteristic determinant of a 'truncated' non-negative matrix of order $n - 1$, and $B_{ki}^{(n)}(\lambda)$ is the algebraic complement of the element $\lambda \delta_{ik} - a_{ik}$ in $B_{nn}(\lambda)$ $(i, k = 1, 2, \ldots, n-1)$. The maximal non-negative root of $B_{nn}(\lambda)$ will be denoted by r_n. Then setting $\lambda = r_n$ in (46) and observing that by the induction hypothesis

$$B_{ki}^{(n)}(r_n) \geqq O \qquad (i, k = 1, 2, \ldots, n-1),$$

we obtain from (46):

$$\Delta(r_n) \leqq 0.$$

On the other hand $\Delta(\lambda) = \lambda^n + \ldots$, so that $\Delta(+\infty) = +\infty$. Therefore r_n either is a root of $\Delta(\lambda)$ or is less than some real root of $\Delta(\lambda)$. In both cases,

---

[22] For since $B(\lambda) \equiv (\lambda E - A)^{-1} \Delta(\lambda)$, we have $B(\lambda) \equiv E$, $\frac{d}{d\lambda} B(\lambda) \equiv O$ for $n = 1$.

$$r_n \leqq r.$$

Since every principal minor $B_{jj}(\lambda)$ of order $n-1$ can be brought into the position of $B_{nn}(\lambda)$ by a permutation of A, we have

$$r_j \leqq r \qquad (j=1, 2, \ldots, n), \tag{47}$$

where r_j denotes the maximal root of the polynomial $B_{jj}(\lambda)$ $(j=1, 2, \ldots, n)$.

Furthermore, $B_{ik}(\lambda)$ may be represented as a minor of order $n-1$ of the characteristic matrix $\lambda E - A$, multiplied by $(-1)^{i+k}$. When we differentiate this determinant with respect to λ, we obtain:

$$\frac{d}{d\lambda} B_{ik}(\lambda) = \sum B_{ik}^{(j)}(\lambda) \qquad (i, k=1, 2, \ldots, n-1), \tag{48}$$

where $B^{(j)}(\lambda) = \| B_{ik}^{(j)} \|$ $(i \neq j,\ k \neq j;\ j=1, 2, \ldots, n)$ is the adjoint matrix of the matrix $\| a_{ik} \|$ $(i, k=1, 2, \ldots, j-1, j+1, \ldots, n)$ of order $n-1$. But, by the induction hypothesis,

$$B^{(j)}(\lambda) \geqq O \quad \text{for} \quad \lambda \geqq r_j \quad (j=1, 2, \ldots, n);$$

and so, by (47) and (48),

$$\frac{d}{d\lambda} B(\lambda) \geqq O \quad \text{for} \quad \lambda \geqq r. \tag{49}$$

From (45) and (49) it follows that

$$B(\lambda) \geqq O \quad \text{for} \quad \lambda \geqq r.$$

The proof of the theorem is now complete.

Note. In the passage to the limit (42) the inequalities (37) are preserved. They hold, therefore, for an arbitrary non-negative matrix. However, the conditions under which the equality sign holds in (37) are not valid for a reducible matrix.

2. A number of important propositions follow from Theorem 3:

1. *If $A = \| a_{ik} \|_1^n$ is a non-negative matrix with maximal characteristic value r and $C(\lambda)$ is its reduced adjoint matrix, then*

$$C(\lambda) \geqq O \quad \textit{for} \quad \lambda \geqq r. \tag{50}$$

For

$$C(\lambda)=\frac{B(\lambda)}{D_{n-1}(\lambda)}, \tag{51}$$

where $D_{n-1}(\lambda)$ is the greatest common divisor of the elements of $B(\lambda)$. Since $D_{n-1}(\lambda)$ divides the characteristic polynomial $\Delta(\lambda)$ and $D_{n-1}(\lambda) = \lambda^{n-1} + \dots$,

$$D_{n-1}(\lambda) > 0 \quad \text{for} \quad \lambda > r. \tag{52}$$

Now (43), (51), and (52) imply (50).

2. *If $A \geqq O$ is an irreducible matrix with maximal characteristic value r, then*

$$B(\lambda) > O,\ C(\lambda) > O \quad \textit{for} \quad \lambda \geqq r. \tag{53}$$

Indeed, by (35) $B(r) > O$. But also (see (43)) $\frac{d}{d\lambda}B(\lambda) \geqq O$ for $\lambda \geqq r$. Therefore

$$B(\lambda) > O \quad \text{for} \quad \lambda \geqq r. \tag{54}$$

The other of the inequalities (53) follows from (51), (52), and (54).

3. *If $A \geqq O$ is an irreducible matrix with maximal characteristic value r, then*

$$(\lambda E - A)^{-1} > O \quad \textit{for} \quad \lambda > r. \tag{55}$$

This inequality follows from the formula

$$(\lambda E - A)^{-1} = \frac{B(\lambda)}{\Delta(\lambda)},$$

since $B(\lambda) > O$ and $\Delta(\lambda) > 0$ for $\lambda > r$.

4. *The maximal characteristic value r' of every principal minor*[23] *(of order less than n) of a non-negative matrix $A = \| a_{ik} \|_1^n$ does not exceed the maximal characteristic value r of A:*

$$r' \leqq r. \tag{56}$$

If A is irreducible, then the equality sign in (56) cannot occur.

If A is reducible, then the equality sign in (56) holds for at least one principal minor.

---

[23] We mean here by a principal minor the matrix formed from the elements of a principal minor.

For the inequality (56) is true for every principal minor of order $n-1$ (see (47)). If A is irreducible, then by (35′) $B_{jj}(r) > 0$ ($j=1, 2, \ldots, n$) and therefore $r' \neq r$.

By descent from $n-1$ to $n-2$, from $n-2$ to $n-3$, etc., we show the truth of (56) for the principal minors of every order.

If A is a reducible matrix, then by means of a permutation it can be put into the form

$$A=\begin{pmatrix} B & O \\ C & D \end{pmatrix}.$$

Then r must be a characteristic value of one of the two principal minors B and D. This proves Proposition 4.

From 4. we deduce:

5. *If $A \geqq O$ and if in the characteristic determinant*

$$\Delta(r)=\begin{vmatrix} r-a_{11} & -a_{12} & \ldots & -a_{1n} \\ -a_{21} & r-a_{22} & \ldots & -a_{2n} \\ \cdot & \cdot & \cdot & \cdot \\ -a_{n1} & -a_{n2} & \ldots & r-a_{nn} \end{vmatrix}$$

any principal minor vanishes (A is reducible!), then every 'augmented' principal minor also vanishes; in particular, so does one of the principal minors of order $n-1$

$$B_{11}(\lambda),\ B_{22}(\lambda),\ \ldots,\ B_{nn}(\lambda).$$

From 4. and 5. we deduce:

6. *A matrix $A \geqq O$ is reducible if and only if in one of the relations*

$$B_{ii}(r) \geqq 0 \qquad (i=1, 2, \ldots, n)$$

the equality sign holds.

From 4. we also deduce:

7. *If r is the maximal characteristic value of a matrix $A \geqq O$, then for every $\lambda > r$ all the principal minors of the characteristic matrix $A_\lambda \equiv \lambda E - A$ are positive:*

$$A_\lambda\begin{pmatrix} i_1 & i_2 & \ldots & i_p \\ i_1 & i_2 & \ldots & i_p \end{pmatrix}>0 \quad (\lambda>r;\ 1\leqq i_1<i_2<\cdots<i_p\leqq n;\ p=1, 2, \ldots, n). \tag{57}$$

It is easy to see that, conversely, (57) implies that $\lambda > r$. For

$$\Delta(\lambda+\mu) = |(\lambda+\mu)E - A| = |A_\lambda + \mu E| = \sum_{k=0}^{n} S_k \mu^{n-k},$$

where S_k is the sum of all the principal minors of order k of the characteristic matrix $A_\lambda \equiv \lambda E - A$ $(k = 1, 2, \ldots, n)$.[24] Therefore, if for some real λ all the principal minors of A_λ are positive, then for $\mu \geqq 0$

$$\Delta(\lambda+\mu) \neq 0,$$

i.e., no number greater than λ is a characteristic value of A. Therefore

$$r < \lambda.$$

Thus, (57) is a necessary and sufficient condition for λ to be an upper bound for the moduli of the characteristic values of A.[25] However, the inequalities (57) are not all independent.

The matrix $\lambda E - A$ is a matrix with non-positive elements outside the main diagonal.[26] D. M. Kotelyanskiĭ has proved that for such matrices, just as for symmetric matrices, all the principal minors are positive, provided the successive principal minors are positive.[27]

LEMMA 3 (Kotelyanskiĭ): *If in a real matrix $G = \| g_{ik} \|_1^n$ all the non-diagonal elements are negative or zero*

$$g_{ik} \leqq 0 \qquad (i \neq k;\, i, k = 1, 2, \ldots, n) \tag{58}$$

and the successive principal minors are positive

$$g_{11} = G\begin{pmatrix}1\\1\end{pmatrix} > 0,\ G\begin{pmatrix}1 & 2\\1 & 2\end{pmatrix} > 0,\ \ldots,\ G\begin{pmatrix}1 & 2 & \ldots & n\\1 & 2 & \ldots & n\end{pmatrix} > 0, \tag{59}$$

then all the principal minors are positive:

$$G\begin{pmatrix}i_1 & i_2 & \ldots & i_p\\i_1 & i_2 & \ldots & i_p\end{pmatrix} > 0 \qquad (1 \leq i_1 < i_2 < \cdots < i_p \leq n;\ p = 1, 2, \ldots, n).$$

---

[24] See Vol. I, p. 70.

[25] See [344].

[26] It is easy to see that, conversely, every matrix with negative or zero non-diagonal elements can be represented in the form $\lambda E - A$, where A is a non-negative matrix and λ is a real number.

[27] See [215]. This paper contains a number of results about matrices in which all the non-diagonal elements are of like sign.

Proof. We shall prove the lemma by induction on the order n of the matrix. For $n=2$ the lemma holds, since it follows from

$$g_{12} \leqq 0 \quad g_{21} \leqq 0, \quad g_{11} > 0, \quad g_{11}g_{22} - g_{12}g_{21} > 0$$

that $g_{22} > 0$. Let us assume now that the lemma is true for matrices of order less than n; we shall then prove it for $G = \| g_{ik} \|_1^n$. We consider the bordered determinants

$$t_{ik} = G\begin{pmatrix} 1 & i \\ 1 & k \end{pmatrix} = g_{11}g_{ik} - g_{1k}g_{i1} \qquad (i, k = 2, \ldots, n).$$

From (58) and (59) it follows that

$$t_{ik} \leqq 0 \qquad (i \neq k;\ i, k = 2, \ldots, n).$$

On the other hand, by applying Sylvester's identity (Vol. I, Chapter II, (30), p. 33) to the matrix $T = \| t_{ik} \|_2^n$, we obtain:

$$T\begin{pmatrix} i_1 & i_2 & \ldots & i_p \\ i_1 & i_2 & \ldots & i_p \end{pmatrix} = (g_{11})^{p-1} G\begin{pmatrix} 1 & i_1 & i_2 & \ldots & i_p \\ 1 & i_1 & i_2 & \ldots & i_p \end{pmatrix} \qquad \begin{pmatrix} 2 \leqq i_1 < i_2 < \cdots < i_p \leqq n, \\ p = 1, 2, \ldots, n-1 \end{pmatrix}. \tag{60}$$

Hence it follows by (59) that the successive principal minors of the matrix $T = \| t_{ik} \|_2^n$ are positive:

$$t_{22} = T\begin{pmatrix} 2 \\ 2 \end{pmatrix} > 0, \ T\begin{pmatrix} 2 & 3 \\ 2 & 3 \end{pmatrix} > 0, \ \ldots, \ T\begin{pmatrix} 2 & 3 & \ldots & n \\ 2 & 3 & \ldots & n \end{pmatrix} > 0.$$

Thus, the matrix $T = \| t_{ik} \|_2^n$ of order $n-1$ satisfies the condition of the lemma. Therefore by the induction hypothesis all the principal minors are positive:

$$T\begin{pmatrix} i_1 & i_2 & \ldots & i_p \\ i_1 & i_2 & \ldots & i_p \end{pmatrix} > 0 \quad (2 \leqq i_1 < i_2 < \cdots < i_p \leqq n;\ p = 1, 2, \ldots, n-1).$$

But then it follows from (60) that all the principal minors of G containing the first row are positive:

$$G\begin{pmatrix} 1 & i_1 & i_2 & \ldots & i_p \\ 1 & i_1 & i_2 & \ldots & i_p \end{pmatrix} > 0 \quad (2 \leqq i_1 < i_2 < \cdots < i_p \leqq n;\ p = 1, 2, \ldots, n-1). \tag{61}$$

Let us choose fixed indices $i_1, i_2, \ldots, i_{n-2}$ (where $1 < i_1 < i_2 < \ldots < i_{n-2} \leqq n$) and form the matrix of order $n - 1$:

$$\| g_{\alpha\beta} \| \qquad (\alpha, \beta = 1, i_1, i_2, \ldots, i_{n-2}). \tag{62}$$

The successive principal minors of this matrix are positive, by (61):

$$g_{11} > 0,\ G\begin{pmatrix} 1 & i_1 \\ 1 & i_1 \end{pmatrix} > 0,\ \ldots,\ G\begin{pmatrix} 1 & i_1 & i_2 & \ldots & i_{n-2} \\ 1 & i_1 & i_2 & \ldots & i_{n-2} \end{pmatrix} > 0;$$

and the non-diagonal elements are non-positive:

$$g_{\alpha\beta} \leqq 0 \qquad (\alpha \neq \beta;\ \alpha, \beta = 1, i_1, i_2, \ldots, i_{n-2}).$$

But the order of (62) is $n - 1$. Therefore, by the induction hypothesis, all the principal minors of this matrix are positive; in particular,

$$G\begin{pmatrix} i_1 & i_2 & \ldots & i_p \\ i_1 & i_2 & \ldots & i_p \end{pmatrix} > 0 \tag{63}$$
$$(2 \leqq i_1 < i_2 < \cdots < i_p \leqq n;\ p = 1, 2, \ldots, n-2).$$

Thus, *all* the minors of G of order not exceeding $n - 2$ are positive.

Since by (63) $g_{22} > 0$, we may now consider the determinants of order two bordering the element g_{22} (and not g_{11} as before):

$$t^*_{ik} = G\begin{pmatrix} 2 & i \\ 2 & k \end{pmatrix} \qquad (i, k = 1, 3, \ldots, n).$$

By operating with the matrix $T^* = \| t^*_{ik} \|$, as we have done above with T, we obtain inequalities analogous to (61):

$$G\begin{pmatrix} 2 & i_1 & \ldots & i_p \\ 2 & i_1 & \ldots & i_p \end{pmatrix} > 0 \tag{64}$$
$$(i_1 < i_2 < \cdots < i_p;\ \ i_1, \ldots, i_p = 1, 3, \ldots, n;\ \ p = 1, 2, \ldots, n-1).$$

Since every principal minor of $G = \| g_{ik} \|_1^n$ contains either the first or the second row or is of order not exceeding $n - 2$, it follows from (61), (63), and (64) that all the principal minors of A are positive. This completes the proof of the lemma.

This lemma allows us to retain only the successive principal minors in the condition (57) and to formulate the following theorem:

---

[28] See [344] and [215]. Since $C = A - \lambda E$ and $A \geqq O$, λ_n is real (this follows from $\lambda_n + \lambda = r$) and the corresponding characteristic vector of C is non-negative: $Cy = \lambda_n y$ $(y \geqq o,\ y \neq o)$.

Theorem 4: *A real number λ is greater than the maximal characteristic value r of the matrix $A = \| a_{ik} \|_1^n \geqq O$*

$$r < \lambda$$

if and only if for this value λ all the successive principal minors of the characteristic matrix $A_\lambda \equiv \lambda E - A$ are positive:

$$\begin{vmatrix} \lambda - a_{11} & -a_{12} \\ -a_{21} & \lambda - a_{22} \end{vmatrix} > 0, \ \ldots, \ \begin{vmatrix} \lambda - a_{11} & -a_{12} & \ldots & -a_{1n} \\ -a_{21} & \lambda - a_{22} & \ldots & -a_{2n} \\ \cdot & \cdot & \cdot & \cdot \\ -a_{n1} & -a_{n2} & \ldots & \lambda - a_{nn} \end{vmatrix} > 0. \qquad (65)$$

Let us consider one application of Theorem 4. Suppose that in the matrix $C = \| c_{ik} \|_1^n$ all the non-diagonal elements are non-negative. Then for some $\lambda > 0$ we have $A = C + \lambda E \geqq O$. We arrange the characteristic values λ_i $(i = 1, 2, \ldots, n)$ of C with their real parts in ascending order:

$$\operatorname{Re} \lambda_1 \leqq \operatorname{Re} \lambda_2 \leqq \ldots \leqq \operatorname{Re} \lambda_n.$$

We denote by r the maximal characteristic value of A. Since the characteristic values of A are the sums $\lambda_i + \lambda$ $(i = 1, 2, \ldots, n)$, we have

$$\lambda_n + \lambda = r.$$

In this case the inequality $r < \lambda$ holds for $\lambda_n < 0$ only, and signifies that all the characteristic values of C have negative real parts. When we write down the inequality (65) for the matrix $-C = \lambda E - A$, we obtain the following theorem:

Theorem 5: *The real parts of all the characteristic values of a real matrix $C = \| c_{ik} \|_1^n$ with non-negative non-diagonal elements*

$$c_{ik} \geqq 0 \quad (i \neq k;\ i, k = 1, 2, \ldots, n)$$

are negative if and only if

$$c_{11} < 0, \ \begin{vmatrix} c_{11} & c_{12} \\ c_{21} & c_{22} \end{vmatrix} > 0, \ \ldots, \ (-1)^n \begin{vmatrix} c_{11} & c_{12} & \ldots & c_{1n} \\ c_{21} & c_{22} & \ldots & c_{2n} \\ \cdot & \cdot & \cdot & \cdot \\ c_{n1} & c_{n2} & \ldots & c_{nn} \end{vmatrix} > 0. \qquad (66)$$

§ 4. The Normal Form of a Reducible Matrix

1. We consider an arbitrary reducible matrix $A = \| a_{ik} \|_1^n$. By means of a permutation we can put it into the form

$$A = \begin{pmatrix} B & O \\ C & D \end{pmatrix}, \tag{67}$$

where B and D are square matrices.

If one of the matrices B or D is reducible, then it can also be represented in a form similar to (67), so that A then assumes the form

$$A = \begin{pmatrix} K & O & O \\ H & L & O \\ F & G & M \end{pmatrix}.$$

If one of the matrices K, L, M is reducible, then the process can be continued. Finally, by a suitable permutation we can reduce A to triangular block form

$$A = \begin{pmatrix} A_{11} & O & \dots & O \\ A_{21} & A_{22} & \dots & O \\ \cdot & \cdot & \cdot & \cdot \\ A_{s1} & A_{s2} & \dots & A_{ss} \end{pmatrix}, \tag{68}$$

where the diagonal blocks are square irreducible matrices.

A diagonal block A_{ii} $(1 \leqq i \leqq s)$ is called *isolated* if

$$A_{ik} = O \quad (k = 1, 2, \dots, i-1, i+1, \dots, s).$$

By a permutation of the blocks (see p. 50) in (68) we can put all the isolated blocks in the first places along the main diagonal, so that A then assumes the form

$$A = \begin{pmatrix} A_1 & O & \dots O & O & \dots O \\ O & A_2 & \dots O & O & \dots O \\ \cdot & \cdot & \cdot & \cdot & \cdot \\ O & O & \dots A_g & O & \dots O \\ A_{g+1,1} & A_{g+1,2} & \dots A_{g+1,g} & A_{g+1} & \dots O \\ \cdot & \cdot & \cdot & \cdot & \cdot \\ A_{s1} & A_{s2} & \dots A_{sg} & A_{s,g+1} & \dots A_s \end{pmatrix}; \tag{69}$$

here $A_1, A_2, \dots, A_s$ are irreducible matrices, and in each row

$$A_{f1}, A_{f2}, \dots, A_{f,f-1} \quad (f = g+1, \dots, s)$$

at least one matrix is different from zero.

We shall call the matrix (69) the *normal form* of the reducible matrix A.

Let us show that *the normal form of a matrix* A *is uniquely determined to within a permutation of the blocks and permutations within the diagonal blocks* (the same for rows and columns).[29] For this purpose we consider the operator $\boldsymbol{A}$ corresponding to A in an n-dimensional vector space $\boldsymbol{R}$. To the representation of A in the form (69) there corresponds a decomposition of $\boldsymbol{R}$ into coordinate subspaces

$$\boldsymbol{R} = \boldsymbol{R}_1 + \boldsymbol{R}_2 + \dots + \boldsymbol{R}_g + \boldsymbol{R}_{g+1} + \dots + \boldsymbol{R}_s; \tag{70}$$

here $\boldsymbol{R}_s, \boldsymbol{R}_{s-1} + \boldsymbol{R}_s, \boldsymbol{R}_{s-2} + \boldsymbol{R}_{s-1} + \boldsymbol{R}_s, \dots$ are invariant coordinate subspaces for $\boldsymbol{A}$, and there is no intermediate invariant subspace between any two adjacent ones in this sequence.

Suppose then that apart from the normal form (69) of the given matrix there is another normal form corresponding to another decomposition of $\boldsymbol{R}$ into coordinate subspaces:

$$\boldsymbol{R} = \widehat{\boldsymbol{R}}_1 + \widehat{\boldsymbol{R}}_2 + \dots + \widehat{\boldsymbol{R}}_g + \widehat{\boldsymbol{R}}_{g+1} + \dots + \widehat{\boldsymbol{R}}_t. \tag{71}$$

The uniqueness of the normal form will be proved if we can show that the decompositions (70) and (71) coincide apart from the order of the terms.

Suppose that the invariant subspace $\widehat{\boldsymbol{R}}_t$ has coordinate vectors in common with $\boldsymbol{R}_k$, but not with $\boldsymbol{R}_{k+1}, \dots, \boldsymbol{R}_s$. Then $\widehat{\boldsymbol{R}}_t$ must be entirely contained in $\boldsymbol{R}_k$, since otherwise $\widehat{\boldsymbol{R}}_t$ would contain a 'smaller' invariant subspace, the intersection of $\widehat{\boldsymbol{R}}_t$ with $\boldsymbol{R}_k + \boldsymbol{R}_{k+1} + \dots + \boldsymbol{R}_s$. Moreover, $\widehat{\boldsymbol{R}}_t$ must coincide with $\boldsymbol{R}_k$, since otherwise the invariant subspace $\widehat{\boldsymbol{R}}_t + \boldsymbol{R}_{k+1} + \dots + \boldsymbol{R}_s$ would be intermediate between $\boldsymbol{R}_k + \boldsymbol{R}_{k+1} + \dots + \boldsymbol{R}_s$ and $\boldsymbol{R}_{k+1} + \dots + \boldsymbol{R}_s$. Since $\boldsymbol{R}_k$ coincides with $\widehat{\boldsymbol{R}}_t$, $\boldsymbol{R}_k$ is an invariant subspace. Therefore, without infringing the normal form of the matrix, $\boldsymbol{R}_k$ can be put in the place of $\boldsymbol{R}_s$. Thus, we may assume that in (70) and (71) $\boldsymbol{R}_s \equiv \widehat{\boldsymbol{R}}_t$.

Let us now consider the coordinate subspace $\widehat{\boldsymbol{R}}_{t-1}$. Suppose that it has coordinate vectors in common with $\boldsymbol{R}_l$ ($l < s$), but not with $\boldsymbol{R}_{l+1}, \dots, \boldsymbol{R}_s$. Then the invariant subspace $\widehat{\boldsymbol{R}}_{t-1} + \widehat{\boldsymbol{R}}_t$ must be entirely contained in $\boldsymbol{R}_l + \boldsymbol{R}_{l+1} + \dots + \boldsymbol{R}_s$, since otherwise there would be an invariant coordinate subspace intermediate between $\widehat{\boldsymbol{R}}_t$ and $\widehat{\boldsymbol{R}}_{t-1} + \widehat{\boldsymbol{R}}_t$. Therefore $\widehat{\boldsymbol{R}}_{t-1} \subset \boldsymbol{R}_l$. Moreover $\widehat{\boldsymbol{R}}_{t-1} \equiv \boldsymbol{R}_l$, since otherwise $\widehat{\boldsymbol{R}}_{t-1} + \boldsymbol{R}_{l+1} + \dots + \boldsymbol{R}_s$ would be an invariant subspace intermediate between $\boldsymbol{R}_l + \boldsymbol{R}_{l+1} + \dots + \boldsymbol{R}_s$ and $\boldsymbol{R}_{l+1} +$

[29] Without violating the normal form we can permute the first g blocks arbitrarily among each other. Moreover, sometimes certain permutations among the last $s - g$ blocks are possible with preservation of the normal form.

$\ldots + \boldsymbol{R}_s$. From $\widehat{\boldsymbol{R}}_{t-1} \equiv \boldsymbol{R}_l$ it follows that $\boldsymbol{R}_l + \boldsymbol{R}_s$ is an invariant subspace. Therefore $\boldsymbol{R}_l$ may be put in the place of $\boldsymbol{R}_{s-1}$ and then we have

$$\widehat{\boldsymbol{R}}_{t-1} \equiv \boldsymbol{R}_{s-1}, \quad \widehat{\boldsymbol{R}}_t \equiv \boldsymbol{R}_s.$$

Continuing this process, we finally reach the conclusion that $s = t$ and that the decompositions (70) and (71) coincide apart from the order of the terms. The corresponding normal forms then coincide to within a permutation of the blocks.

From the uniqueness of the normal form it follows that the numbers g and s are invariants of the non-negative matrix A.[30]

2. Making use of the normal form, we shall now prove the following theorem:

THEOREM 6: *To the maximal characteristic value r of the matrix $A \geqq O$ there belongs a positive characteristic vector if and only if in the normal form* (69) *of A:* 1) *each of the matrices $A_1, A_2, \ldots, A_g$ has r as a characteristic value; and* (*in case $g < s$*) 2) *none of the matrices $A_{g+1}, \ldots, A_s$ has this property.*

Proof. 1. Let $z > o$ be a positive characteristic vector belonging to the maximal characteristic value r. In accordance with the dissection into blocks in (69) we dissect the column z into parts z^k ($k = 1, 2, \ldots, s$). Then the equation

$$Az = rz \qquad (z > o) \tag{72}$$

is replaced by two systems of equations

$$A_i z^i = r z^i \qquad (i = 1, 2, \ldots, g), \tag{72'}$$

$$\sum_{h=1}^{j-1} A_{jh} z^h + A_j z^j = r z^j \qquad (j = g+1, \ldots, s). \tag{72''}$$

From (72′) it follows that r is a characteristic value of each of the matrices $A_1, A_2, \ldots, A_g$. From (72″) we find:

$$A_j z^j \leqq r z^j, \quad A_j z^j \neq r z^j \qquad (j = g+1, \ldots, s). \tag{73}$$

We denote by r_j the maximal characteristic value of A_j ($j = g+1, \ldots, s$). Then (see (41) on p. 65) we find from (73):

$$r_j \leqq \max_i \frac{(A_j z^j)_i}{z_i^j} \leqq r \qquad (j = g+1, \ldots, s).$$

[30] For an irreducible matrix, $g = s = 1$.

On the other hand, the equation $r_j = r$ would contradict the second of the relations (73) (see Note 5 on p. 65). Therefore

$$r_j < r \quad (j = g+1, \ldots, s). \tag{74}$$

2. Suppose now, conversely, that the maximal characteristic values of the matrices A_i $(i = 1, 2, \ldots, g)$ are equal to r and that (74) holds for the matrices A_j $(j = g+1, \ldots, s)$. Then by replacing the required equation (72) by the systems (72′), (72″) we can define positive characteristic columns z^i of the matrices A_i $(i = 1, 2, \ldots, g)$ by means of (72′). Next we find columns z^j $(j = g+1, \ldots, s)$ from (72″):

$$z^j = (rE_j - A_j)^{-1} \sum_{h=1}^{j-1} A_{jh} z^h \quad (j = g+1, \ldots, s), \tag{75}$$

where E_j is the unit matrix of the same order as A_j $(j = g+1, \ldots, s)$.

Since $r_j < r$ $(j = g+1, \ldots, s)$, we have (see (55) on p. 69)

$$(rE_j - A_j)^{-1} > O \quad (j = g+1, \ldots, s). \tag{76}$$

Let us prove by induction that the columns $z^{g+1}, \ldots, z^s$ defined by (75) are positive. We shall show that for every j $(g+1 \leqq j \leqq s)$ the fact that $z^1, z^2, \ldots, z^{j-1}$ are positive implies that $z^j > o$. Indeed, in this case,

$$\sum_{h=1}^{j-1} A_{jh} z^h \geqq o, \quad \sum_{h=1}^{j-1} A_{jh} z^h \neq o,$$

which in conjunction with (76) yields, by (75):

$$z^j > o.$$

Thus, the positive column $z = (z^1, \ldots, z^s)$ is a characteristic vector of A for the characteristic value r. This completes the proof of the theorem.

3. The following theorem gives a characterization of a matrix $A \geqq O$ which together with its transpose A^{T} has the property that a positive characteristic vector belongs to the maximal characteristic value.

THEOREM 7:[31] *To the maximal characteristic value r of a matrix $A \geqq O$ there belongs a positive characteristic vector both of A and of A^{T} if and only if A can be represented by a permutation in quasi-diagonal form*

$$A = \{A_1, A_2, \ldots, A_s\}, \tag{77}$$

where $A_1, A_2, \ldots, A_s$ are irreducible matrices each of which has r as its maximal characteristic value.

---

[31] See [166].

Proof. Suppose that A and A^{T} have positive characteristic vectors for $\lambda = r$. Then, by Theorem 6, A is representable in the normal form (69), where $A_1, A_2, \ldots, A_g$ have r as maximal characteristic value and (for $g < s$) the maximal characteristic values of $A_{g+1}, \ldots, A_s$ are less than r. Then

$$A^{\mathsf{T}} = \begin{pmatrix} A_1^{\mathsf{T}} & \ldots & 0 & A_{g+1,1}^{\mathsf{T}} & \ldots & A_{s1}^{\mathsf{T}} \\ \vdots & \ddots & & & & \\ 0 & \ldots & A_g^{\mathsf{T}} & A_{g+1,g}^{\mathsf{T}} & \ldots & A_{sg}^{\mathsf{T}} \\ 0 & \ldots & 0 & A_{g+1}^{\mathsf{T}} & & \\ \vdots & & \vdots & & \ddots & \\ 0 & \ldots & 0 & 0 & \ldots & A_s^{\mathsf{T}} \end{pmatrix}.$$

Let us reverse here the order of the blocks in this matrix:

$$\begin{pmatrix} A_s^{\mathsf{T}} & 0 & 0 \ldots 0 \\ A_{s,s-1}^{\mathsf{T}} & A_{s-1}^{\mathsf{T}} & 0 \ldots 0 \\ \vdots & \ddots & \vdots \\ A_{s1}^{\mathsf{T}} & A_{s-1,1}^{\mathsf{T}} & \ldots A_1^{\mathsf{T}} \end{pmatrix}. \tag{78}$$

Since $A_s^{\mathsf{T}}, A_{s-1}^{\mathsf{T}}, \ldots, A_1^{\mathsf{T}}$ are irreducible, we obtain a normal form for (78) by a permutation of the blocks, placing the isolated blocks first along the main diagonal. One of these isolated blocks is A_s^{T}. Since the normal form of A^{T} must satisfy the conditions of the preceding theorem, the maximal characteristic value of A_s^{T} must be equal to r. This is only possible when $g = s$. But then the normal form (69) goes over into (77).

If, conversely, a representation (77) of A is given, then

$$A^{\mathsf{T}} = \{A_1^{\mathsf{T}}, A_2^{\mathsf{T}}, \ldots, A_s^{\mathsf{T}}\}. \tag{79}$$

We then deduce from (77) and (79), by the preceding theorem, that A and A^{T} have positive characteristic vectors for the maximal characteristic value r.

This proves the theorem.

COROLLARY. *If the maximal characteristic value r of a matrix $A \geqq O$ is simple and if positive characteristic vectors belong to r both in A and A^{T}, then A is irreducible.*

Since, conversely, every irreducible matrix has the properties of this corollary, these properties provide a spectral characterization of an irreducible non-negative matrix.

§ 5. Primitive and Imprimitive Matrices

1. We begin with a classification of irreducible matrices.

DEFINITION 3: *If an irreducible matrix $A \geqq O$ has h characteristic values $\lambda_1, \lambda_2, \ldots, \lambda_h$ of maximal modulus r $(\lambda_1 = |\lambda_2| = \ldots = |\lambda_h| = r)$, then A is called primitive if $h = 1$ and imprimitive if $h > 1$. h is called the index of imprimitivity of A.*

The index of imprimitivity h is easily determined if the coefficients of the characteristic equation of the matrix are known

$$\Delta(\lambda) \equiv \lambda^n + a_1 \lambda^{n_1} + a_2 \lambda^{n_2} + \cdots + a_t \lambda^{n_t} = 0$$
$$(n > n_1 > \cdots > n_t;\ a_1 \neq 0,\ a_2 \neq 0, \ldots, a_t \neq 0)\,;$$

namely: *h is the greatest common divisor of the differences*

$$n - n_1,\ n_1 - n_2, \ldots, n_{t-1} - n_t. \tag{80}$$

For by Frobenius' theorem the spectrum of A in the complex λ-plane goes over into itself under a rotation through $2\pi/h$ around the point $\lambda = 0$. Therefore the polynomial $\Delta(\lambda)$ must be obtained from some polynomial $g(\mu)$ by the formula

$$\Delta(\lambda) = g(\lambda^h)\,\lambda^{n'}.$$

Hence it follows that h is a common divisor of the differences (80). But then h is the greatest common divisor d of these differences, since the spectrum does not change under a rotation by $2\pi/d$, which is impossible for $h < d$.

The following theorem establishes an important property of a primitive matrix:

THEOREM 8: *A matrix $A \geqq O$ is primitive if and only if some power of A is positive:*

$$A^p > O \qquad (p \geqq 1). \tag{81}$$

Proof. If $A^p > O$, then A is irreducible, since the reducibility of A would imply that of A^p. Moreover, for A we have $h = 1$, since otherwise the positive matrix A^p would have h (> 1) characteristic values

$$\lambda_1^p, \lambda_2^p, \ldots, \lambda_h^p$$

of maximal modulus r^p, and this contradicts Perron's theorem.

Suppose now, conversely, that A is primitive. We apply the formula (23) of Chapter V (Vol. I, p. 107) to A^p

$$A^p = \sum_{k=1}^{s} \frac{1}{(m_k-1)!} \left[\frac{C(\lambda)\lambda^p}{\overset{k}{\psi}(\lambda)} \right]^{(m_k-1)}_{\lambda=\lambda_k}, \tag{82}$$

where

$$\psi(\lambda) = (\lambda-\lambda_1)^{m_1}(\lambda-\lambda_2)^{m_2}\cdots(\lambda-\lambda_s)^{m_s} \quad (\lambda_j \neq \lambda_f \text{ for } j \neq f)$$

is the minimal polynomial of A, $\overset{k}{\psi}(\lambda) = \frac{\psi(\lambda)}{(\lambda-\lambda_k)^{m_k}}$ $(k=1, 2, \ldots, s)$ and $C(\lambda) = (\lambda E - A)^{-1}\psi(\lambda)$ is the reduced adjoint matrix.

In this case, we can set:

$$\lambda_1 = r > |\lambda_2| \geqq \cdots \geqq |\lambda_s| \quad \text{and} \quad m_1 = 1. \tag{83}$$

Then (82) assumes the form

$$A^p = \frac{C(r)}{\psi'(r)} r^p + \sum_{k=2}^{s} \frac{1}{(m_k-1)!} \left[\frac{C(\lambda)\lambda^p}{\overset{k}{\psi}(\lambda)} \right]^{(m_k-1)}_{\lambda=\lambda_k}.$$

Hence it is easy to deduce by (83) that

$$\lim_{p\to\infty} \frac{A^p}{r^p} = \frac{C(r)}{\psi'(r)}. \tag{84}$$

On the other hand, $C(r) > O$ (see (53)) and $\psi'(r) > 0$ by (83). Therefore

$$\lim_{p\to\infty} \frac{A^p}{r^p} > O,$$

and so (73) must hold from some p onwards.[32] This completes the proof.

We shall now prove the following theorem:

THEOREM 9: *If $A \geqq O$ is an irreducible matrix and some power A^q of A is reducible, then A^q is completely reducible,* i.e., *A^q can be represented by means of a permutation in the form*

$$A^q = \{A_1, A_2, \ldots, A_d\}, \tag{85}$$

where $A_1, A_2, \ldots, A_d$ are irreducible matrices having one and the same maximal characteristic value. Here d is the greatest common divisor of q and h, where h is the index of imprimitivity of A.

---

[32] As regards a lower bound for the exponent p in (81), see [384].

Proof. Since A is irreducible, we know by Frobenius' theorem that positive characteristic vectors belong to the maximal characteristic value r, both in A and in A^{T}. But then these positive vectors are also characteristic vectors of the non-negative matrices A^q and $(A^q)^{\mathsf{T}}$ for the characteristic value $\lambda = r^q$. Therefore by applying Theorem 7 to A^q, we represent this matrix (after a suitable permutation) in the form (85), where $A_1, A_2, \ldots, A_d$ are irreducible matrices with the same maximal characteristic value r^q. But A has h characteristic values of maximal modulus r:

$$r, r\varepsilon, \ldots, r\varepsilon^{h-1} \quad \left(\varepsilon = e^{\frac{2\pi i}{h}}\right).$$

Therefore A^q also has h characteristic values of maximal modulus

$$r^q, r^q\varepsilon^q, \ldots, r^q\varepsilon^{q(h-1)},$$

among which d are equal to r^q. This is only possible when d is the greatest common divisor of q and h. This proves the theorem.

For $h = 1$, we obtain:

COROLLARY 1: *A power of a primitive matrix is irreducible and primitive.*

If we set $q = h$ in the theorem, then we obtain:

COROLLARY 2: *If A is an imprimitive matrix with index of imprimitivity h, then A^h splits into h primitive matrices with the same maximal characteristic value.*

§ 6. Stochastic Matrices

1. We consider n possible states of a certain system

$$S_1, S_2, \ldots, S_n \tag{86}$$

and a sequence of instants

$$t_0, t_1, t_2, \ldots.$$

Suppose that at each of these instants the system is in one and only one of the states (86) and that p_{ij} denotes the probability of finding the system in the state S_j at the instant t_k if it is known that at the preceding instant t_{k-1} the system is in the state S_i ($i, j = 1, 2, \ldots, n$; $k = 1, 2, \ldots$). We shall assume that the *transition probability* p_{ij} ($i, j = 1, 2, \ldots, n$) does not depend on the index k (of the instant t_k).

If the matrix of transition probabilities is given,

$$P = \| p_{ij} \|_1^n,$$

then we say that we have a *homogeneous Markov chain with a finite number of states.*[33] It is obvious that

$$p_{ij} \geqq 0, \quad \sum_{j=1}^{n} p_{ij} = 1 \qquad (i, j = 1, 2, \ldots, n). \tag{87}$$

DEFINITION 4: *A square matrix* $P = \| p_{ij} \|_1^n$ *is called stochastic if* P *is non-negative and if the sum of the elements of each row of* P *is* 1, i.e., *if the relations* (87) *hold.*[34]

Thus, for every homogeneous Markov chain the matrix of transition probabilities is stochastic and, conversely, every stochastic matrix can be regarded as the matrix of transition probabilities of some homogeneous Markov chain. This is the basis of the matrix method of investigating homogeneous Markov chains.[35]

A stochastic matrix is a special form of a non-negative matrix. Therefore all the concepts and propositions of the preceding sections are applicable to it.

We mention some specific properties of a stochastic matrix. From the definition of a stochastic matrix it follows that it has the characteristic value 1 with the positive characteristic vector $z = (1, 1, \ldots, 1)$. It is easy to see that, conversely, every matrix $P \geqq O$ having the characteristic vector $(1, 1, \ldots, 1)$ for the characteristic value 1 is stochastic. Moreover, 1 is the maximal characteristic value of a stochastic matrix, since the maximal characteristic value is always included between the largest and the smallest of the row sums[36] and in a stochastic matrix all the row sums are 1. Thus, we have proved the proposition:

1. *A non-negative matrix* $P \geqq O$ *is stochastic if and only if it has the characteristic vector* $(1, 1, \ldots, 1)$ *for the characteristic value* 1. *For a stochastic matrix the maximal characteristic value is* 1.

Now let $A = \| a_{ik} \|_1^n$ be a non-negative matrix with a positive maximal characteristic value $r > 0$ and a corresponding positive characteristic vector $z = (z_1, z_2, \ldots, z_n) > o$:

---

[33] See [212] and [46], pp. 9-12.

[34] Sometimes the additional condition $\sum_{i=1}^{n} p_{ij} \neq 0$ $(j = 1, 2, \ldots, n)$ is included in the definition of a stochastic matrix. See [46], p. 13.

[35] The theory of homogeneous Markov chains with a finite (and a countable) number of states was introduced by Kolmogorov (see [212]). The reader can find an account of the later introduction and development of the matrix method with applications to homogeneous Markov chains in the memoir [329] and in the monograph [46] by V. I. Romanovskiĭ (see also [4], Appendix 5).

[36] See (37) and the note on p. 68.

$$\sum_{j=1}^{n} a_{ij} z_j = r z_i \qquad (i = 1, 2, \ldots, n). \tag{88}$$

We introduce the diagonal matrix $Z = \{z_1, z_2, \ldots, z_n\}$ and the matrix $P = \| p_{ij} \|_1^n$

$$P = \frac{1}{r} Z^{-1} A Z.$$

Then

$$p_{ij} = \frac{1}{r} z_i^{-1} a_{ij} z_j \geqq 0 \qquad (i, j = 1, 2, \ldots, n),$$

and by (88)

$$\sum_{j=1}^{n} p_{ij} = 1 \qquad (i = 1, 2, \ldots, n).$$

Thus:

2. *A non-negative matrix $A \geqq O$ with the maximal positive characteristic value $r > 0$ and with a corresponding positive characteristic vector $z = (z_1, z_2, \ldots z_n) > o$ is similar to the product of r and a stochastic matrix:*[37]

$$A = Z r P Z^{-1} \qquad (Z = \{z_1, z_2, \ldots, z_n\} > O). \tag{89}$$

In a preceding section we have given (see Theorem 6, § 4) a characterization of the class of non-negative matrices having a positive characteristic vector for $\lambda = r$. The formula (89) establishes a close connection between this class and the class of stochastic matrices.

2. We shall now prove the following theorem:

THEOREM 10: *To the characteristic value 1 of a stochastic matrix there always correspond only elementary divisors of the first degree.*

Proof. We apply the decomposition (69), § 4, to the stochastic matrix $P = \| p_{ij} \|_1^n$.

$$P = \begin{pmatrix} A_1 & O & \cdots & & & & O \\ O & A_2 & \cdots & & & & O \\ \vdots & & & & & & \vdots \\ O & \cdots & & A_g & O & \cdots & O \\ A_{g+1,1} & \cdots & & A_{g+1,g} & A_{g+1} & \cdots & O \\ \vdots & & & & & \ddots & \vdots \\ A_{s1} & \cdots & & A_{sg} & \cdots & & A_s \end{pmatrix},$$

where $A_1, A_2, \ldots, A_s$ are irreducible and

---

[37] Proposition 2. also holds for $r = 0$, since $A \geqq O$, $z > o$ implies that $A = O$.

$$A_{f1}+A_{f2}+\cdots+A_{f,f-1}\neq O \qquad (f=g+1,\ldots,s).$$

Here $A_1, A_2, \ldots, A_g$ are *stochastic* matrices, so that each has the simple characteristic value 1. As regards the remaining irreducible matrices $A_{g+1}, \ldots, A_s$, by the Remark 2 on p. 63 their maximal characteristic values are less than 1, since in each of these matrices at least one row sum is less than 1.[38]

Thus, the matrix P is representable in the form

$$P=\begin{pmatrix} Q_1 & O \\ S & Q_2 \end{pmatrix},$$

where in Q_1 to the value 1 there correspond elementary divisors of the first degree, and where 1 is not a characteristic value of Q_2. The theorem now follows immediately from the following lemma:

LEMMA 4: *If a matrix A has the form*

$$A=\begin{pmatrix} Q_1 & O \\ S & Q_2 \end{pmatrix}, \tag{90}$$

where Q_1 and Q_2 are square matrices, and if the characteristic value λ_0 of A is also a characteristic value of Q_1, but not of Q_2,

$$|Q_1-\lambda_0 E|=0, \qquad |Q_2-\lambda_0 E|\neq 0,$$

then the elementary divisors of A and Q_1 corresponding to the characteristic value λ_0 are the same.

Proof. 1. To begin with, we consider the case where Q_1 and Q_2 do not have characteristic values in common. Let us show that in this case the elementary divisors of Q_1 and Q_2 together form the system of elementary divisors of A, i.e., for some matrix T ($|T|\neq 0$)

$$TAT^{-1}=\begin{pmatrix} Q_1 & O \\ O & Q_2 \end{pmatrix}. \tag{91}$$

We shall look for the matrix T in the form

$$T=\begin{pmatrix} E_1 & O \\ U & E_2 \end{pmatrix}$$

[38] These properties of the matrices $A_1, \ldots, A_s$ also follow from Theorem 6.

(the dissection of T into blocks corresponds to that of A; E_1 and E_2 are unit matrices). Then

$$TAT^{-1}=\begin{pmatrix}E_1 & O\\ U & E_2\end{pmatrix}\begin{pmatrix}Q_1 & O\\ S & Q_2\end{pmatrix}\begin{pmatrix}E_1 & O\\ -U & E_2\end{pmatrix}=\begin{pmatrix}Q_1 & O\\ UQ_1-Q_2U+S & Q_2\end{pmatrix}. \quad (91')$$

The equation (91′) reduces to (91) if we choose the rectangular matrix U so that it satisfies the matrix equation

$$Q_2U-UQ_1=S.$$

If Q_1 and Q_2 have no characteristic values in common, then this equation always has a unique solution for every right-hand side S (see Vol. I, Chapter VIII, § 3).

2. In the case where Q_1 and Q_2 have characteristic values in common, we replace Q_1 in (90) by its Jordan form J (as a result, A is replaced by a similar matrix). Let $J=\{J_1 J_2\}$, where all the Jordan blocks with the characteristic value λ_0 are combined in J_1. Then

$$A=\begin{pmatrix}J_1 & O & O & O\\ O & J_2 & O & O\\ S_{11} & S_{12} & \multicolumn{2}{c}{\multirow{2}{*}{Q_2}}\\ S_{21} & S_{22} & & \end{pmatrix}=\begin{pmatrix}J_1 & O\ O\ O\\ O & \\ S_{11} & \widehat{Q}_2\\ S_{21} & \end{pmatrix}.$$

This matrix falls under the preceding case, since the matrices J_1 and $\widehat{Q}_2$ have no characteristic values in common. Hence it follows that the elementary divisors of the form $(\lambda-\lambda_0)^p$ are the same for A and J_1 and therefore also for A and Q_1. This proves the lemma.

If an irreducible stochastic matrix P has a complex characteristic value λ_0 with $|\lambda_0|=1$, then $\lambda_0 P$ is similar to P (see (16)) and so it follows from Theorem 10 that to λ_0 there correspond only elementary divisors of the first degree. With the help of the normal form and of Lemma 4 it is easy to extend this statement to reducible stochastic matrices. Thus we obtain:

COROLLARY 1. *If λ_0 is a characteristic value of a stochastic matrix P and $|\lambda_0|=1$, then the elementary divisors corresponding to λ_0 are of the first degree.*

From Theorem 10 we also deduce by 2. (p. 84):

COROLLARY 2. *If a positive characteristic vector belongs to the maximal characteristic value r of a non-negative matrix A, then all the elementary divisors of A that belong to a characteristic value λ_0 with $|\lambda_0|=r$ are of the first degree.*

We shall now mention some papers that deal with the distribution of the characteristic values of stochastic matrices.

A characteristic value of a stochastic matrix P always lies in the disc $|\lambda| \leqq 1$ of the λ-plane. The set of all points of this disc that are characteristic values of any stochastic matrices of order n will be denoted by M_n.

3. In 1938, in connection with investigation on Markov chains A. N. Kolmogorov raised the problem of determining the structure of the domain M_n. This problem was partially solved in 1945 by N. A. Dmitriev and E. B. Dynkin [133], [133a] and completely in 1951 in a paper by F. I. Karpelevich [209]. It turned out that the boundary of M_n consists of a finite number of points on the circle $|\lambda| = 1$ and certain curvilinear arcs joining these points in cyclic order.

We note that by Proposition 2. (p. 84) the characteristic values of the matrices $A = \| a_{ik} \|_1^n \geqq O$ having a positive characteristic vector for $\lambda = r$ with a fixed r form the set $r \cdot M_n$.[39] Since every matrix $A = \| a_{ik} \|_1^n \geqq O$ can be regarded as the limit of a sequence of non-negative matrices of that type and the set $r \cdot M_n$ is closed, the characteristic values of arbitrary matrices $A = \| a_{ik} \|_1^n \geqq O$ with a given maximal characteristic value r fill out the set $r \cdot M_n$.[40]

A paper by H. R. Suleimanova [359] is relevant in this context; it contains sufficiency criteria for n given real numbers $\lambda_1, \lambda_2, \ldots, \lambda_n$ to be the characteristic values of a stochastic matrix $P = \| p_{ij} \|_1^n$.[41]

§ 7. Limiting Probabilities for a Homogeneous Markov Chain with a Finite Number of States

1. Let

$$S_1, S_2, \ldots, S_n$$

be all the possible states of a system in a homogeneous Markov chain and let $P = \| p_{ij} \|_1^n$ be the stochastic matrix determined by this chain that is formed from the transition probabilities p_{ij} $(i, j = 1, 2, \ldots, n)$ (see p. 82).

We denote by $p_{ij}^{(q)}$ the probability of finding the system in the state S_j at the instant t_k if it is known that at the instant t_{k-q} it is in the state S_i $(i, j = 1, 2, \ldots, n;\ q = 1, 2, \ldots)$. Clearly, $p_{ij}^{(1)} = p_{ij}$ $(i, j = 1, 2, \ldots, n)$.

---

[39] $r \cdot M_n$ is the set of points in the λ-plane of the form $r\mu$, where $\mu \in M_n$.

[40] Kolmogorov has shown (see [133a (1946)], Appendix) that this problem for an arbitrary matrix $A \geqq O$ can be reduced to the analogous problem for a stochastic matrix.

[41] See also [312].

Making use of the theorems on the addition and multiplication of probabilities, we find easily:

$$p_{ij}^{(q+1)} = \sum_{h=1}^{n} p_{ih}^{(q)} p_{hj} \qquad (i, j = 1, 2, \ldots, n)$$

or, in matrix notation,

$$\| p_{ij}^{(q+1)} \| = \| p_{ij}^{(q)} \|_1^n \, \| p_{ij} \|_1^n .$$

Hence, by giving to q in succession the values 1, 2, ..., we obtain the important formula[42]

$$\| p_{ij}^{(q)} \| = P^q \qquad (q = 1, 2, \ldots).$$

If the limits

$$\lim_{q \to \infty} p_{ij}^{(q)} = p_{ij}^{\infty} \qquad (i, j = 1, 2, \ldots, n)$$

or, in matrix notation,

$$\lim_{q \to \infty} P^q = P^{\infty} = \| p_{ij}^{\infty} \|_1^n ,$$

exist, then the values p_{ij}^{∞} $(i, j = 1, 2, \ldots, n)$ are called the *limiting* or *final transition probabilities.*[42]

In order to investigate under what conditions limiting transition probabilities exist and to derive the corresponding formulas, we introduce the following terminology.

We shall call a stochastic matrix P and the corresponding homogeneous Markov chain *regular* if P has no characteristic values of modulus 1 other than 1 itself and *fully regular* if, in addition, 1 is a simple root of the characteristic equation of P.

A regular matrix P is characterized by the fact that in its normal form (69) (p. 75) the matrices $A_1, A_2, \ldots, A_g$ are primitive. For a fully regular matrix we have, in addition, $g = 1$.

Furthermore, a homogeneous Markov chain is *irreducible, reducible, acyclic* or *cyclic* if the stochastic matrix P of the chain is irreducible, reducible, primitive, or imprimitive, respectively. Just as a primitive stochastic matrix is a special form of a regular matrix, so an acyclic Markov chain is a special form of a regular chain.

We shall prove that: *Limiting transition probabilities exist for regular homogeneous Markov chains only.*

---

[42] It follows from this formula that the probabilities $p_{ij}^{(q)}$ as well as p_{ij} $(i, j = 1, 2, 3, \ldots, n;\ q = 1, 2, \ldots)$ do not depend on the index k of the original instant t_k.

[43] The matrix P^{∞}, as the limit of stochastic matrices, is itself stochastic.

For let $\psi(\lambda)$ be the minimal polynomial of the regular matrix $P = \| p_{ij} \|_1^n$. Then

$$\psi(\lambda) = (\lambda - \lambda_1)^{m_1} (\lambda - \lambda_2)^{m_2} \cdots (\lambda - \lambda_u)^{m_u} \quad (\lambda_i \neq \lambda_k;\ i, k = 1, 2, \ldots, u). \tag{92}$$

By Theorem 10 we may assume that

$$\lambda_1 = 1, \quad m_1 = 1. \tag{93}$$

By the formula (23) of Chapter V (Vol. I, p. 107),

$$P^q = \frac{C(1)}{\overset{1}{\psi}(1)} + \sum_{k=2}^{u} \frac{1}{(m_k - 1)!} \left[\frac{C(\lambda)}{\overset{k}{\psi}(\lambda)} \lambda^q \right]_{\lambda = \lambda_k}^{(m_k - 1)}, \tag{94}$$

where $C(\lambda) = (\lambda E - P)^{-1} \psi(\lambda)$ is the reduced adjoint matrix and

$$\overset{k}{\psi}(\lambda) = \frac{\psi(\lambda)}{(\lambda - \lambda_k)^{m_k}} \qquad (k = 1, 2, \ldots, u);$$

moreover

$$\overset{1}{\psi}(\lambda) = \frac{\psi(\lambda)}{\lambda - 1} \quad \text{and} \quad \overset{1}{\psi}(1) = \psi'(1).$$

If P is a regular matrix, then

$$|\lambda_k| < 1 \qquad (k = 2, 3, \ldots, u),$$

and therefore all the terms on the right-hand side of (94), except the first, tend to zero for $q \to \infty$. Therefore, for a regular matrix P the matrix P^∞ formed from the limiting transition probabilities exists, and

$$P^\infty = \frac{C(1)}{\psi'(1)}. \tag{95}$$

The converse proposition is obvious. If the limit

$$P^\infty = \lim_{q \to \infty} P^q \tag{96}$$

exists, then the matrix P cannot have any characteristic value λ_k for which $\lambda_k \neq 1$ and $|\lambda_k| = 1$, since then the limit $\lim_{q \to \infty} \lambda_k^q$ would not exist. (This limit must exist, since the limit (96) exists.)

We have proved that the matrix P^∞ exists for a regular homogeneous Markov chain (and for such a regular chain only). This matrix is determined by (95).

We shall now show that P^∞ can be expressed by the characteristic polynomial

$$\Delta(\lambda) = (\lambda - \lambda_1)^{n_1} (\lambda - \lambda_2)^{n_2} \cdots (\lambda - \lambda_u)^{n_u} \tag{97}$$

and the adjoint matrix $B(\lambda) = (\lambda E - P)^{-1} \Delta(\lambda)$.

From the identity

$$\frac{B(\lambda)}{\Delta(\lambda)} = \frac{C(\lambda)}{\psi(\lambda)}$$

it follows by (92), (93), and (97) that

$$\frac{n_1 B^{(n_1-1)}(1)}{\Delta^{(n_1)}(1)} = \frac{C(1)}{\psi'(1)}.$$

Therefore (95) may be replaced by the formula

$$P^\infty = \frac{n_1 B^{(n_1-1)}(1)}{\Delta^{(n_1)}(1)}. \tag{98}$$

For a fully regular Markov chain, inasmuch as it is a special form of a regular chain, the matrix P^∞ exists and is determined by (95) or (98). In this case $n_1 = 1$, and (98) assumes the form

$$P^\infty = \frac{B(1)}{\Delta'(1)}. \tag{99}$$

2. Let us consider a regular chain of general type (not fully regular). We write the corresponding matrix P in the normal form

$$P = \begin{pmatrix} Q_1 & \cdots & O & O & \cdots & O \\ \vdots & \ddots & \vdots & & \ddots & \vdots \\ O & & Q_g & O & & O \\ U_{g+1,1} & \cdots & U_{g+1,g} & Q_{g+1} & & \vdots \\ \vdots & & \vdots & & \ddots & \vdots \\ U_{s1} & \cdots & U_{sg} & \cdots & U_{s,s-1} & Q_s \end{pmatrix}, \tag{100}$$

where $Q_1, \ldots, Q_g$ are primitive stochastic matrices and the maximal values of the irreducible matrices $Q_{g+1}, \ldots, Q_s$ are less than 1. Setting

$$U = \begin{pmatrix} U_{g+1,1} & \cdots & U_{g+1,g} \\ \cdot & \cdots & \cdot \\ U_{s1} & \cdots & U_{sg} \end{pmatrix}, \quad W = \begin{pmatrix} Q_{g+1} & \cdots & O \\ \cdot & \cdots & \cdot \\ U_{s,g+1} & \cdots & Q_s \end{pmatrix},$$

we write P in the form

$$P = \begin{pmatrix} Q_1 & \cdots & O & O \\ \vdots & \ddots & \vdots & \vdots \\ O & \cdots & Q_g & O \\ & U & & W \end{pmatrix}.$$

Then

$$P^q = \begin{pmatrix} Q_1^q & \cdots & O & O \\ \vdots & \ddots & \vdots & \vdots \\ O & \cdots & Q_g^q & O \\ & U^q & & W^q \end{pmatrix} \tag{101}$$

and

$$P^\infty = \lim_{q \to \infty} P^q = \begin{pmatrix} Q_1^\infty & \cdots & O & O \\ \vdots & \ddots & \vdots & \vdots \\ O & \cdots & Q_g^\infty & O \\ & U_\infty & & W^\infty \end{pmatrix}.$$

But $W^\infty = \lim_{q \to \infty} W^q = O$, because all the characteristic values of W are of modulus less than 1. Therefore

$$P^\infty = \begin{pmatrix} Q_1^\infty & \cdots & O & O \\ \vdots & \ddots & \vdots & \vdots \\ O & \cdots & Q_g^\infty & O \\ & U_\infty & & O \end{pmatrix}. \tag{102}$$

Since $Q_1, \ldots, Q_g$ are primitive stochastic matrices, the matrices $Q_1^\infty, \ldots, Q_g^\infty$ are positive, by (99) and (35) (p. 62)

$$Q_1^\infty > O, \ldots, Q_g^\infty > O,$$

and in each of these matrices all the elements belonging to any one column are equal:

$$Q_h^\infty = \| q_{*j}^{(h)} \|_{i,j=1}^n \quad (h=1, 2, \ldots, g).$$

We note that the states $S_1, S_2, \ldots, S_n$ of the system fall into groups corresponding to the normal form (100) of P:

$$\Sigma_1, \Sigma_2, \ldots, \Sigma_g, \Sigma_{g+1}, \ldots, \Sigma_s. \tag{103}$$

To each group Σ in (103) there corresponds a group of rows in (100). In the terminology of Kolmogorov the states of the system that occur in $\Sigma_1, \Sigma_2, \ldots, \Sigma_g$ are called *essential* and the states that occur in the remaining groups $\Sigma_{g+1}, \ldots, \Sigma_s$ *non-essential.*

From the form (101) of P^q it follows that in any finite number q of steps (from the instant t_{k-q} to t_k) only the following transitions of the system are possible: a) from an essential state to an essential state of the same group; b) from a non-essential state to an essential state; and c) from a non-essential state to a non essential state of the same or a preceding group.

From the form (102) of P^∞ it follows that: *A limiting transition can only lead from an arbitrary state to an essential state,* i.e., the probability of transition to any non-essential state tends to zero when the number of steps q tends to infinity. The essential states are therefore sometimes also called *limiting states.*

3. From (95) it follows that[45]

$$(E-P)\,P^\infty = O.$$

Hence it is clear that: *Every column of P^∞ is a characteristic vector of the stochastic matrix P for the characteristic value $\lambda=1$.*

For a fully regular matrix P, 1 is a simple root of the characteristic equation and (apart from scalar factors) only one characteristic vector $(1, 1, \ldots, 1)$ of P belongs to it. Therefore all the elements of the j-th column of P^∞ are equal to one and the same non-negative number p_{*j}^∞:

$$p_{ij}^\infty = p_{*j}^\infty \geqq 0 \quad (j=1, 2, \ldots, n;\ \sum_{j=1}^n p_{*j} = 1). \tag{104}$$

---

[44] See [212] and [46], pp. 37-39.

[45] This formula holds for an arbitrary regular chain and can be obtained from the obvious equation $P^q - P \cdot P^{q-1} = O$ by passing to the limit $q \to \infty$.

Thus, in a fully regular chain the limiting transition probabilities do not depend on the initial state.

Conversely, if in a regular homogeneous Markov chain the limiting transition probabilities do not depend on the initial state, i.e., if (104) holds, then obviously in the scheme (102) for P^∞ we have $g=1$. But then $n_1=1$ and the chain is fully regular.

For an acyclic chain, which is a special case of a fully regular chain, P is a primitive matrix. Therefore $P^q > O$ (see Theorem 8 on p. 80) for some $q > 0$. But then also $P^\infty = P^\infty P^q > O$.[46]

Conversely, it follows from $P^\infty > O$ that $P^q > O$ for some $q > 0$, and this means by Theorem 8 that P is primitive and hence that the given homogeneous Markov chain is acyclic.

We formulate these results in the following theorem:

THEOREM 11: 1. *In a homogeneous Markov chain all the limiting transition probabilities exist if and only if the chain is regular. In that case the matrix P^∞ formed from the limiting transition probabilities is determined by* (95) *or* (98).

2. *In a regular homogeneous Markov chain the limiting transition probabilities are independent of the initial state if and only if the chain is fully regular. In that case the matrix P^∞ is determined by* (99).

3. *In a regular homogeneous Markov chain all the limiting transition probabilities are different from zero if and only if the chain is acyclic.*[47]

4. We now consider the columns of *absolute probabilities*

$$\overset{k}{p} = (\overset{k}{p}_1, \overset{k}{p}_2, \ldots, \overset{k}{p}_n) \qquad (k=0, 1, 2, \ldots), \tag{105}$$

where $\overset{k}{p}_i$ is the probability of finding the system in the state S_i ($i=1, 2, \ldots, n$; $k=0, 1, 2, \ldots$) at the instant t_k. Making use of the theorems on the addition and multiplication of probabilities, we find:

$$\overset{k}{p}_i = \sum_{h=1}^{n} \overset{0}{p}_h p_{hi}^{(k)} \qquad (i=1, 2, \ldots, n;\ k=1, 2, \ldots)$$

or, in matrix notation,

---

[46] This matrix equation is obtained by passing to the limit $m \to \infty$ from the equation $P^m = P^{m-q} \cdot P^q$ $(m > q)$. P^∞ is a stochastic matrix; therefore $P^\infty \geqq O$ and there are non-zero elements in every row of P^∞. Hence $P^\infty P^q > O$. Instead of Theorem 8 we can use here the formula (99) and the inequality (35) (p. 62).

[47] Note that $P^\infty > O$ implies that the chain is acyclic and therefore regular. Hence it follows automatically from $P^\infty > O$ that the limiting transition probabilities do not depend on the initial state, i.e., that the formulas (104) hold.

$$\overset{k}{p} = (P^{\mathsf{T}})^k \overset{0}{p} \qquad (k = 1, 2, \ldots), \tag{106}$$

where P^{T} is the transpose of P.

All the absolute probabilities (105) can be determined from (106) if the initial probabilities $\overset{0}{p}_1, \overset{0}{p}_2, \ldots, \overset{0}{p}_n$ and the matrix of transition probabilities $P = \| p_{ij} \|_1^n$ are known.

We introduce the *limiting absolute probabilities*

$$\overset{\infty}{p}_i = \lim_{k \to \infty} \overset{k}{p}_i \qquad (i = 1, 2, \ldots, n)$$

or

$$\overset{\infty}{p} = (\overset{\infty}{p}_1, \overset{\infty}{p}_2, \ldots, \overset{\infty}{p}_n) = \lim_{k \to \infty} \overset{k}{p}.$$

When we take the limit $k \to \infty$ on both sides of (106), we obtain:

$$\overset{\infty}{p} = (P^{\infty})^{\mathsf{T}} \overset{0}{p}. \tag{107}$$

Note that the existence of the matrix of limiting transition probabilities P^{∞} implies the existence of the limiting absolute probabilities

$$\overset{\infty}{p} = (\overset{\infty}{p}_1, \overset{\infty}{p}_2, \ldots, \overset{\infty}{p}_n)$$

for arbitrary initial probabilities $\overset{0}{p} = (\overset{0}{p}_1, \overset{0}{p}_2, \ldots, \overset{0}{p}_n)$, and vice versa.

From the formula (107) and the form (102) of P^{∞} it follows that: *The limiting absolute probabilities corresponding to non-essential states are zero.*

Multiplying both sides of the matrix equation

$$P^{\mathsf{T}} \cdot (P^{\infty})^{\mathsf{T}} = (P^{\infty})^{\mathsf{T}}$$

by $\overset{0}{p}$ on the right, we obtain by (107):

$$P^{\mathsf{T}} \overset{\infty}{p} = \overset{\infty}{p}, \tag{108}$$

i.e.: *The column of limiting absolute probabilities $\overset{\infty}{p}$ is a characteristic vector of P^{T} for the characteristic value $\lambda = 1$.*

If a fully regular Markov chain is given, then $\lambda = 1$ is a simple root of the characteristic equation of P^{T}. In this case, the column of limiting absolute probabilities is uniquely determined by (108) (because $\overset{\infty}{p}_i \geqq 0$ $(j = 1, 2, \ldots, n)$ and $\sum_{j=1}^{n} \overset{\infty}{p}_j = 1$).

Suppose that a fully regular Markov chain is given. Then it follows from (104) and (107) that:

$$\overset{\infty}{p} - \sum_{h=1}^{n} \overset{0}{p}_h p_{hj}^{\infty} = p_{*j}^{\infty} \sum_{h=1}^{n} \overset{0}{p}_h = p_{*j}^{\infty} \qquad (j = 1, 2, \ldots, n). \tag{109}$$

In this case the limiting absolute probabilities $\overset{\infty}{p}_1, \overset{\infty}{p}_2, \ldots, \overset{\infty}{p}_n$ do not depend on the initial probabilities $\overset{0}{p}_1, \overset{0}{p}_2, \ldots, \overset{0}{p}_n$.

Conversely, $\overset{\infty}{p}$ is independent of $\overset{0}{p}$ on account of (107) if and only if all the rows of P^{∞} are equal, i.e.,

$$p_{hj}^{\infty} = p_{*j}^{\infty} \quad (h, j = 1, 2, \ldots, n)$$

so that (by Theorem 11) P is a fully regular matrix.

If P is primitive, then $P^{\infty} > O$ and hence, by (109),

$$\overset{\infty}{p}_j > 0 \qquad (j = 1, 2, \ldots, n).$$

Conversely, if all the $\overset{\infty}{p}_j$ $(j = 1, 2, \ldots, n)$ are positive and do not depend on the initial probabilities, then all the elements in every column of P^{∞} are equal and by (109) $P^{\infty} > O$, and this means by Theorem 11 that P is primitive, i.e., that the given chain is acyclic.

From these remarks it follows that Theorem 11 can also be formulated as follows:

THEOREM 11′: 1. *In a homogeneous Markov chain all the limiting absolute probabilities exist for arbitrary initial probabilities if and only if the chain is regular.*

2. *In a homogeneous Markov chain the limiting absolute probabilities exist for arbitrary initial probabilities and are independent of them if and only if the chain is fully regular.*

3. *In a homogeneous Markov chain positive limiting absolute probabilities exist for arbitrary initial probabilities and are independent of them if and only if the chain is acyclic.*[48]

5. We now consider a homogeneous Markov chain of general type with a matrix P of transition probabilities.

---

[48] The second part of Theorem 11′ is sometimes called the *ergodic theorem* and the first part the *general quasi-ergodic theorem* for homogeneous Markov chains (see [4], pp. 473 and 476).

We choose the normal form (69) for P and denote by $h_1, h_2, \ldots, h_g$ the indices of imprimitivity of the matrices $A_1, A_2, \ldots, A_g$ in (69). Let h be the least common multiple of the integers $h_1, h_2, \ldots, h_g$. Then the matrix P^h has no characteristic values, other than 1, of modulus 1, i.e., P^h is regular; here h is the least exponent for which P^h is regular. We shall call h the *period* of the given homogeneous Markov chain.

Since P^h is regular, the limit

$$\lim_{q\to\infty} P^{hq} = (P^h)^\infty$$

exists and hence the limits

$$P_r = \lim_{q\to\infty} P^{r+qh} = P^r (P^h)^\infty \qquad (r = 0, 1, \ldots, h-1)$$

also exist.

Thus, in general, the sequence of matrices

$$P, P^2, P^3, \ldots$$

splits into h subsequences with the limits $P_r = P^r(P^h)^\infty$ $(r=0, 1, \ldots, h-1)$.

When we go from the transition probabilities to the absolute probabilities by means of (106), we find that the sequence

$$\overset{1}{p}, \overset{2}{p}, \overset{3}{p}, \ldots$$

splits into h subsequences with the limits

$$\lim_{q\to\infty} \overset{r+qh}{p} = (P^{\mathsf{T}h})^\infty \overset{r}{p} \quad (r = 0, 1, 2, \ldots, h-1).$$

For an arbitrary homogeneous Markov chain with a finite number of states the limits of the arithmetic means always exist:

$$\tilde{P} = \lim_{N\to\infty} \frac{1}{N} \sum_{k=1}^{N} P^k = \frac{1}{h} (E + P + \cdots + P^{h-1}) (P^h)^\infty \tag{110}$$

and

$$\tilde{p} = \lim_{N\to\infty} \frac{1}{N} \sum_{k=1}^{N} \overset{k}{p} = P^{\mathsf{T}} \overset{0}{p}. \tag{110'}$$

Here $\tilde{P} = \| \tilde{p}_{ij} \|_1^n$ and $\tilde{p} = (\tilde{p}_1, \tilde{p}_2, \ldots, \tilde{p}_n)$. The values $\tilde{p}_{ij}$ $(i, j = 1, 2, 3, \ldots, n)$ and $\tilde{p}_j$ $(j = 1, 2, \ldots, n)$ are called the *mean limiting transition probabilities* and *mean limiting absolute probabilities*, respectively.

Since

$$\lim_{N\to\infty} \frac{1}{N}\sum_{k=2}^{N+1} P^k = \lim_{N\to\infty} \frac{1}{N}\sum_{k=1}^{N} \tilde{P}^k,$$

we have

$$\tilde{P}P = \tilde{P}$$

and therefore, by (110′),

$$P^{\mathsf{T}}\tilde{p} = \tilde{p}\,; \tag{111}$$

i.e., $\tilde{p}$ is a characteristic vector of P^{T} for $\lambda = 1$.

Note that by (69) and (110) we may represent $\tilde{P}$ in the form

$$\tilde{P} = \begin{pmatrix} \tilde{A}_1 & O & \cdot & \cdot & \cdot & O & \\ O & \tilde{A}_2 & \cdot & \cdot & \cdot & O & O \\ \cdot & \cdot & \cdot & \cdot & \cdot & \cdot & \cdot \\ O & O & \cdot & \cdot & \cdot & \tilde{A}_g & \\ & & & \tilde{U} & & & \tilde{W} \end{pmatrix},$$

where

$$\tilde{A}_i = \lim_{N\to\infty}\frac{1}{N}\sum_{k=1}^{N} A_i^k \; (i = 1, 2, \ldots, g) \quad \tilde{W} = \lim_{N\to\infty}\frac{1}{N}\sum_{k=1}^{N} W^k,$$

$$W = \begin{pmatrix} A_{g+1} & O & \cdot & \cdot & \cdot & O \\ * & A_{g+2} & \cdot & \cdot & \cdot & O \\ \cdot & \cdot & \cdot & \cdot & \cdot & \cdot \\ * & * & & \cdot & \cdot & A_s \end{pmatrix}.$$

Since all the characteristic values of W are of modulus less than 1, we have

$$\lim_{k\to\infty} W^k = O,$$

and therefore $\tilde{W} = O$.

Hence

$$\tilde{P} = \begin{pmatrix} \tilde{A}_1 & O & \cdot & \cdot & \cdot & O & \\ O & \tilde{A}_2 & \cdot & \cdot & \cdot & O & O \\ \cdot & \cdot & \cdot & \cdot & \cdot & \cdot & \cdot \\ O & O & \cdot & \cdot & \cdot & \tilde{A}_g & \\ & & & \tilde{U} & & & O \end{pmatrix}. \tag{112}$$

Since $\tilde{P}$ is a stochastic matrix, the matrices $\tilde{A}_1, \tilde{A}_2, \ldots, \tilde{A}_g$ are also stochastic.

From this representation of $\tilde{P}$ and from (107) is follows that: *The mean limiting absolute probabilities corresponding to non-essential states are always zero.*

If $g=1$ in the normal form of P, then $\lambda=1$ is a simple characteristic value of P^{T}.

In this case $\tilde{p}$ is uniquely determined by (111), and the mean limiting probabilities $\tilde{p}_1, \tilde{p}_2, \ldots, \tilde{p}_n$ do not depend on the initial probabilities $\overset{0}{p}_1, \overset{0}{p}_2, \ldots, \overset{0}{p}_n$. Conversely, if $\tilde{p}$ does not depend on $\overset{0}{p}$, then P is of rank 1 by (110′). But the rank of (112) can be 1 only if $g=1$.

We formulate these results in the following theorem:[49]

THEOREM 12: *For an arbitrary homogeneous Markov chain with period h the probability matrices P^k and $\overset{k}{p}$ tend to a periodic repetition with period h for $k \to \infty$; moreover, the mean limiting transition probabilities and the absolute probabilities $\tilde{P} = \| \tilde{p}_{ij} \|_1^n$ and $\tilde{p} = (\tilde{p}_1, \tilde{p}_2, \ldots, \tilde{p}_n)$ defined by* (110) *and* (110′) *always exist.*

The mean absolute probabilities corresponding to non-essential states are always zero.

If $g=1$ in the normal form of P (and only in this case), the mean limiting absolute probabilities $\tilde{p}_1, \tilde{p}_2, \ldots, \tilde{p}_n$ are independent of the initial probabilities $\overset{0}{p}_1, \overset{0}{p}_2, \ldots, \overset{0}{p}_n$ and are uniquely determined by (111).

§ 8. Totally Non-negative Matrices

In this and the following sections we consider real matrices in which not only the elements, but also all the minors of every order are non-negative. Such matrices have important applications in the theory of small oscillations of elastic systems. The reader will find a detailed study of these matrices and their applications in the book [17]. Here we shall only deal with some of their basic properties.

1. We begin with a definition:

DEFINITION 5: *A rectangular matrix*

$$A = \| a_{ik} \| \quad (i=1, 2, \ldots, m;\ k=1, 2, \ldots, n)$$

is called totally non-negative (totally positive) if all its minors of any order are non-negative (positive):

---

[49] This theorem is sometimes called the *asymptotic theorem* for homogeneous Markov chains. See [4], pp. 479-82.

$$A\begin{pmatrix} i_1 & i_2 & \dots & i_p \\ k_1 & k_2 & \dots & k_p \end{pmatrix} \geqq 0 \qquad (>0)$$

$$\left(1 \leqq \begin{matrix} i_1 < i_2 < \cdots < i_p \\ k_1 < k_2 < \cdots < k_p \end{matrix} \leqq n;\ p = 1, 2, \ldots, \min(m, n)\right).$$

In what follows we shall only consider square totally non-negative and totally positive matrices.

Example 1. The *generalized Vandermonde matrix*

$$V = \| a_i^{\alpha_k} \|_1^n \qquad (0 < a_1 < a_2 < \ldots < a_n;\ \alpha_1 < \alpha_2 < \ldots < \alpha_n)$$

is totally positive. Let us show first that $| V | \neq 0$. Indeed, from $| V | = 0$ it would follow that we could determine real numbers $c_1, c_2, \ldots, c_n$, not all equal to zero, such that the function

$$f(x) = \sum_{k=1}^{n} c_k x^{\alpha_k} \qquad (\alpha_i \neq \alpha_j \text{ for } i \neq j)$$

has the n zeros $x_i = a_i$ $(i = 1, 2, \ldots, n)$, where n is the number of terms in the above summand. For $n = 1$ this is impossible. Let us make the induction hypothesis that it is impossible for a sum of n_1 terms, where $n_1 < n$, and show that it is then also impossible for the given function $f(x)$. Assume the contrary. Then by Rolle's Theorem the function $f_1(x) = [x^{-\alpha_1} f(x)]'$ consisting of $n - 1$ terms would have $n - 1$ positive zeros, and this contradicts the induction hypothesis.

Thus, $| V | \neq 0$. But for $\alpha_1 = 0, \alpha_2 = 1, \ldots, \alpha_n = n - 1$ the determinant $| V |$ goes over into the ordinary Vandermonde determinant $| a_i^{k-1} |_1^n$, which is positive. Since the transition from this to the generalized Vandermonde determinant can be carried out by means of a continuous change of the exponents $\alpha_1, \alpha_2, \ldots, \alpha_n$ with preservation of the inequalities $\alpha_1 < \alpha_2 < \ldots < \alpha_n$, and since, by what we have shown, the determinant does not vanish in this process, we have $| V | > 0$ for arbitrary $0 < \alpha_1 < \alpha_2 < \ldots < \alpha_n$.

Since every minor of V can be regarded as the determinant of some generalized Vandermonde matrix, all the minors of V are positive.

Example 2. We consider a *Jacobi matrix*

$$J = \begin{Vmatrix} a_1 & b_1 & 0 & \dots & 0 & 0 \\ c_1 & a_2 & b_2 & \dots & 0 & 0 \\ 0 & c_2 & a_3 & \dots & 0 & 0 \\ \cdot & \cdot & \cdot & \cdot & \cdot & \cdot \\ \cdot & \cdot & \cdot & \cdot & \cdot & \cdot \\ 0 & 0 & 0 & \dots & c_{n-1} & a_n \end{Vmatrix}, \tag{113}$$

in which all the elements are zero outside the main diagonal and the first super-diagonal and sub-diagonal. Let us set up a formula that expresses an arbitrary minor of the matrix in terms of principal minors and the elements *b, c.* Suppose that

$$1 \leq \begin{matrix} i_1 < i_2 < \cdots < i_p \\ k_1 < k_2 < \cdots < k_p \end{matrix} \leq n$$

and

$$i_1 = k_1,\ i_2 = k_2, \ldots, i_{\nu_1} = k_{\nu_1}\ ;\ i_{\nu_1+1} \neq k_{\nu_1+1}, \ldots, i_{\nu_2} \neq k_{\nu_2};\ i_{\nu_2+1} = k_{\nu_2+1}, \ldots, i_{\nu_3} = k_{\nu_3}; \ldots;$$

then

$$J\begin{pmatrix} i_1 & i_2 & \ldots & i_p \\ k_1 & k_2 & \ldots & k_p \end{pmatrix} = J\begin{pmatrix} i_1 & \ldots & i_{\nu_1} \\ k_1 & \ldots & k_{\nu_1} \end{pmatrix} J\begin{pmatrix} i_{\nu_1+1} \\ k_{\nu_1+1} \end{pmatrix} \cdots J\begin{pmatrix} i_{\nu_2} \\ k_{\nu_2} \end{pmatrix} J\begin{pmatrix} i_{\nu_2+1} & \ldots & i_{\nu_3} \\ k_{\nu_2+1} & \ldots & k_{\nu_3} \end{pmatrix} \cdots. \tag{114}$$

This formula is a consequence of the easily verifiable equation:

$$J\begin{pmatrix} i_1 & \ldots & i_p \\ k_1 & \ldots & k_p \end{pmatrix} = J\begin{pmatrix} i_1 & \ldots & i_{\nu-1} \\ k_1 & \ldots & k_{\nu-1} \end{pmatrix} J\begin{pmatrix} i_\nu \\ k_\nu \end{pmatrix} J\begin{pmatrix} i_{\nu+1} & \ldots & i_p \\ k_{\nu+1} & \ldots & k_p \end{pmatrix} \quad (\text{for } i_\nu \neq k_\nu). \tag{115}$$

From (114) it follows that every minor is the product of certain principal minors and certain elements of J. Thus: *For J to be totally non-negative it is necessary and sufficient that all the principal minors and the elements b, c should be non-negative.*

2. A totally non-negative matrix $A = \| a_{ik} \|_1^n$ always satisfies the following important determinantal inequality:[50]

$$A\begin{pmatrix} 1 & 2 & \ldots & n \\ 1 & 2 & \ldots & n \end{pmatrix} \leq A\begin{pmatrix} 1 & 2 & \ldots & p \\ 1 & 2 & \ldots & p \end{pmatrix} A\begin{pmatrix} p+1 & \ldots & n \\ p+1 & \ldots & n \end{pmatrix} \quad (p < n). \tag{116}$$

Before deriving this inequality, we prove the following lemma:

LEMMA 5: *If in a totally non-negative matrix $A = \| a_{ik} \|_1^n$ any principal minor vanishes, then every principal minor 'bordering' it also vanishes.*

Proof. The lemma will be proved if we can show that for a totally non-negative matrix $A = \| a_{ik} \|_1^n$ it follows from

---

[50] See [172] and [17], pp. 111ff, where it is also shown that the equality sign in (116) can only hold in the following obvious cases:

1) One of the factors on the right-hand side of (116) is zero;

2) All the elements a_{ik} $(i = 1, 2, \ldots, p;\ k = p+1, \ldots, n)$ or a_{ik} $(i = p+1, \ldots, n;\ k = 1, 2, \ldots, p)$ are zero.

The inequality (116) has the same outward form as the generalized Hadamard inequality (see (33), Vol. I, p. 255) for a positive-definite hermitian or quadratic form.

$$A\begin{pmatrix}1 & 2 & \dots & q\\ 1 & 2 & \dots & q\end{pmatrix}=0 \quad (q<n) \tag{117}$$

that

$$A\begin{pmatrix}1 & 2 & \dots & n\\ 1 & 2 & \dots & n\end{pmatrix}=0. \tag{118}$$

For this purpose we consider two cases:

1) $a_{11}=0$. Since $\begin{vmatrix}a_{11} & a_{1k}\\ a_{i1} & a_{ik}\end{vmatrix}=-a_{i1}a_{1k}\geqq 0$, $a_{i1}\geqq 0$, $a_{1k}\geqq 0$ $(i, k=2, \dots, n)$, either all the $a_{i1}=0$ $(i=2, \dots, n)$ or all the $a_{1k}=0$ $(k=2, \dots, n)$. These equations and $a_{11}=0$ imply (118).

2) $a_{11}\neq 0$. Then for some p $(1\leqq p\leqq q)$

$$A\begin{pmatrix}1 & 2 & \dots & p-1\\ 1 & 2 & \dots & p-1\end{pmatrix}\neq 0, \quad A\begin{pmatrix}1 & 2 & \dots & p-1 & p\\ 1 & 2 & \dots & p-1 & p\end{pmatrix}=0. \tag{119}$$

We introduce bordered determinants

$$d_{ik}=A\begin{pmatrix}1 & 2 & \dots & p-1 & i\\ 1 & 2 & \dots & p-1 & k\end{pmatrix} \quad (i, k=p, p+1, \dots, n) \tag{120}$$

and form from them a matrix $D=\| d_{ik} \|_p^n$.

By Sylvester's identity (Vol. I, Chapter II, § 3),

$$\begin{gathered}D\begin{pmatrix}i_1 & i_2 & \dots & i_g\\ k_1 & k_2 & \dots & k_g\end{pmatrix}\\ =\left[A\begin{pmatrix}1 & 2 & \dots & p-1\\ 1 & 2 & \dots & p-1\end{pmatrix}\right]^{g-1} A\begin{pmatrix}1 & 2 & \dots & p-1 & i_1 & i_2 & \dots & i_g\\ 1 & 2 & \dots & p-1 & k_1 & k_2 & \dots & k_g\end{pmatrix}\geqq 0 \\ \left(p\leqq \begin{matrix} i_1<i_2<\dots<i_g\\ k_1<k_2<\dots<k_g\end{matrix}\leqq n;\quad g=1, 2, \dots, n-p+1\right),\end{gathered} \tag{121}$$

so that D is a totally non-negative matrix.

Since by (119)

$$d_{pp}=A\begin{pmatrix}1 & 2 & \dots & p\\ 1 & 2 & \dots & p\end{pmatrix}=0,$$

the matrix D falls under the case 1) and

$$D\begin{pmatrix}p & p+1 & \dots & n\\ p & p+1 & \dots & n\end{pmatrix}=\left[A\begin{pmatrix}1 & 2 & \dots & p-1\\ 1 & 2 & \dots & p-1\end{pmatrix}\right]^{n-p} A\begin{pmatrix}1 & 2 & \dots & n\\ 1 & 2 & \dots & n\end{pmatrix}=0.$$

Since $A\begin{pmatrix}1 & 2 & \dots & p-1\\ 1 & 2 & \dots & p-1\end{pmatrix} \neq 0$, (118) follows, and the lemma is proved.

3. We may now assume in the derivation of the inequality (116) that all the principal minors of A are different from zero, since by Lemma 5 one of the principal minors can only be zero when $|A| = 0$, and in this case the inequality (116) is obvious.

For $n = 2$, (116) can be verified immediately:

$$A\begin{pmatrix}1 & 2\\ 1 & 2\end{pmatrix} = a_{11}a_{22} - a_{12}a_{21} \leqq a_{11}a_{22},$$

since $a_{12} \geqq 0$, $a_{21} \geqq 0$. We shall establish (116) for $n > 2$ under the assumption that it is true for matrices of order less than n. Moreover, without loss of generality, we may assume that $p > 1$, since otherwise by reversing the numbering of the rows and columns we could interchange the roles of p and $n - p$.

We now consider again the matrix $D = \| d_{ik} \|_p^n$, where the d_{ik} $(i, k = p, p+1, \dots, n)$ are defined by (120); we use Sylvester's identity twice as well as the basic inequality (116) for matrices of order less than n and obtain

$$A\begin{pmatrix}1 & 2 & \dots & n\\ 1 & 2 & \dots & n\end{pmatrix} = \frac{D\begin{pmatrix}p & p+1 & \dots & n\\ p & p+1 & \dots & n\end{pmatrix}}{\left[A\begin{pmatrix}1 & 2 & \dots & p-1\\ 1 & 2 & \dots & p-1\end{pmatrix}\right]^{n-p}} \leqq \frac{d_{pp}D\begin{pmatrix}p+1 & \dots & n\\ p+1 & \dots & n\end{pmatrix}}{\left[A\begin{pmatrix}1 & 2 & \dots & p-1\\ 1 & 2 & \dots & p-1\end{pmatrix}\right]^{n-p}}$$

$$= \frac{A\begin{pmatrix}1 & 2 & \dots & p\\ 1 & 2 & \dots & p\end{pmatrix} A\begin{pmatrix}1 & 2 & \dots & p-1 & p+1 & \dots & n\\ 1 & 2 & \dots & p-1 & p+1 & \dots & n\end{pmatrix}}{A\begin{pmatrix}1 & 2 & \dots & p-1\\ 1 & 2 & \dots & p-1\end{pmatrix}}$$

$$\leqq A\begin{pmatrix}1 & 2 & \dots & p\\ 1 & 2 & \dots & p\end{pmatrix} A\begin{pmatrix}p+1 & \dots & n\\ p+1 & \dots & n\end{pmatrix}. \tag{122}$$

Thus, the inequality (116) has been established.

Let us make the following definition:

DEFINITION 6. *A minor*

$$A\begin{pmatrix}i_1 & i_2 & \dots & i_p\\ k_1 & k_2 & \dots & k_p\end{pmatrix} \quad \left(1 \leqq \begin{matrix}i_1 < i_2 < \dots < i_p\\ k_1 < k_2 < \dots < k_p\end{matrix} \leqq n\right) \tag{123}$$

of the matrix $A = \| a_{ik} \|_1^n$ *will be called almost principal if of the differences* $i_1 - k_1, i_2 - k_2, \dots, i_p - k_p$ *only one is not zero.*

We can then point out that the whole derivation of (116) (and the proof of the auxiliary lemma) remain valid if the condition 'A is totally non-negative' is replaced by the weaker condition 'all the principal and almost principal minors of A are non-negative.'[51]

§ 9. Oscillatory Matrices

1. The characteristic values and characteristic vectors of totally positive matrices have a number of remarkable properties. However, the class of totally positive matrices is not wide enough from the point of view of applications to small oscillations of elastic systems. In this respect, the class of totally non-negative matrices is suffiently extensive. But the spectral properties we need do not hold for all totally non-negative matrices. Now there exists an intermediate class (between that of totally positive and that of totally non-negative matrices) in which the spectral properties of totally positive matrices are preserved and which is of sufficiently wide scope for the applications. The matrices of this intermediate class have been called 'oscillatory.' The name is due to the fact that oscillatory matrices form the mathematical apparatus for the study of oscillatory properties of small vibrations of elastic systems.[52]

DEFINITION 7. *A matrix* $A = \| a_{ik} \|_1^n$ *is called oscillatory if* A *is totally non-negative and if there exists an integer* $q > 0$ *such that* A^q *is totally positive.*

Example. A Jacobi matrix J (see (113)) is oscillatory if and only if 1. all the numbers b, c are positive and 2. the successive principal minors are positive:

---

[51] See [214]. We take this opportunity of mentioning that in the second edition of the book [17] by F. R. Gantmacher and M. G. Kreĭn a mistake crept in which was first pointed out to the authors by D. M. Kotelyanskiĭ. On p. 111 of that book an almost principal minor (123) was defined by the equation

$$\sum_{\nu=1}^{p} | i_\nu - k_\nu | = 1 .$$

With this definition, the inequality (116) does not follow from the fact that the principal and the almost principal minors are non-negative. However, all the statements and proofs of § 6, Chapter II in [17] that refer to the fundamental inequality remain valid if an almost principal minor is defined as above and as we have done in the paper [214].

[52] See [17], Introduction, Chapter III, and Chapter IV.

$$a_1>0,\quad \begin{vmatrix} a_1 & b_1 \\ c_1 & a_2 \end{vmatrix}>0,\quad \begin{vmatrix} a_1 & b_1 & 0 \\ c_1 & a_2 & b_2 \\ 0 & c_2 & a_3 \end{vmatrix}>0,\ \dots,\ \begin{vmatrix} a_1 & b_1 & 0 & \dots & 0 & 0 \\ c_1 & a_2 & b_2 & \dots & 0 & 0 \\ 0 & c_2 & a_3 & \dots & 0 & 0 \\ \cdot & \cdot & \cdot & \cdot & \cdot & \cdot \\ 0 & 0 & 0 & \dots & c_{n-1} & a_n \end{vmatrix}>0. \qquad (124)$$

Necessity of 1., 2. The numbers b, c are non-negative, because $J \geqq O$. But none of the numbers b, c may be zero, since otherwise the matrix would be reducible and then the inequality $J^q > O$ could not hold for any $q > 0$. Hence, all the numbers b, c are positive. All the principal minors of (124) are positive, by Lemma 5, since it follows from $|J| \geqq 0$ and $|J^q| > 0$ that $|J| > 0$.

Sufficiency of 1., 2. When we expand $|J|$ we easily see that the numbers b, c occur in $|J|$ only as products $b_1c_1, b_2c_2, \dots, b_{n-1}c_{n-1}$. The same applies to every principal minor of 'zero density,' i.e., a minor formed from successive rows and columns (without gaps). But every principal minor of J is a product of principal minors of zero density. Therefore: *In every principal minor of J the numbers b and c occur only as products $b_1c_1, b_2c_2, \dots, b_{n-1}c_{n-1}$.*

We now form the symmetrical Jacobi matrix

$$\tilde{J} = \begin{Vmatrix} a_1 & \tilde{b}_1 & & & 0 \\ \tilde{b}_1 & a_2 & \tilde{b}_2 & & \\ & \tilde{b}_2 & \ddots & \ddots & \\ & & \ddots & \ddots & \tilde{b}_{n-1} \\ 0 & & & \tilde{b}_{n-1} & a_n \end{Vmatrix}, \quad \tilde{b}_i = \sqrt{b_i c_i} > 0 \quad (i = 1, 2, \dots, n). \qquad (125)$$

From the above properties of the principal minors of a Jacobi matrix it follows that the corresponding principal minors of J and $\tilde{J}$ are equal. But then (124) means that the quadratic form

$$\tilde{J}(x, x)$$

is positive definite (see Vol. I, Chapter X, Theorem 3, p. 306). But in a positive-definite quadratic form all the principal minors are positive. Therefore in J too all the principal minors are positive. Since by 1. all the numbers b, c are positive, by (114) all the minors of J are non-negative; i.e., J is totally non-negative.

That a totally non-negative matrix J for which 1. and 2. are satisfied is oscillatory follows immediately from the following *criterion for an oscillatory matrix.*

A totally non-negative matrix $A = \| a_{ik} \|_1^n$ is oscillatory if and only if:

1) A is non-singular ($|A| > 0$);

2) All the elements of A in the principal diagonal and the first super-diagonals and sub-diagonals are different from zero ($a_{ik} > 0$ for $|i-k| \leqq 1$).

The reader can find a proof of this proposition in [17], Chapter II, § 7.

2. In order to formulate properties of the characteristic values and characteristic vectors of oscillatory matrices, we introduce some preliminary concepts and notations.

We consider a vector (column)

$$u = (u_1, u_2, \ldots, u_n).$$

Let us count the number of variations of sign in the sequence of coordinates $u_1, u_2, \ldots, u_n$ of u, attributing arbitrary signs to the zero coordinates (if any such exist). Depending on what signs we give to the zero coordinates the number of variations of sign will vary within certain limits. The *maximal* and *minimal* number of variations of sign so obtained will be denoted by S_u^+ and S_u^-, respectively. If $S_u^- = S_u^+$, we shall speak of the *exact* number of sign changes and denote it by S_u. Obviously $S_u^- = S_u^+$ if and only if 1. the extreme coordinates u_1 and u_n of u are different from zero, and 2. $u_i = 0$ $(1 < i < n)$ always implies that $u_{i-1}u_{i+1} < 0$.

We shall now prove the following fundamental theorem:

THEOREM 13: 1. *An oscillatory matrix $A = \| a_{ik} \|_1^n$ always has n distinct positive characteristic values*

$$\lambda_1 > \lambda_2 > \cdots > \lambda_n > 0. \tag{126}$$

2. *The characteristic vector $\overset{1}{u} = (u_{11}, u_{21}, \ldots, u_{n1})$ of A that belongs to the largest characteristic value λ_1 has only non-zero coordinates of like sign; the characteristic vector $\overset{2}{u} = (u_{12}, u_{22}, \ldots, u_{n2})$ that belongs to the second largest characteristic value λ_2 has exactly one variation of sign in its coordinates; more generally, the characteristic vector $\overset{k}{u} = (u_{1k}, u_{2k}, \ldots, u_{nk})$ that belongs to the characteristic value λ_k has exactly $k-1$ variations of sign* $(k = 1, 2, \ldots, n)$.

3. *For arbitrary real numbers $c_g, c_{g+1}, \ldots, c_h$ $(1 \leqq g \leqq h \leqq n;$ $\sum_{k=g}^{h} c_k^2 > 0)$ the number of variations of sign in the coordinates of the vector*

$$u = \sum_{k=g}^{h} c_k \overset{k}{u} \tag{127}$$

lies between $g-1$ and $h-1$:

$$g-1 \leqq S_u^- \leqq S_u^+ \leqq h-1. \tag{128}$$

Proof. 1. We number the characteristic values $\lambda_1, \lambda_2, \ldots, \lambda_n$ of A so that

$$|\lambda_1| \geqq |\lambda_2| \geqq \cdots \geqq |\lambda_n|$$

and consider the p-th compound matrix $\mathfrak{A}_p$ $(p=1, 2, \ldots, n)$ (see Chapter I, § 4). The characteristic values of $\mathfrak{A}_p$ are all the possible products of p characteristic values of A (see Vol. I, p. 75), i.e., the products

$$\lambda_1\lambda_2\cdots\lambda_p, \quad \lambda_1\lambda_2\cdots\lambda_{p-1}\lambda_{p+1}, \quad \ldots .$$

From the conditions of the theorem it follows that for some integer q A^q is totally positive. But then $\mathfrak{A}_p \geqq O$, $\mathfrak{A}_p^q > O$;[53] i.e., $\mathfrak{A}_p$ is irreducible, non-negative, and primitive. Applying Frobenius' theorem (see § 2, p. 40) to the primitive matrix $\mathfrak{A}_p$ $(p=1, 2, \ldots, n)$, we obtain

$$\lambda_1\lambda_2\cdots\lambda_p > 0 \quad (p=1, 2, \ldots, n),$$
$$\lambda_1\lambda_2\cdots\lambda_p > \lambda_1\lambda_2\cdots\lambda_{p-1}\lambda_{p+1} \quad (p=1, 2, \ldots, n-1).$$

Hence (126) follows.

2. From this inequality (126) it follows that $A = \| a_{ik} \|_1^n$ is a matrix of simple structure. Then all the compound matrices $\mathfrak{A}_p$ $(p=1, 2, \ldots, n)$ are also of simple structure (see Vol. I, p. 74).

We consider the fundamental matrix $U = \| u_{ik} \|_1^n$ of A (the k-th column of U contains the coordinates of the k-th characteristic vector $\overset{k}{u}$ of A; $k = 1, 2, \ldots, n$). Then (see Vol. I, Chapter III, p. 74), the characteristic vector of $\mathfrak{A}_p$ belonging to the characteristic value $\lambda_1\lambda_2 \ldots \lambda_p$ has the coordinates

$$U\begin{pmatrix} i_1 & i_2 & \ldots & i_p \\ 1 & 2 & \ldots & p \end{pmatrix} \quad (1 \leqq i_1 < i_2 < \cdots < i_p \leqq n) \tag{129}$$

By Frobenius' theorem all the numbers (129) are different from zero and are of like sign. Multiplying the vectors $\overset{1}{u}, \overset{2}{u}, \ldots, \overset{n}{u}$ by ± 1, we can make all the minors of (129) positive:

$$U\begin{pmatrix} i_1 & i_2 & \ldots & i_p \\ 1 & 2 & \ldots & p \end{pmatrix} > 0 \quad \begin{pmatrix} 1 \leqq i_1 < i_2 < \cdots < i_p \leqq n \\ p = 1, 2, \ldots, n \end{pmatrix}. \tag{130}$$

---

[53] The matrix $\mathfrak{A}_p^q$ is the p-th compound matrix A^q (see Vol. I, Chapter I, p. 20.)

The fundamental matrix $U = \| u_{ik} \|_1^n$ is connected with A by the equation

$$A = U\{\lambda_1, \lambda_2, \ldots, \lambda_n\}\, U^{-1}. \tag{131}$$

But then

$$A^{\mathsf{T}} = (U^{\mathsf{T}})^{-1}\{\lambda_1, \lambda_2, \ldots, \lambda_n\}\, U^{\mathsf{T}}. \tag{132}$$

Comparing (131) with (132), we see that

$$V = (U^{\mathsf{T}})^{-1} \tag{133}$$

is the fundamental matrix of A^{T} with the same characteristic values $\lambda_1, \lambda_2, \ldots, \lambda_n$. But since A is oscillatory, so is A^{T}. Therefore in V as well for every $p = 1, 2, \ldots, n$ all the minors

$$V\begin{pmatrix} i_1 & i_2 & \ldots & i_p \\ 1 & 2 & \ldots & p \end{pmatrix} \quad (1 \leq i_1 < i_2 < \cdots < i_p \leq n) \tag{134}$$

are different from zero and are of the same sign.

On the other hand, by (133) U and V are connected by the equation

$$U^{\mathsf{T}} V = E.$$

Going over to the p-th compound matrices (see Vol. I, Chapter I, § 4), we have:

$$\mathfrak{U}_p \mathfrak{B}_p = \mathfrak{E}_p.$$

Hence, in particular, noting that the diagonal elements of $\mathfrak{E}_p$ are 1, we obtain:

$$\sum_{1 \leq i_1 < i_2 < \cdots < i_p \leq n} U\begin{pmatrix} i_1 & i_2 & \ldots & i_p \\ 1 & 2 & \ldots & p \end{pmatrix} V\begin{pmatrix} i_1 & i_2 & \ldots & i_p \\ 1 & 2 & \ldots & p \end{pmatrix} = 1. \tag{135}$$

On the left-hand side of this equation, the first factor in each of the summands is positive and the second factors are different from zero and are of like sign. It is then obvious that the second factors as well are positive; i.e.,

$$V\begin{pmatrix} i_1 & i_2 & \ldots & i_p \\ 1 & 2 & \ldots & p \end{pmatrix} > 0 \quad \begin{pmatrix} 1 \leq i_1 < i_2 < \cdots < i_p \leq n \\ p = 1, 2, \ldots, n \end{pmatrix}. \tag{136}$$

Thus, the inequalities (130) and (136) hold for $U = \| u_{ik} \|_1^n$ and $V = (U^{\mathsf{T}})^{-1}$ simultaneously.

When we express the minors of V in terms of those of the inverse matrix $V^{-1}=U^{\mathsf{T}}$ by the well-known formulas (see Vol. I, pp. 21-22), we obtain

$$V\begin{pmatrix} j_1 & j_2 & \dots & j_{n-p} \\ 1 & 2 & \dots & n-p \end{pmatrix} = \frac{(-1)^{np+\sum_{\nu=1}^{p} i_\nu}}{|U|} U\begin{pmatrix} i_1 & i_2 & & i_p \\ n & n-1 & \dots & n-p+1 \end{pmatrix}, \tag{137}$$

where $i_1 < i_2 < \dots < i_p$ and $j_1 < j_2 < \dots < j_{n-p}$ together give the complete system of indices $1, 2, \dots, n$. Since, by (130), $|U| > 0$ it follows from (136) and (137) that

$$(-1)^{np+\sum_{\nu=1}^{p} i_\nu} \quad U\begin{pmatrix} i_1 & i_2 & \dots & i_p \\ 1 & 2 & \dots & p \end{pmatrix} > 0 \quad \begin{pmatrix} 1 \leqq i_1 < i_2 < \cdots < i_p \leqq n \\ p=1, 2, \dots, n \end{pmatrix}. \tag{138}$$

Now let $u = \sum_{k=g}^{h} c_k \overset{k}{u}$ $\left(\sum_{k=g}^{h} c_k^2 > 0\right)$. We shall show that the inequalities (130) imply the second part of (128):

$$S_u^+ \leqq h-1, \tag{139}$$

and the inequalities (138), the first part:

$$S_u^- \geqq g-1. \tag{140}$$

Suppose that $S_u^+ > h-1$. Then we can find $h+1$ coordinates of u

$$u_{i_1}, u_{i_2}, \dots, u_{i_{h+1}} \quad (1 \leqq i_1 < i_2 < \cdots < i_{h+1} \leqq n) \tag{141}$$

such that

$$u_{i_\alpha} u_{i_{\alpha+1}} \leqq 0 \quad (\alpha = 1, 2, \dots, h).$$

Furthermore, the coordinates (141) cannot all be zero; for then we could equate the corresponding coordinates of the vector $u = \sum_{k=1}^{h} c_k \overset{k}{u}$ $(c_1 = \dots = c_{g-1} = 0;\ \sum_{k=1}^{h} c_k^2 > 0)$ to zero and thus obtain a system of homogeneous equations

$$\sum_{k=1}^{h} c_k u_{i_\alpha k} = 0 \quad (\alpha = 1, 2, \dots, h)$$

with the non-zero solution $c_1, c_2, \dots, c_h$, whereas the determinant of the system

$$U\begin{pmatrix} i_1 & i_2 & \dots & i_h \\ 1 & 2 & \dots & h \end{pmatrix}$$

is different from zero, by (130).

We now consider the vanishing determinan.

$$\begin{vmatrix} u_{i_1 1} & \dots & u_{i_1 h} & u_{i_1} \\ u_{i_2 1} & \dots & u_{i_2 h} & u_{i_2} \\ \cdot & \cdot & \cdot & \cdot \\ \cdot & \cdot & \cdot & \cdot \\ u_{i_{h+1} 1} & \dots & u_{i_{h+1} h} & u_{i_{h+1}} \end{vmatrix} = 0.$$

We expand it with respect to the elements of the last column:

$$\sum_{\alpha=1}^{h+1} (-1)^{h+\alpha+1} u_{i_\alpha} \; U\begin{pmatrix} i_1 \dots i_{\alpha-1} \; i_{\alpha+1} \dots i_{h+1} \\ 1 \; . \; . \; . \; . \; . \; . \; . \; . \; h \end{pmatrix} = 0.$$

But such an equation cannot hold, since on the left-hand side all the terms are of like sign and at least one term is different from zero. Hence the assumption that $S_u^+ > h-1$ has led to a contradiction, and (139) can be regarded as proved.

We consider the vector

$$\overset{k*}{u} = (u_{1k}^*, u_{2k}^*, \dots, u_{nk}^*) \qquad (k=1, 2, \dots, n),$$

where

$$u_{ik}^* = (-1)^{n+i+k} u_{ik} \qquad (i, k = 1, 2, \dots, n);$$

then for the matrix $U^* = \| u_{ik}^* \|_1^n$ we have, by (138):

$$U^* \begin{pmatrix} i_1 & i_2 & \dots & i_p \\ n & n-1 & \dots & n-p+1 \end{pmatrix} > 0 \quad \begin{pmatrix} 1 \leqq i_1 < i_2 < \dots < i_p \leqq n \\ p = 1, 2, \dots, n \end{pmatrix}. \tag{142}$$

But the inequalities (142) are analogous to (130). Therefore, by setting

$$\overset{*}{u} = \sum_{k \neq g}^{h} (-1)^k c_k \overset{k*}{u}, \tag{143}$$

we have the inequality analogous to (139):[54]

$$S_{u^*}^+ \leqq n-g. \tag{144}$$

Let $u = (u_1, u_2, \dots, u_n)$ and $u^* = (\overset{*}{u}_1, \overset{*}{u}_2, \dots, \overset{*}{u}_n)$. It is easy to see that

$$\overset{*}{u}_i = (-1)^i u_i \qquad (i = 1, 2, \dots, n).$$

Therefore

---

[54] In the inequalities (142), the vectors $\overset{k}{u}$ $(k = 1, 2, \dots, n)$ occur in the inverse order $\overset{n}{u}, \overset{n-1}{u}, \dots$. The vector $\overset{g}{u}$ is preceded by $n-g$ vectors of this kind.

$$S_{u*}^{+} + S_{u}^{-} = n - 1,$$

and so the relation (140) holds, by (144).

This establishes the inequality (128). Since the second statement of the theorem is obtained from (128) by setting $g = h = k$, the theorem is now completely proved.

3. As an application of this theorem, let us study the small oscillations of n masses $m_1, m_2, \ldots, m_n$ concentrated at n movable points $x_1 < x_2 < \ldots < x_n$ of a segmentary elastic continuum (a string or a rod of finite length), stretched (in a state of equilibrium) along the segment $0 \leqq x \leqq l$ of the x-axis.

We denote by $K(x, s)$ $(0 \leqq x, s \leqq l)$ the function of influence of this continuum ($K(x, s)$ is the displacement at the point x under the action of a unit force applied at the point s) and by k_{ij} the coefficients of influence for the given n masses:

$$k_{ij} = K(x_i, x_j) \quad (i, j = 1, 2, \ldots, n).$$

If at the points $x_1, x_2, \ldots, x_n$ n forces $F_1, F_2, \ldots, F_n$ are applied, then the corresponding static displacement $y(x)$ $(0 \leqq x \leqq l)$, is given, by virtue of the linear superposition of displacements, by the formula

$$y(x) = \sum_{j=1}^{n} K(x, x_j) F_j.$$

When we here replace the forces F_j by the inertial forces $-m_j \frac{\partial^2}{\partial t^2} y(x_j, t)$ $(j = 1, 2, \ldots, n)$, we obtain the equation of free oscillations

$$y(x) = -\sum_{j=1}^{n} m_j K(x, x_j) \frac{\partial^2}{\partial t^2} y(x_j, t). \tag{145}$$

We shall seek harmonic oscillations of the continuum in the form

$$y(x) = u(x) \sin(\omega t + \alpha) \quad (0 \leqq x \leqq l). \tag{146}$$

Here $u(x)$ is the amplitude function, ω the frequency, and α the initial phase. Substituting this expression for $y(x)$ in (145) and cancelling $\sin(\omega t + \alpha)$, we obtain

$$u(x) = \omega^2 \sum_{j=1}^{n} m_j K(x, x_j) u(x_j). \tag{147}$$

Let us introduce a notation for the variable displacements and the displacements in amplitude at the points of distribution of mass:

$$y_i = y(x_i, t), \quad u_i = u(x_i) \qquad (i = 1, 2, \ldots, n).$$

Then

$$y_i = u_i \sin(\omega t + \alpha) \qquad (i = 1, 2, \ldots, n).$$

We also introduce the *reduced amplitude displacements* and the *reduced coefficients of influence*

$$\tilde{u}_i = \sqrt{m_i}\, u_i, \quad a_{ij} = \sqrt{m_i m_j}\, k_{ij} \qquad (i, j = 1, 2, \ldots, n). \tag{148}$$

Replacing x in (147) by x_i $(i = 1, 2, \ldots, n)$ successively, we obtain a system of equations for the amplitude displacements:

$$\sum_{j=1}^{n} a_{ij} \tilde{u}_j = \lambda \tilde{u}_i \qquad \left(\lambda = \frac{1}{\omega^2};\quad i = 1, 2, \ldots, n\right). \tag{149}$$

Hence it is clear that the amplitude vector $\tilde{u} = (\tilde{u}_1, \tilde{u}_2, \ldots, \tilde{u}_n)$ is a characteristic vector of $A = \| a_{ij} \|_1^n = \| \sqrt{m_i m_j} k_{ij} \|_1^n$ for $\lambda = 1/\omega^2$ (see Vol. I, Chapter X, § 8).

It can be established, as the result of a detailed analysis,[55] that *the matrix of the coefficients of influence* $\| k_{ij} \|_1^n$ *of a segmentary continuum is always oscillatory.* But then the matrix $A = \| a_{ij} \|_1^n = \| \sqrt{m_i m_j} k_{ij} \|_1^n$ is also oscillatory! Therefore (by Theorem 13) A has n positive characteristic values

$$\lambda_1 > \lambda_2 > \cdots > \lambda_n > 0;$$

i.e., there exist n harmonic oscillations of the continuum with *distinct* frequencies:

$$(0 <)\, \omega_1 < \omega_2 < \cdots < \omega_n \qquad \left(\lambda_i = \frac{1}{\omega_i^2};\quad i = 1, 2, \ldots, n\right).$$

By the same theorem to the fundamental frequency ω_1 there correspond amplitude displacements different from zero and of like sign. Among the displacements in amplitude corresponding to the first overtone with the frequency ω_2 there is exactly one variation of sign and, in general, among the displacements in amplitude for the overtone with the frequency ω_j there are exactly $j - 1$ variations of sign $(j = 1, 2, \ldots, n)$.

---

[55] See [239], [240], and [17], Chapter III.

From the fact that the matrix of the coefficients of influence $\| k_{ij} \|_1^n$ is oscillatory there follow other oscillatory properties of the continuum: 1) For $\omega = \omega_1$ the amplitude function $u(x)$, which is connected with the amplitude displacements by (147), has no nodes; and, in general, for $\omega = \omega_j$ the function has $j - 1$ nodes ($j = 1, 2, \ldots, n$); 2) The nodes of two adjacent harmonics alternate, etc.

We cannot dwell here on the justification of these properties.[56]

[56] See [17], Chapters III and IV.

CHAPTER XIV

APPLICATIONS OF THE THEORY OF MATRICES TO THE INVESTIGATION OF SYSTEMS OF LINEAR DIFFERENTIAL EQUATIONS

§ 1. Systems of Linear Differential Equations with Variable Coefficients. General Concepts

1. Suppose given a system of linear homogeneous differential equations of the first order:

$$\frac{dx_i}{dt}=\sum_{k=1}^{n} p_{ik}(t)\, x_k \qquad (i=1, 2, \ldots, n)\,, \tag{1}$$

where $p_{ik}(t)$ $(i, k=1, 2, \ldots, n)$ are complex functions of a real argument t, continuous in some interval, finite or infinite, of the variable t.[1]

Setting $P(t)=\| p_{ik}(t) \|_1^n$ and $x=(x_1, x_2, \ldots, x_n)$, we write (1) as

$$\frac{dx}{dt}=P(t)\, x\,. \tag{2}$$

An *integral matrix* of the system (1) shall be defined as a square matrix $X(t)=\| x_{ik}(t) \|_1^n$ whose columns are n linearly independent solutions of the system.

Since every column of X satisfies (2), the integral matrix X satisfies the equation

$$\frac{dX}{dt}=P(t)\, X\,. \tag{3}$$

In what follows, we shall consider the matrix equation (3) instead of the system (1).

From the theorem on the existence and uniqueness of the solution of a system of differential equations[2] it follows that the integral matrix $X(t)$ is uniquely determined when the value of the matrix for some ('initial')

---

[1] In this section, all the relations that involve functions of t refer to the given interval.

[2] A proof of this theorem will be given in § 5. See also I. G. Petrowski (Petrovskiĭ), *Vorlesungen über die Theorie der gewöhnlichen Differentialgleichungen*, Leipzig, 1954 (translated from the Russian: Moscow, 1952).

value $t=t_0$ is known,[3] $X(t_0)=X_0$. For X_0 we can take an arbitrary non-singular square matrix of order n. In the particular case where $X(t_0)=E$, the integral matrix $X(t)$ will be called *normalized.*

Let us differentiate the determinant of X by differentiating its rows in succession and let us then use the differential relations

$$\frac{dx_{ij}}{dt}=\sum_{k=1}^{n} p_{ik}x_{kj} \quad (i, j=1, 2, \ldots, n).$$

We obtain:

$$\frac{d\,|X|}{dt}=(p_{11}+p_{22}+\cdots+p_{nn})\,|X|.$$

Hence there follows the well-known *Jacobi identity*

$$|X|=ce^{\int_{t_0}^{t}\operatorname{tr} P\,dt}, \tag{4}$$

where c is a constant and

$$\operatorname{tr} P=p_{11}+p_{22}+\ldots+p_{nn}$$

is the trace of $P(t)$.

Since the determinant $|X|$ cannot vanish identically, we have $c\neq 0$. But then it follows from the Jacobi identity that $|X|$ is different from zero for every value of the argument

$$|X|\neq 0;$$

i.e., *an integral matrix is non-singular for every value of the argument.*

If $\tilde{X}(t)$ is a non-singular ($|\tilde{X}(t)|\neq 0$) particular solution of (3), then the general solution is determined by the formula

$$X=\tilde{X}C, \tag{5}$$

where C is an arbitrary constant matrix.

For, by multiplying both sides of the equation

$$\frac{d\tilde{X}}{dt}=P\tilde{X} \tag{6}$$

by C on the right, we see that the matrix $\tilde{X}C$ also satisfies (3). On the other hand, if X is an arbitrary solution of (3), then (6) implies:

---

[3] It is assumed that t_0 belongs to the given interval of t.

$$\frac{dX}{dt}=\frac{d}{dt}(\tilde{X}\cdot\tilde{X}^{-1}X)=\frac{d\tilde{X}}{dt}\tilde{X}^{-1}X+\tilde{X}\frac{d}{dt}(\tilde{X}^{-1}X)=PX+\tilde{X}\frac{d}{dt}(\tilde{X}^{-1}X),$$

and hence by (3)

$$\frac{d}{dt}(\tilde{X}^{-1}X)=0$$

and

$$\tilde{X}^{-1}X=\text{const.}=C\,;$$

i.e., (5) holds.

All the integral matrices X of the system (1) are obtained by the formula (5) with $|\,C\,|\neq 0$.

2. Let us consider the special case:

$$\frac{dX}{dt}=AX, \tag{7}$$

where A is a constant matrix. Here $\tilde{X}=e^{At}$ is a particular non-singular solution of (7),[4] so that the general solution is of the form

$$X=e^{At}C \tag{8}$$

where C is an arbitrary constant matrix.

Setting $t=t_0$ in (8) we find: $X_0=e^{At_0}C$. Hence $C=e^{-At_0}X_0$ and therefore (8) can be represented in the form

$$X=e^{A(t-t_0)}X_0. \tag{9}$$

This formula is equivalent to our earlier formula (46) of Chapter V (Vol. I, p. 118).

Let us now consider the so-called *Cauchy system*:

$$\frac{dX}{dt}=\frac{A}{t-a}X \qquad (A \text{ is a constant matrix}). \tag{10}$$

This case reduces to the preceding one by a change of argument:

$$u=\ln(t-a).$$

Therefore the general solution of (10) looks as follows:

$$X=e^{A\ln(t-a)}C=(t-a)^{A}C. \tag{11}$$

The functions e^{At} and $(t-a)^A$ that occur in (8) and (11) may be represented in the form (Vol. I, p. 117)

---

[4] By term-by-term differentiation of the series $e^{At}=\sum_{k=0}^{\infty}\frac{A^k}{k!}t^k$ we find $\frac{d}{dt}e^{At}=Ae^{At}$.

$$e^{At} = \sum_{k=1}^{s} (Z_{k1} + Z_{k2}t + \cdots + Z_{km_k}t^{m_k-1})\, e^{\lambda_k t}, \tag{12}$$

$$(t-a)^A = \sum_{k=1}^{s} (Z_{k1} + Z_{k2} \ln(t-a) + \cdots + Z_{km_k} [\ln(t-a)]^{m_k-1})\,(t-a)^{\lambda_k}. \tag{13}$$

Here

$$\psi(\lambda) = (\lambda-\lambda_1)^{m_1}(\lambda-\lambda_2)^{m_2}\cdots(\lambda-\lambda_s)^{m_s}$$
$$(\lambda_i \neq \lambda_k \text{ for } i \neq k;\ i, k = 1, 2, \ldots, s)$$

is the minimal polynomial of A, and Z_{kj} $(j=1, 2, \ldots, m_k;\ k=1, 2, \ldots, s)$ are linearly independent constant matrices that are polynomials in A.[5]

Note. Sometimes an integral matrix of the system of differential equations (1) is taken to be a matrix W in which the *rows* are linearly independent solutions of the system. It is obvious that W is the transpose of X:

$$W = X^{\mathsf{T}}.$$

When we go over to the transposed matrices on both sides of (3), we obtain instead of (3) the following equation for W:

$$\frac{dW}{dt} = WP(t). \tag{3'}$$

Here W is the first factor on the right-hand side, not the second, as X was in (3).

§ 2. Lyapunov Transformations

1. Let us now assume that in the system (1) (and in the equation (3)) the coefficient matrix $P(t) = \| p_{ik}(t) \|_1^n$ is a continuous bounded function of t in the interval $[t_0, \infty)$.[6]

In place of the unknown functions $x_1, x_2, \ldots, x_n$ we introduce the new unknown functions $y_1, y_2, \ldots, y_n$ by means of the transformation

$$x_i = \sum_{k=1}^{n} l_{ik}(t)\, y_k \qquad (i = 1, 2, \ldots, n). \tag{14}$$

---

[5] Every term $X_k = (Z_{k1} + Z_{k2}t + \cdots + Z_{km_k}t^{m_k-1})\, e^{\lambda_k t}$ $(k=1, 2, \ldots, s)$ on the right-hand side of (12) is a solution of (7). For the product $g(A)e^{At}$, with an arbitrary function $g(\lambda)$, satisfies this equation. But $X_k = f(A) = g(A)e^{At}$ if $f(\lambda) = g(\lambda)e^{\lambda t}$ and $g(\lambda^k) = 1$, and all the remaining $m-1$ values of $g(\lambda)$ on the spectrum of A are zero (see Vol. I, Chapter V, formula (17), on p. 104).

[6] This means that each function $p_{ik}(t)$ $(i, k = 1, 2, \ldots, n)$ is continuous and bounded in the interval $[t_0, \infty)$, i.e., $t \geqq t_0$.

We impose the following restrictions on the matrix $L(t) = \| l_{ik}(t) \|_1^n$ of the transformation:

1. $L(t)$ has a continuous derivative $\frac{dL}{dt}$ in the interval $[t_0, \infty)$;
2. $L(t)$ and $\frac{dL}{dt}$ are bounded in the interval $[t_0, \infty)$;
3. There exists a constant m such that

$$0 < m < \text{absolute value of } | L(t) | \qquad (t \geqq t_0),$$

i.e., the determinant $| L(t) |$ is bounded in modulus from below by the positive constant m.

A transformation (14) in which the coefficient matrix $L(t) = \| l_{ik}(t) \|_1^n$ satisfies 1.-3. will be called a *Lyapunov transformation* and the corresponding matrix $L(t)$ a *Lyapunov matrix.*

Such transformations were investigated by A. M. Lyapunov in his famous memoir 'The General Problem of Stability of Motion' [32].

Examples. 1. If $L =$ const. and $| L | \neq 0$, then L satisfies the conditions 1.-3. Therefore a non-singular transformation with constant coefficients is always a Lyapunov transformation.

2. If $D = \| d_{ik} \|_1^n$ is a matrix of simple structure with pure imaginary characteristic values, then the matrix

$$L(t) = e^{Dt}$$

satisfies the conditions 1.-3. and is therefore a Lyapunov matrix.[7]

2. It is easy to verify that the conditions 1.-3. of a matrix $L(t)$ imply the existence of the inverse matrix $L^{-1}(t)$ also satisfying the conditions 1.-3.; i.e., the inverse of a Lyapunov transformation is itself a Lyapunov transformation. In the same way it can be verified that two Lyapunov transformations in succession yield a Lyapunov transformation. Thus, the Lyapunov transformations form a group. They have the following important property:

If under the transformation (14) *the system* (1) *goes over into*

$$\frac{dy_i}{dt} = \sum_{k=1}^{n} q_{ik}(t)\, y_k \tag{15}$$

and if the zero solution of this system is stable, asymptotically stable, or unstable in the sense of Lyapunov (see Vol. I, Chapter V, § 6), *then the zero solution of the original system* (1) *has the same property.*

---

[7] Here all the $m_k = 1$ in (12) and $\lambda_k = i\varphi_k$ (φ_k real, $k = 1, 2, \ldots, s$).

In other words, Lyapunov transformations do not alter the character of the zero solution (as regards stability). This is the reason why these transformations can be used in the investigation of stability in order to simplify the original system of equations.

A Lyapunov transformation establishes a one-to-one correspondence between the solutions of the systems (1) and (15); moreover, linearly independent solutions remain so after the transformation. Therefore a Lyapunov transformation carries an integral matrix X of (1) into some integral matrix Y of (15) such that

$$X = L(t)\,Y. \tag{16}$$

In matrix notation, the system (15) has the form

$$\frac{dY}{dt} = Q(t)\,Y, \tag{17}$$

where $Q(t) = \| q_{ik}(t) \|_1^n$ is the coefficient matrix of (15).

Substituting LY for X in (3) and comparing the equation so obtained with (17), we easily find the following formula which expresses Q in terms of P and L:

$$Q = L^{-1}PL - L^{-1}\frac{dL}{dt}. \tag{18}$$

Two systems (1) and (15) or, what is the same, (3) and (17) will be called *equivalent* (in the sense of Lyapunov) if they can be carried into one another by a Lyapunov transformation. The coefficient matrices P and Q of equivalent systems are always connected by the formula (18) in which L satisfies the conditions 1.-3.

§ 3. Reducible Systems

1. Among the systems of linear differential equations of the first order the simplest and best known are those with constant coefficients. It is, therefore, of interest to study systems that can be carried by a Lyapunov transformation into systems with constant coefficients. Lyapunov has called such systems *reducible*.

Suppose given a reducible system

$$\frac{dX}{dt} = PX. \tag{19}$$

Then some Lyapunov transformation

$$X = L(t)\,Y \tag{20}$$

carries it into a system

$$\frac{dY}{dt} = AY, \tag{21}$$

where A is a constant matrix. Therefore (19) has the particular solution

$$\tilde{X} = L(t)\, e^{At}. \tag{22}$$

It is easy to see that, conversely, every system (19) with a particular solution of the form (22), where $L(t)$ is a Lyapunov matrix and A a constant matrix, is reducible and is reduced to the form (21) by means of the Lyapunov transformation (20).

Following Lyapunov, we shall show that: *Every system* (19) *with periodic coefficients is reducible.*[8]

Let $P(t)$ in (19) be a continuous function in $(-\infty, +\infty)$ with period τ:

$$P(t+\tau) = P(t). \tag{23}$$

Replacing t in (19) by $t+\tau$ and using (23), we obtain:

$$\frac{dX(t+\tau)}{dt} = P(t)\, X(t+\tau).$$

Thus, $X(t+\tau)$ is an integral matrix of (19) if $X(t)$ is. Therefore

$$X(t+\tau) = X(t)\, V,$$

where V is a constant non-singular matrix. Since $|V| \neq 0$, we can determine[9]

$$V^{\frac{t}{\tau}} = e^{\frac{t}{\tau}\ln V}.$$

This matrix function of t, just like $X(t)$, is multiplied on the right by V when the argument is increased by τ. Therefore the 'quotient'

$$L(t) = X(t)\, V^{-\frac{t}{\tau}} = X(t)\, e^{-\frac{t}{\tau}\ln V}$$

is continuous and periodic with period τ:

$$L(t+\tau) = L(t),$$

and with $|L| \neq 0$. The matrix $L(t)$ satisfies the conditions 1.-3. of the preceding section and is therefore a Lyapunov matrix.

[8] See [32], § 47.

[9] Here $\ln V = f(V)$, where $f(\lambda)$ is any single-valued branch of $\ln \lambda$ in the simply-connected domain G containing all the characteristic values of V, but not containing 0. See Vol. I, Chapter V.

On the other hand, since the solution X of (19) can be represented in the form

$$X = L(t)\, e^{\frac{\ln V}{\tau} t},$$

the system (19) is reducible.

In this case the Lyapunov transformation

$$X = L(t)\, Y,$$

which carries (19) into the form

$$\frac{dY}{dt} = \frac{1}{\tau} \ln V \cdot Y$$

has periodic coefficients with period τ.

Lyapunov has established[10] a very important criterion for stability and instability of a first linear approximation to a non-linear system of differential equations

$$\frac{dx_i}{dt} = \sum_{k=1}^{n} a_{ik} x_k + (**) \qquad (i = 1, 2, \ldots, n), \tag{24}$$

where we have convergent power series in $x_1, x_2, \ldots, x_n$ on the right-hand side and where (**) denotes the sum of the terms of second and higher orders in $x_1, x_2, \ldots, x_n$; the coefficients a_{ik} $(i, k = 1, 2, \ldots, n)$ of the linear terms are constant.[11]

LYAPUNOV'S CRITERION: *The zero solution of* (24) *is stable (and even asymptotically stable) if all the characteristic values of the coefficient matrix* $A = \| a_{ik} \|_1^n$ *of the first linear approximation have negative real parts, and unstable if at least one characteristic value has a positive real part.*

2. The arguments used above enable us to apply this criterion to a system whose linear terms have periodic coefficients:

$$\frac{dx_i}{dt} = \sum_{k=1}^{n} p_{ik}(t)\, x_k + (**). \tag{25}$$

For on the basis of the preceding arguments we reduce the system (25) to the form (24) by means of a Lyapunov transformation, where

---

[10] See [32], § 24.

[11] The coefficients in the non-linear terms may depend on t. These functional coefficients are subject to certain restrictions (see [32], § 11).

$$A = \| a_{ik} \|_1^n = \frac{1}{\tau} \ln V$$

and where V is the constant matrix by which an integral matrix of the corresponding linear system (19) is multiplied when the argument is changed by τ. Without loss of generality, we may assume that $\tau > 0$. By the properties of Lyapunov transformations the zero solutions of the original and of the transformed systems are simultaneously stable, asymptotically stable, or unstable. But the characteristic values λ_i and ν_i $(i = 1, 2, \ldots, n)$ of A and V are connected by the formula

$$\lambda_i = \frac{1}{\tau} \ln \nu_i \quad (i = 1, 2, \ldots, n).$$

Therefore, by applying Lyapunov's criterion to the reduced systems we find:[12]

The zero solution of (25) *is asymptotically stable if all the characteristic values* $\nu_1, \nu_2, \ldots, \nu_n$ *of* V *are of modulus less than* 1 *and unstable if at least one characteristic value is of modulus greater than* 1.

Lyapunov has established his criterion for the stability of a linear approximation for a considerably wider class of systems, namely those of the form (24) in which the linear approximation is not necessarily a system with constant coefficients, but belongs to a class of systems that he has called regular.[13]

The class of regular linear systems contains all the reducible systems.

A criterion for instability in the case when the first linear approximation is a regular system was set up by N. G. Chetaev.[14]

§ 4. The Canonical Form of a Reducible System. Erugin's Theorem

1. Suppose that a reducible system (19) and an equivalent system

$$\frac{dY}{dt} = AY$$

(in the sense of Lyapunov) are given, where A is a constant matrix.

We shall be interested in the question: *To what extent is the matrix* A *determined by the given system* (19)? This question can also be formulated as follows:

---

[12] *Loc. cit.*, § 55.

[13] *Loc. cit.*, § 9.

[14] See [9], p. 181.

When are two systems

$$\frac{dY}{dt}=AY \quad \text{and} \quad \frac{dZ}{dt}=BZ,$$

where A and B are constant matrices, equivalent in the sense of Lyapunov; i.e., when can they be carried into one another by a Lyapunov transformation?

In order to answer this question we introduce the notion of matrices with one and the same real part of the spectrum.

We shall say that two matrices A and B of order n *have one and the same real part of the spectrum* if and only if the elementary divisors of A and B are of the form

$$(\lambda-\lambda_1)^{m_1}, (\lambda-\lambda_2)^{m_2}, \ldots, (\lambda-\lambda_s)^{m_s}; (\lambda-\mu_1)^{m_1}, (\lambda-\mu_2)^{m_2}, \ldots, (\lambda-\mu_s)^{m_s},$$

where

$$\operatorname{Re}\lambda_k=\operatorname{Re}\mu_k \quad (k=1, 2, \ldots, s).$$

Then the following theorem due to N. P. Erugin holds:[15]

THEOREM 1 (Erugin): *Two systems*

$$\frac{dY}{dt}=AY \quad \text{and} \quad \frac{dZ}{dt}=BZ \tag{26}$$

(A and B are constant matrices of order n) are equivalent in the sense of Lyapunov if and only if the matrices A and B have one and the same real part of the spectrum.

Proof. Suppose that the systems (26) are given. We reduce A to the normal Jordan form[16] (see Vol. I, Chapter VI, § 7)

$$A=T\{\lambda_1E_1+H_1, \lambda_2E_2+H_2, \ldots, \lambda_sE_s+H_s\}T^{-1}, \tag{27}$$

where

$$\lambda_k=\alpha_k+i\beta_k \qquad (\alpha_k, \beta_k \text{ are real numbers}; k=1, 2, \ldots, s). \tag{28}$$

In accordance with (27) and (28) we set

$$\left.\begin{aligned} A_1&=T\{\alpha_1E_1+H_1, \alpha_2E_2+H_2, \ldots, \alpha_sE_s+H_s\}T^{-1},\\ A_2&=T\{i\beta_1E_1, i\beta_2E_2, \ldots, i\beta_sE_s\}T^{-1}.\end{aligned}\right\} \tag{29}$$

---

[15] Our proof of the theorem differs from that of Erugin.

[16] E_k is the unit matrix; in H_k the elements of the first superdiagonal are 1, and the remaining elements are zero; the orders of E_k, H_k are the degrees of the k-th elementary divisor of A, i.e., m_k $(k=1, 2, \ldots, s)$.

Then

$$A = A_1 + A_2, \quad A_1 A_2 = A_2 A_1. \tag{30}$$

We define a matrix $L(t)$ by the equation

$$L(t) = e^{A_2 t}.$$

$L(t)$ is a Lyapunov matrix (see Example 2 on p. 117).

But by (30) a particular solution of the first of the systems (26) is of the form

$$e^{At} = e^{A_2 t} e^{A_1 t} = L(t)\, e^{A_1 (t)}.$$

Hence it follows that the first of the systems (26) is equivalent to

$$\frac{dU}{dt} = A_1 U, \tag{31}$$

where, by (29), the matrix A_1 has real characteristic values and its spectrum coincides with the real part of the spectrum of A.

Similarly, we replace the second of the systems (26) by the equivalent system

$$\frac{dV}{dt} = B_1 V, \tag{32}$$

where the matrix B_1 has real characteristic values and its spectrum coincides with the real part of the spectrum of B.

Our theorem will be proved if we can show that *the two systems* (31) *and* (32) *in which* A_1 *and* B_1 *are constant matrices with real characteristic values are equivalent if and only if* A_1 *and* B_1 *are similar.*[17]

Suppose that the Lyapunov transformation

$$U = L_1 V$$

carries (31) into (32). Then the matrix L_1 satisfies the equation

$$\frac{dL_1}{dt} = A_1 L_1 - L_1 B_1. \tag{33}$$

This matrix equation for L_1 is equivalent to a system of n^2 differential equations in the n^2 elements of L_1. The right-hand side of (33) is a linear operation on the 'vector' L_1 in an n^2-dimensional space

---

[17] This proposition implies Theorem 1, since the equivalence of the systems (31) and (32) means that the systems (26) are equivalent, and the similarity of A_1 and B_1 means that these matrices have the same elementary divisors, so that the matrices A and B have one and the same real part of the spectrum.

$$\frac{dL_1}{dt} = \widehat{F}(L_1), \quad [\widehat{F}(L_1) = A_1L_1 - L_1B_1]. \tag{33'}$$

Every characteristic value of the linear operator $\widehat{F}$ (and of the corresponding matrix of order n^2) can be represented in the form of a difference $\gamma - \delta$, where γ is a characteristic value of A_1 and δ a characteristic value of B_1.[18] Hence it follows that the operator $\widehat{F}$ has only real characteristic values.

We denote by

$$\widehat{\psi}(\lambda) = (\lambda - \widehat{\lambda}_1)^{\widehat{m}_1} (\lambda - \widehat{\lambda}_2)^{\widehat{m}_2} \cdots (\lambda - \widehat{\lambda}_u)^{\widehat{m}_u}$$

(the $\widehat{\lambda}_i$ are real; $\widehat{\lambda}_i \neq \widehat{\lambda}_j$ for $i \neq j$; $i, j = 1, 2, \ldots, u$) the minimal polynomial of $\widehat{F}$. Then the solution $L_1(t) = e^{\widehat{F}t}L^{(0)}$ of (33′) can, by formula (12) (p. 116), be written as follows:

$$L_1(t) = \sum_{k=1}^{u} \sum_{j=0}^{\widehat{m}_k - 1} L_{kj} t^j e^{\widehat{\lambda}_k t}, \tag{34}$$

where the L_{kj} are constant matrices of order n. Since the matrix $L_1(t)$ is bounded in the interval (t_0, ∞), both for every $\widehat{\lambda}_k > 0$ and for $\widehat{\lambda}_k = 0$ and $j > 0$, the corresponding matrices $L_{kj} = O$. We denote by $L_-(t)$ the sum of all the terms in (34) for which $\widehat{\lambda}_k < 0$. Then

$$L_1(t) = L_-(t) + L_0, \tag{35}$$

where

$$\lim_{t \to +\infty} L_-(t) = O, \quad \lim_{t \to +\infty} \frac{dL_-(t)}{dt} = O, \quad L_0 = \text{const.} \tag{35'}$$

Then, by (35) and (35′),

$$\lim_{t \to +\infty} L_1(t) = L_0,$$

[18] For let Λ_0 be any characteristic value of the operator $\widehat{F}$. Then there exists a matrix $L \neq O$ such that $\widehat{F}(L) = \Lambda_0 L$, or

$$(A_1 - \Lambda_0 E)L = LB_1. \tag{*}$$

The matrices $A_1 - \Lambda_0 E$ and B_1 have at least one characteristic value in common, since otherwise there would exist a polynomial $g(\lambda)$ such that

$$g(A_1 - \Lambda_0 E) = O, \; g(B_1) = E,$$

and this is impossible, because it follows from (*) that $g(A_1 - \Lambda_0 E) \cdot L = L \cdot g(B_1)$ and $L \neq O$. But if $A_1 - \Lambda_0 E$ and B_1 have a common characteristic value, then $\Lambda_0 = \gamma - \delta$, where γ and δ are characteristic values of A_1 and B_1, respectively. A detailed study of the operator $\widehat{F}$ can be found in the paper [179] by F. Golubchikov.

from which it follows that

$$|L_0| \neq 0,$$

because the determinant $|L_1(t)|$ is bounded in modulus from below.

When we substitute for $L_1(t)$ in (33) the sum $L_-(t) + L_0$, we obtain:

$$\frac{dL_-(t)}{dt} - A_1 L_-(t) + B_1 L_-(t) = A_1 L_0 - B_1 L_0;$$

hence by (35′)

$$A_1 L_0 - L_0 B_1 = O$$

and therefore

$$B_1 = L_0^{-1} A_1 L_0. \tag{36}$$

Conversely, if (36) holds, then the Lyapunov transformation

$$U = L_0 V$$

carries (31) into (32). This completes the proof of the theorem.

2. From this theorem it follows that: *Every reducible system* (19) *can be carried by the Lyapunov transformation* $X = LY$ *into the form*

$$\frac{dY}{dt} = JY,$$

where J *is a Jordan matrix with real characteristic values.* This canonical form of the system is uniquely determined by the given matrix $P(t)$ to within the order of the diagonal blocks of J.

§ 5. The Matricant

1. We consider a system of differential equations

$$\frac{dX}{dt} = P(t)\,X, \tag{37}$$

where $P(t) = \| p_{ik}(t) \|_1^n$ is a continuous matrix function of the argument t in some interval (a, b).[19]

[19] (a, b) is an arbitrary interval (finite or infinite). All the elements $p_{ik}(t)$ $(i, k = 1, 2, \dots, n)$ of $P(t)$ are complex functions of the real argument t, continuous in (a, b). Everything that follows remains valid if, instead of continuity, we require (in every finite subinterval of (a, b)) only boundedness and Riemann integrability of all the functions $p_{ik}(t)$.

We use the method of successive approximations to determine a normalized solution of (37), i.e., a solution that for $t = t_0$ becomes the unit matrix (t_0 is a fixed number of the interval (a, b)). The successive approximations X_k $(k = 0, 1, 2, \ldots)$ are found from the recurrence relations

$$\frac{dX_k}{dt} = P(t)\, X_{k-1} \qquad (k = 1, 2, \ldots),$$

when X_0 is taken to be the unit matrix E.

Setting $X_k(t_0) = E$ $(k = 0, 1, 2, \ldots)$ we may represent X_k in the form

$$X_k = E + \int_{t_0}^{t} P(\tau)\, X_{k-1}\, d\tau.$$

Thus

$$X_0 = E, \quad X_1 = E + \int_{t_0}^{t} P(\tau)\, d\tau, \quad X_2 = E + \int_{t_0}^{t} P(\tau)\, d\tau + \int_{t_0}^{t} P(\tau) \int_{t_0}^{\tau} P(\sigma)\, d\sigma\, d\tau, \ldots,$$

i.e., X_k $(k = 0, 1, 2, \ldots)$ is the sum of the first $k + 1$ terms of the matrix series

$$E + \int_{t_0}^{t} P(\tau)\, d\tau + \int_{t_0}^{t} P(\tau) \int_{t_0}^{\tau} P(\sigma)\, d\sigma\, d\tau + \cdots. \tag{38}$$

In order to prove that this series is absolutely and uniformly convergent in every closed subinterval of the interval (a, b) and determines the required solution of (37), we construct a majorant.

We define non-negative functions $g(t)$ and $h(t)$ in (a, b) by the equations[20]

$$g(t) = \max\left[\,|p_{11}(t)|, |p_{12}(t)|, \ldots, |p_{nn}(t)|\,\right], \quad h(t) = \left|\int_{t_0}^{t} g(\tau)\, d\tau\right|.$$

It is easy to verify that $g(t)$, and consequently $h(t)$ as well, is continuous in (a, b).[21]

Each of the n^2 scalar series into which the matrix series (38) splits is majorized by the series

$$1 + h(t) + \frac{nh^2(t)}{2!} + \frac{n^2h^3(t)}{3!} + \cdots. \tag{39}$$

---

[20] By definition, the value of $g(t)$ for any value of t is the largest of the n^2 moduli of the values of $p_{ik}(t)$ $(i, k = 1, 2, \ldots, n)$ for that value of t.

[21] The continuity of $g(t)$ at any point t_1 of the interval (a, b) follows from the fact that the difference $g(t) - g(t_1)$ for t sufficiently near t_1 always coincides with one of the n^2 differences $|p_{ik}(t)| - |p_{ik}(t_1)|$ $(i, k = 1, 2, \ldots, n)$.

For

$$\left|\left(\int_{t_0}^{t} P(\tau)\,d\tau\right)_{i,k}\right| = \left|\int_{t_0}^{t} p_{ik}(\tau)\,d\tau\right| \leqq \left|\int_{t_0}^{t} g(\tau)\,d\tau\right| = h(t),$$

$$\left|\left(\int_{t_0}^{t} P(\tau)\int_{t_0}^{\tau} P(\sigma)\,d\sigma\,d\tau\right)_{i,k}\right| = \left|\sum_{j=1}^{n}\int_{t_0}^{t} p_{ij}(\tau)\int_{t_0}^{\tau} p_{jk}(\sigma)\,d\sigma\,d\tau\right| \leqq n\left|\int_{t_0}^{t} g(\tau)\int_{t_0}^{\tau} g(\sigma)\,d\sigma\,d\tau\right| = \frac{n\,h^2(t)}{2},$$

etc.

The series (39) converges in (a, b) and converges uniformly in every closed part of this interval. Hence it follows that the matrix series (38) also converges in (a, b) and does so absolutely and uniformly in every closed interval contained in (a, b).

By term-by-term differentiation we verify that the sum of (38) is a solution of (37); this solution becomes E for $t = t_0$. The term-by-term differentiation of (38) is permissible, because the series obtained after differentiation differs from (38) by the factor P and therefore, like (38), is uniformly convergent in every closed interval contained in (a, b).

Thus we have proved the theorem on the existence of a normal solution of (37). This solution will be denoted by $\Omega_{t_0}^{t}(P)$ or simply $\Omega_{t_0}^{t}$. Every other solution, as we have shown in § 1, is of the form

$$X = \Omega_{t_0}^{t} C,$$

where C is an arbitrary constant matrix. From this formula it follows that every solution, in particular the normalized one, is uniquely determined by its value for $t = t_0$.

This normalized solution $\Omega_{t_0}^{t}$ of (37) is often called the *matricant*.

We have seen that the matricant can be represented in the form of a series[22]

$$\Omega_{t_0}^{t} = E + \int_{t_0}^{t} P(\tau)\,d\tau + \int_{t_0}^{t} P(\tau)\int_{t_0}^{\tau} P(\sigma)\,d\sigma\,d\tau + \cdots, \tag{40}$$

which converges absolutely and uniformly in every closed interval in which $P(t)$ is continuous.

2. We mention a few formulas involving the matricant.

1. $$\Omega_{t_0}^{t} = \Omega_{t_1}^{t}\Omega_{t_0}^{t_1} \quad (t_0, t_1, t \in (a, b)).$$

For since $\Omega_{t_0}^{t}$ and $\Omega_{t_1}^{t}$ are two solutions of (37), we have

[22] The representation of the matricant in the form of such a series was first obtained by Peano [308].

$$\Omega_{t_0}^t = \Omega_{t_1}^t C \quad (C \text{ is a constant matrix}).$$

Setting $t = t_1$ in this equation, we obtain $C = \Omega_{t_0}^{t_1}$.

2. $\Omega_{t_0}^t (P+Q) = \Omega_{t_0}^t (P)\, \Omega_{t_0}^t (S)$ *with* $S = [\Omega_{t_0}^t (P)]^{-1} Q \Omega_{t_0}^t (P)$.

To derive this formula we set:

$$X = \Omega_{t_0}^t (P), \quad Y = \Omega_{t_0}^t (P+Q),$$

and

$$Y = XZ. \tag{41}$$

Differentiating (41) term by term, we find:

$$(P+Q)\, XZ = PXZ + X\frac{dZ}{dt}.$$

Hence

$$\frac{dZ}{dt} = X^{-1}QXZ$$

and since it follows from (41) that $Z(t_0) = E$,

$$Z = \Omega_{t_0}^t (X^{-1}QX).$$

When we substitute their respective matricants for X, Y, Z in (41), we obtain the formula 2.

3. $$\ln | \Omega_{t_0}^t (P) | = \int_{t_0}^{t} \operatorname{tr} P \, d\tau.$$

This formula follows from the Jacobi identity (4) (p. 114) when we substitute $\Omega_{t_0}^t (P)$ for $X(t)$ in that identity.

4. *If* $A = \| a_{ik} \|_1^n = \text{const.}$, *then*

$$\Omega_{t_0}^t (A) = e^{A(t-t_0)}.$$

We introduce the following notation. If $P = \| p_{ik} \|_1^n$, then we shall mean by mod P the matrix

$$\operatorname{mod} P = \| \, |p_{ik}| \, \|_1^n .$$

Furthermore, if $A = \| a_{ik} \|_1^n$ and $B = \| b_{ik} \|_1^n$ are two real matrices and

$$a_{ik} \leqq b_{ik} \qquad (i, k = 1, 2, \ldots, n),$$

then we shall write

$$A \leqq B.$$

Then it follows from the representation (40) that:

5. *If* $\operatorname{mod} P(t) \leqq \operatorname{mod} Q(t)$ $(t \geqq t_0)$, *then the series* (40) *for* $\Omega_{t_0}^t(P)$ *is majorized, beginning with the first term, by the same series for* $\Omega_{t_0}^t(Q)$, *so that for all* $t \geqq t_0$

$$\operatorname{mod} \Omega_{t_0}^t(P) \leqq \Omega_{t_0}^t(Q), \quad \operatorname{mod}[\Omega_{t_0}^t(P) - E] \leqq \Omega_{t_0}^t(Q) - E,$$

$$\operatorname{mod}[\Omega_{t_0}^t(P) - E - \int_{t_0}^t P d\tau] \leqq \Omega_{t_0}^t(Q) - E - \int_{t_0}^t Q d\tau, \qquad \text{etc.}$$

In what follows we shall denote the matrix of order n in which all the elements are 1 by I:

$$I = \| 1 \|.$$

We consider the function $g(t)$ defined on p. 126. Then we have

$$\operatorname{mod} P(t) \leqq g(t) I.$$

But $\Omega_{t_0}^t(g(t)I)$ is the normalized solution of the equation

$$\frac{dX}{dt} = g(t) IX.$$

Therefore, by 4.,[23]

$$\Omega_{t_0}^t(g(t)I) = e^{h(t)I} = E + \left(h(t) + \frac{nh^2(t)}{2!} + \frac{n^2h^3(t)}{3!} + \cdots\right) I, \tag{42}$$

where

$$h(t) = \int_{t_0}^t g(\tau)\, d\tau, \quad g(t) = \max_{1 \leqq i, k \leqq n} |p_{ik}(t)|.$$

Therefore it follows from 5. and (42) that:

6.
$$\operatorname{mod} \Omega_{t_0}^t(P) \leqq E + \frac{1}{n}(e^{nh(t)} - 1) I,$$

$$\operatorname{mod}[\Omega_{t_0}^t(P) - E] \leqq \frac{1}{n}(e^{nh(t)} - 1) I,$$

$$\operatorname{mod}[\Omega_{t_0}^t(P) - E - \int_{t_0}^t P d\tau] \leqq \frac{1}{n}(e^{nh(t)} - 1 - nh(t)) I, \quad \text{etc.}$$

We shall now derive an important formula giving an estimate for the modulus of the difference between two matricants:

---

[23] By replacing the independent variable t by $h = \int_{t_0}^t g(t) dt$.

7. $\operatorname{mod}[\Omega_{t_0}^t(P) - \Omega_{t_0}^t(Q)] \leqq \frac{1}{n} e^{nq(t-t_0)}(e^{nd(t-t_0)} - 1) I \qquad (t \geqq t_0),$

if

$$\operatorname{mod} Q \leqq qI, \quad \operatorname{mod}(P - Q) \leqq d \cdot I, \quad I = \| 1 \|$$

(q, d are non-negative numbers; n is the order of P and Q).

We denote the difference $P - Q$ by D. Then

$$P = Q + D, \qquad \operatorname{mod} D \leqq d \cdot I.$$

Using the expansion (40) of the matricant in a series, we find:

$$\Omega_{t_0}^t(Q + D) - \Omega_{t_0}^t(Q)$$
$$= \int_{t_0}^{t} D(\tau)\,d\tau + \int_{t_0}^{t} D(\tau) \int_{t_0}^{\tau} Q(\sigma)\,d\sigma\,d\tau + \int_{t_0}^{t} Q(\tau) \int_{t_0}^{\tau} D(\sigma)\,d\sigma\,d\tau + \int_{t_0}^{t} D(\tau) \int_{t_0}^{\tau} D(\sigma)\,d\sigma\,d\tau + \cdots.$$

From this expression it is clear that, for $t \geqq t_0$,

$$\begin{aligned}
\operatorname{mod}[\Omega_{t_0}^t(Q + D) - \Omega_{t_0}^t(Q)] &\leqq \Omega_{t_0}^t(\operatorname{mod} Q + \operatorname{mod} D) - \Omega_{t_0}^t(\operatorname{mod} Q) \\
&\leqq \Omega_{t_0}^t((q + d) I) - \Omega_{t_0}^t(qI) = e^{(q+d) I (t-t_0)} - e^{qI(t-t_0)} \\
&= e^{qI(t-t_0)}(e^{dI(t-t_0)} - E) \\
&= \frac{1}{n}\left[E + \frac{1}{n}(e^{nq(t-t_0)}) - 1) I\right](e^{nd(t-t_0)} - 1) I \\
&= \frac{1}{n}\left[I + \frac{1}{n}(e^{nq(t-t_0)} - 1) I^2\right](e^{nd(t-t_0)} - 1) \\
&= \frac{1}{n} e^{nq(t-t_0)}(e^{nd(t-t_0)} - 1) I.
\end{aligned}$$

We shall now show how to express by means of the matricant the general solution of a system of linear differential equations with right-hand sides:

$$\frac{dx_i}{dt} = \sum_{k=1}^{n} p_{ik}(t) x_k + f_i(t) \qquad (i = 1, 2, \ldots, n); \tag{43}$$

$p_{ik}(t)$ and $f_i(t)$ ($i, k = 1, 2, \ldots, n$) are continuous functions of t in some interval.

By introducing the column matrices ('vectors') $x = (x_1, x_2, \ldots, x_n)$ and $f = (f_1, f_2, \ldots, f_n)$ and the square matrix $P = \| p_{ik} \|_1^n$, we write the system as follows:

$$\frac{dx}{dt} = P(t)x + f(t). \tag{43'}$$

We shall look for a solution of this equation in the form

$$x = \Omega_{t_0}^{t}(P)\, z, \tag{44}$$

where z is an unknown column depending on t. We substitute this expression for x in (43′) and obtain:

$$P\Omega_{t_0}^{t}(P)\, z + \Omega_{t_0}^{t}(P)\frac{dz}{dt} = P\Omega_{t_0}^{t}(P)\, z + f(t);$$

hence

$$\frac{dz}{dt} = [\Omega_{t_0}^{t}(P)]^{-1} f(t).$$

Integrating this, we find:

$$z = \int_{t_0}^{t} [\Omega_{t_0}^{\tau}(P)]^{-1} f(\tau)\, d\tau + c,$$

where c is an arbitrary constant vector. Substituting this expression in (44), we obtain:

$$x = \Omega_{t_0}^{t}(P) \int_{t_0}^{t} [\Omega_{t_0}^{\tau}(P)]^{-1} f(\tau)\, d\tau + \Omega_{t_0}^{t}(P)\, c. \tag{45}$$

When we give to t the value t_0, we find: $x(t_0) = c$. Therefore (45) assumes the form

$$x = \Omega_{t_0}^{t}(P)\, x(t_0) + \int_{t_0}^{t} K(t, \tau) f(\tau)\, d\tau, \tag{45′}$$

where

$$K(t, \tau) = \Omega_{t_0}^{t}(P)\, [\Omega_{t_0}^{\tau}(P)]^{-1}$$

is the so-called Cauchy matrix.

§ 6. The Multiplicative Integral. The Infinitesimal Calculus of Volterra

1. Let us consider the matricant $\Omega_{t_0}^{t}(P)$. We divide the basic interval (t_0, t) into n parts by introducing intermediate points $t_1, t_2, \dots, t_{n-1}$ and set $\Delta t_k = t_k - t_{k-1}$ ($k = 1, 2, \dots, n$; $t_n = t$). Then by property 1. of the matricant (see the preceding section),

$$\Omega_{t_0}^{t} = \Omega_{t_{n-1}}^{t} \cdots \Omega_{t_1}^{t_2}\Omega_{t_0}^{t_1}. \tag{46}$$

In the interval (t_{k-1}, t_k) we choose an intermediate point τ_k $(k=1, 2, \ldots, n)$. By regarding the Δt_k as small quantities of the first order we can take, for the computation of $\Omega_{t_{k-1}}^{t_k}$ to within small quantities of the second order, $P(t) \approx \text{const.} = P(\tau_k)$. Then

$$\Omega_{t_{k-1}}^{t_k} = e^{P(\tau_k)\Delta t_k} + (**) = E + P(\tau_k)\,\Delta t_k + (**); \tag{47}$$

here we denote by the symbol $(**)$ the sum of terms beginning with terms of the second order.

From (46) and (47) we find:

$$\Omega_{t_0}^{t} = e^{P(\tau_n)\Delta t_n} \cdots e^{P(\tau_2)\Delta t_2} e^{P(\tau_1)\Delta t_1} + (*) \tag{48}$$

and

$$\Omega_{t_0}^{t} = [E + P(\tau_n)\,\Delta t_n] \cdots [E + P(\tau_2)\,\Delta t_2]\,[E + P(\tau_1)\,\Delta t_1] + (*). \tag{49}$$

When we pass to the limit by increasing the number of intervals indefinitely and letting the length of these intervals tend to zero (the small terms $(*)$ disappear in the limit),[24] we obtain the exact limit formulas

$$\Omega_{t_0}^{t}(P) = \lim_{\Delta t_k \to 0} [e^{P(\tau_n)\Delta t_n} \cdots e^{P(\tau_2)\Delta t_2} e^{P(\tau_1)\Delta t_1}] \tag{48'}$$

and

$$\Omega_{t_0}^{t}(P) = \lim_{\Delta t_k \to 0} [E + P(\tau_n)\,\Delta t_n] \cdots [E + P(\tau_2)\,\Delta t_2]\,[E + P(\tau_1)\,\Delta t_1]. \tag{49'}$$

The expression under the limit sign on the right-hand side of the latter equation is the *product integral*.[25] We shall call its limit the *multiplicative integral* and denote it by the symbol

$$\widehat{\int_{t_0}^{t}} [E + P(t)\,dt] = \lim_{\Delta t_k \to 0} [E + P(\tau_n)\,\Delta t_n] \cdots [E + P(\tau_1)\,\Delta t_1]. \tag{50}$$

The formula (49′) gives a representation of the matricant in the form of a multiplicative integral

$$\Omega_{t_0}^{t}(P) = \widehat{\int_{t_0}^{t}} (E + P\,dt), \tag{51}$$

and the formulas (48) and (49) may be used for the approximative computation of the matricant.

---

24 These arguments can be made more precise by an estimate of the terms we have denoted by $(*)$. For a rigorous deduction of (48′) we have to use formula 7. of § 5 in which the matricant $Q(t)$ must be replaced by a piece-wise constant matrix

$$Q(t) = P(t_k) \qquad (t_{k-1} \leqq t \leqq t_k;\ k = 1, 2, \ldots, n).$$

25 An analogue to the sum integral for the ordinary integral.

The multiplicative integral was first introduced by Volterra in 1887. On the basis of this concept Volterra developed an original infinitesimal calculus for matrix functions (see [63]).[26]

The whole peculiarity of the multiplicative integral is tied up with the fact that the various values of the matrix function $P(t)$ in subintervals are not permutable. In the very special case when all these values are permutable

$$P(t')\,P(t'')=P(t'')\,P(t') \qquad (t', t'' \in (t_0, t)),$$

the multiplicative integral, as is clear from (48′) and (51), reduces to the matrix

$$e^{\int_{t_0}^{t} P(t)\,dt}$$

2. We now introduce the *multiplicative derivative*

$$D_t X=\frac{dX}{dt}X^{-1}. \tag{52}$$

The operations D_t and $\widehat{\int}_{t_0}^{t}$ are mutually inverse:

If

$$D_t X=P,$$

then[27]

$$X=\widehat{\int}_{t_0}^{t}(E+P\,dt)\cdot C \qquad (C=X(t_0)),$$

and vice versa. The last formula can also be written as follows: [28]

$$\widehat{\int}_{t_0}^{t}(E+P\,dt)=X(t)\,X(t_0)^{-1}. \tag{53}$$

We leave it to the reader to verify the following differential and integral formulas:[29]

---

[26] The multiplicative integral (in German, *Produkt-Integral*) was used by Schlesinger in investigating systems of linear differential equations with analytic coefficients [49] and [50]; see also [321].

The multiplicative integral (50) exists not only for a function $P(t)$ that is continuous in the interval of integration, but also under considerably more general conditions (see [116]).

[27] Here the arbitrary constant matrix C is an analogue to the arbitrary additive constant in the ordinary indefinite integral.

[28] An analogue to the formula $\int_{t_0}^{t} P\,dt=X(t)-X(t_0)$, where $\frac{dX}{dt}=P$.

[29] These formulas can be deduced immediately from the definitions of the multiplicative derivative and multiplicative integral (see [63]). However, the integral formulas are obtained more quickly and simply if the multiplicative integral is regarded as a matricant and the properties of the matricant that were expounded in the preceding section are used (see [49]).

DIFFERENTIAL FORMULAS

$$\text{I. } D_t(XY) = D_t(X) + X D_t(Y) X^{-1},$$

$$D_t(XC) = D_t(X), \qquad D_t(CY) = C D_t(Y) C^{-1}. \qquad (C \text{ is a constant matrix})$$

$$\text{II. } D_t(X^{\mathsf{T}}) = X^{\mathsf{T}} (D_t X)^{\mathsf{T}} X^{\mathsf{T}-1}.$$

$$\text{III. } D_t(X^{-1}) = -X^{-1} D_t(X) X = -(D_t(X^{\mathsf{T}}))^{\mathsf{T}},$$

$$D_t((X^{\mathsf{T}})^{-1}) = -(D_t(X))^{\mathsf{T}}.$$

INTEGRAL FORMULAS

$$\text{IV. } \widehat{\int}_{t_0}^{t} (E + P\,d\tau) = \widehat{\int}_{t_1}^{t} (E + P\,d\tau) \widehat{\int}_{t_0}^{t_1} (E + P\,d\tau).$$

$$\text{V. } \widehat{\int}_{t_0}^{t} (E + P\,d\tau) = \left[\widehat{\int}_{t}^{t_0} (E + P\,d\tau)\right]^{-1}.$$

$$\text{VI. } \widehat{\int}_{t_0}^{t} (E + CPC^{-1}\,d\tau) = C \widehat{\int}_{t_0}^{t} (D + P\,d\tau)\, C^{-1} \quad (C \text{ is a constant matrix})$$

$$\text{VII. } \widehat{\int}_{t_0}^{t} [E + (Q + D_t X)\,d\tau] = X(t) \widehat{\int}_{t_0}^{t} (E + X^{-1} Q X\,d\tau)\, X(t_0)^{-1}.$$ [37]

$$\text{VIII. } \operatorname{mod}\left[\widehat{\int}_{t_0}^{t} (E + P\,d\tau) - \widehat{\int}_{t_0}^{t} (E + Q\,d\tau)\right] \leqq \frac{1}{n} e^{nq(t-t_0)} \left(e^{nd(t-t_0)} - 1\right) I \ (t > t_0),$$

if

$$\operatorname{mod} Q \leqq q \cdot I, \qquad \operatorname{mod}(P - Q) \leqq d \cdot I, \qquad I = \| 1 \|$$

(q and d are non-negative numbers; n is the order of P and Q).

Suppose now that the matrices P and Q depend on the same parameter α

$$P = P(\tau, \alpha), \qquad Q = Q(\tau, \alpha)$$

and that

$$\lim_{\alpha \to \alpha_0} P(\tau, \alpha) = \lim_{\alpha \to \alpha_0} Q(\tau, \alpha) = P_0(\tau),$$

where the limit is approached uniformly with respect to τ in the interval (t_0, t) in question. Furthermore, let us assume that for $\alpha \to \alpha_0$ the matrix $Q(\tau, \alpha)$ is bounded in modulus by qI, where q is a positive constant. Then, setting

$$\lim_{\alpha \to \alpha_0} d(\alpha) = 0,$$

we have:

$$d(\alpha) = \max_{\substack{1 \leqq i, k \leqq n \\ t_0 \leqq \tau \leqq t}} | p_{ik}(\tau, \alpha) - q_{ik}(\tau, \alpha) |.$$

---

[31] The formula VII can be regarded in a certain sense as a analogue to the formula for integration by parts in ordinary (non-multiplicative) integrals. VII follows from 2. of § 5).

Therefore it follows from formula VIII that:

$$\lim_{\alpha\to\alpha_0}\left[\widehat{\int}_{t_0}^{t}(E+P\,dt)-\widehat{\int}_{t_0}^{t}(E+Q\,dt)\right]=O.$$

In particular, if Q does not depend on α $(Q(t,\alpha)=P_0(t))$, we obtain:

$$\lim_{\alpha\to\alpha_0}\widehat{\int}_{t_0}^{t}[E+P(t,\alpha)\,dt]=\widehat{\int}_{t_0}^{t}[E+P_0(t)\,dt],$$

where

$$P_0(t)=\lim_{\alpha\to\alpha_0}P(t,\alpha).$$

§ 7. Differential Systems in a Complex Domain. General Properties

1. We consider a system of differential equations

$$\frac{dx_i}{dz}=\sum_{k=1}^{n}p_{ik}(z)\,x_k. \tag{54}$$

Here the given function $p_{ik}(z)$ and the unknown functions $x_i(z)$ $(i,k=1,2,\ldots,n)$ are supposed to be single-valued analytic functions of a complex argument z, regular in a domain G of the complex z-plane.

Introducing the square matrix $P(z)=\|\,p_{ik}(z)\,\|_1^n$ and the column matrix $x=(x_1,x_2,\ldots,x_n)$, we can write the system (54), as in the case of a real argument (§ 1), in the form

$$\frac{dx}{dz}=P(z)\,x \tag{54'}$$

Denoting an integral matrix, i.e., a matrix whose columns are n linearly independent solutions of (54), by X, we can write instead of (54′):

$$\frac{dX}{dz}=P(z)\,X \tag{55}$$

Jacobi's formula holds also for a complex argument z:

$$|X|=ce^{\int_{z_0}^{z}\operatorname{tr}P\,dz} \tag{56}$$

Here it is assumed that z_0 and all the points of the path along which $\int_{z_0}^{z}$ is taken are regular points for the single-valued analytic function $\operatorname{tr}P(z)=p_{11}(z)+p_{22}(z)+\ldots+p_{nn}(z)$.[32]

[32] Here, and in what follows, the path of integration is taken as a sectionally smooth curve.

2. A peculiar feature of the case of a complex argument is the fact that for a single-valued function $P(z)$ the integral matrix $X(z)$ may well be a many-valued function of z.

As an example, we consider the Cauchy system

$$\frac{dX}{dz}=\frac{U}{z-a}X \qquad (U \text{ is a constant matrix}). \tag{57}$$

One of the solutions of this system, as in the case of a real argument (see p. 115), is the integral matrix

$$X=e^{U\ln(z-a)}=(z-a)^U. \tag{58}$$

For the domain G we take the whole z-plane except the point $z=a$. All the points of this domain are regular points of the coefficient matrix

$$P(z)=\frac{U}{z-a}\cdot$$

If $U\neq O$, then $z=a$ is a singular point (a pole of the first order) of the matrix function $P(z)=U/(z-a)$.

An element of the integral matrix (58) after going around the point $z=a$ once in the positive direction returns with a new value which is obtained from the old one by multiplication on the right by the constant matrix

$$V=e^{2\pi iU}$$

In the general case of a system (55) we see, by the same reasoning as in the case of a real argument, that two single-valued solutions X and $\tilde{X}$ are always connected in some part of the domain G by the formula

$$X=\tilde{X}C,$$

where C is a constant matrix. This formula remains valid under any analytic continuation of the functions $X(z)$ and $\tilde{X}(z)$ in G.

The proof of the theorem on the existence and (for given initial values) uniqueness of the solution of (54) is similar to that of the real case.

Let us consider a simply-connected star domain G_1 (relative to z_0)[33] forming part of G and let the matrix function $P(z)$ be regular[34] in G_1. We form the series

$$E+\int_{z_0}^{z}P(\zeta)\,d\zeta+\int_{z_0}^{z}P(\zeta)\int_{z_0}^{z}P(\zeta')\,d\zeta'\,d\zeta+\cdots. \tag{59}$$

---

[33] A domain is called a *star domain relative to a point* z_0 if every segment joining z_0 to an arbitrary point z of the domain lies entirely in the given domain.

[34] I.e., all the elements $p_{ik}(z)$ $(i, k=1, 2, \ldots, n)$ of $P(z)$ are regular functions in G_1.

Since G_1 is simply-connected, it follows that every integral that occurs in (59) is independent of the path of integration and is a regular function in G_1. Since G_1 is a star domain relative to z_0, we may assume for the purpose of an estimate of the moduli of these integrals that they are all taken along the straight-line segment joining z_0 and z.

That the series (59) converges absolutely and uniformly in every closed part of G_1 containing z_0 follows from the convergence of the majorant

$$1 + lM + \frac{n}{2!} l^2 M^2 + \frac{n^2}{3!} l^3 M^3 + \cdots.$$

Here M is an upper bound for the modulus of $P(z)$ and l an upper bound for the distance of z from z_0, and both bounds refer to the closed part of G_1 in question.

By differentiating term by term we verify that the sum of the series (59) is a solution of (55). This solution is normalized, because for $z = z_0$ it reduces to the unit matrix E. The single-valued normalized solution of (55) will be called, as in the real case, a matricant and will be denoted by $\Omega^t_{t_0}(P)$. Thus we have obtained a representation of the matricant in G_1 in the form of a series[35]

$$\Omega^z_{z_0}(P) = E + \int_{t_0}^{t} P(\zeta)\, d\zeta + \int_{t_0}^{t} P(\zeta) \int_{t_0}^{\zeta} P(\zeta')\, d\zeta'\, d\zeta + \cdots. \tag{60}$$

The properties 1.-4. *of the matricant that were set up in* § 5 *automatically carry over to the case of a complex argument.*

Any solution of (55) that is regular in G and reduces to the matrix X_0 for $z = z_0$ can be represented in the form

$$X = \Omega^z_{z_0}(P) \cdot C \qquad (C = X_0). \tag{61}$$

The formula (61) comprises all single-valued solutions that are regular in a neighborhood of z_0 (z_0 is a regular point of the coefficient matrix $P(z)$). These solutions when continued analytically in G give all the solutions of (55); i.e., the equation (55) cannot have any solutions for which z_0 would be a singular point.

For the analytic continuation of the matricant in G it is convenient to use the multiplicative integral.

---

[35] Our proof for the existence of a normalized solution and its representation in G_1 by the series (60) remains valid if instead of the assumption that the domain is a star domain we make a wider assumption, namely, that for every closed part of G_1 there exists a positive number l such that every point z of this closed part can be joined to z_0 by a path of length not exceeding l.

§ 8. The Multiplicative Integral in a Complex Domain

1. The multiplicative integral along a curve in the complex plane is defined in the following way.

Suppose that L is some path and $P(z)$ a matrix function, continuous on L. We divide the path L into n parts (z_0, z_1), (z_1, z_2), ..., (z_{n-1}, z_n); here z_0 is the beginning, and $z_n = z$ the end of the path, and $z_1, z_2, \ldots, z_{n-1}$ are intermediate points of division. On the segment $z_{k-1}z_k$ we take an arbitrary point ζ_k and we use the notation $\Delta z_k = z_k - z_{k-1}$ $(k = 1, 2, \ldots, n)$. We then define

$$\overset{\frown}{\int_L} [E + P(z)\,dz] = \lim_{\Delta z_k \to 0} (E + P(\zeta_n)\,\Delta z_n] \cdots [E + P(\zeta_1)\,\Delta z_1].$$

When we compare this definition with that on p. 132, we see that they coincide in the special case where L is a segment of the real axis. However, even in the general case, where L is located anywhere in the complex plane, the new definition may be reduced to the old one by a change of the variable of integration.

If

$$z = z(t)$$

is a parametric equation of the path, where $z(t)$ is a continuous function in the interval (t_0, t) with a piece-wise continuous derivative $\frac{dz}{dt}$, then it is easy to see that

$$\overset{\frown}{\int_L} [E + P(z)\,dz] = \overset{\frown}{\int_{t_0}^{t}} \left\{E + P[z(t)]\,\frac{dz}{dt}\,dt\right\}.$$

This formula shows that the multiplicative integral along an arbitrary path exists if the matrix $P(z)$ under the integral sign is continuous along this path.[36]

2. The multiplicative derivative is defined by the previous formula

$$D_z X = \frac{dX}{dz} X^{-1}.$$

Here it is assumed that $X(z)$ is an analytic function.

All the differential formulas (I-III) of the preceding section carry over without change to the case of a complex argument. As regards the integral formulas IV-VI, their outward form has to be modified somewhat:

---

[36] See footnote 26. Even when $P(z)$ is continuous along L, the function $P[z(t)]\frac{dz}{dt}$ may only be sectionally continuous. In this case we can split the interval (t_0, t) into partial intervals in each of which the derivative $\frac{dz}{dt}$ is continuous and can interpret the integral from t_0 to t as the sum of the integrals along these partial intervals.

IV'. $\widehat{\int\limits_{(L'+L'')}} (E + P\,dz) = \widehat{\int\limits_{L''}} (E + P\,dz) \widehat{\int\limits_{L'}} (E + P\,dz).$

V'. $\widehat{\int\limits_{-L}} (E + P\,dz) = \left[\widehat{\int\limits_{L}} (E + P\,dz)\right]^{-1}.$

VI'. $\widehat{\int\limits_{L}} (E + CPC^{-1}\,dz) = C \widehat{\int\limits_{L}} (E + P\,dz)\, C^{-1}$ (C is a constant matrix).

In IV' we have denoted by $L' + L''$ the composite path that is obtained by traversing first L' and then L''. In V', $-L$ denotes the path that differs from L only in direction.

The formula VII now assumes the form

VII'. $\widehat{\int\limits_{L}} [E + (Q + D_z X)\,dz] = X(z) \widehat{\int\limits_{L}} (E + X^{-1}QX\,dz)\, X(z_0)^{-1}.$

Here $X(z_0)$ and $X(z)$ on the right-hand side denote the values of $X(z)$ at the beginning and at the end of L, respectively.

Formula VIII is now replaced by the formula

VIII'. $\operatorname{mod} \left[\widehat{\int\limits_{L}} (E + P\,dz) - \widehat{\int\limits_{L}} (E + Q\,dz)\right] \leqq \frac{1}{n} e^{nql} (e^{nd \cdot l} - 1)\, I,$

where $\operatorname{mod} Q \leqq qI$, $\operatorname{mod} (P - Q) \leqq d \cdot I$, $I = \| 1 \|$, and l is the length of L.

VIII' is easily obtained from VIII if we make a change of variable in the latter and take as the new variable of integration the arc-length s along L (with $\left|\frac{dz}{ds}\right| = 1$).

3. As in the case of a real argument, there exists a close connection between the multiplicative integral and the matricant.

Suppose that $P(z)$ is a single-valued analytic matrix function, regular in G, and that G_0 is a simply-connected domain containing z_0 and forming part of G. Then the matricant $\Omega_{z_0}^{z}(P)$ is a regular function of z in G_0.

We join the points z_0 and z by an arbitrary path L lying entirely in G_0 and we choose on L intermediate points $z_1, z_2, \ldots, z_{n-1}$. Then, using the equation

$$\Omega_{z_0}^{z} = \Omega_{z_{n-1}}^{z} \cdots \Omega_{z_1}^{z_2} \Omega_{z_0}^{z_1},$$

and proceeding to the limit exactly as in § 6 (p. 132), we obtain:

$$\Omega_{z_0}^{z}(P) = \widehat{\int_L} (E + P\,dz) = \widehat{\int_{z_0}^{z}} (E + P\,dz). \tag{62}$$

From this formula it is clear that the multiplicative integral depends not on the form of the path, but only on the initial point and the end point if the whole path of integration lies in the simply-connected domain G_0 within which the integrand $P(z)$ is regular. In particular, for a closed contour L in G_0, we have:

$$\widehat{\oint} (E + P\,dz) = E. \tag{63}$$

This formula is an analogue to Cauchy's well-known theorem according to which the ordinary (non-multiplicative) integral along a closed contour is zero if the contour lies in a simply-connected domain within which the integrand is regular.

4. The representation of the matricant in the form of the multiplicative integral (62) can be used for the analytic continuation of the matricant along an arbitrary path L in G. In this case the formula

$$X = \widehat{\int_{z_0}^{z}} (E + P\,dz)\, X_0 \tag{64}$$

gives all those branches of the many-valued integral matrix X of the differential equation $\frac{dX}{dz} = PX$ that for $z = z_0$ reduce to X_0 on one of the branches. The various branches are obtained by taking account of the various paths joining z_0 and z.

By Jacobi's formula (56)

$$|X| = |X_0|\, e^{\int_{z_0}^{z} \operatorname{tr} P\,dz}$$

and, in particular, for $X_0 = E$,

$$\left| \widehat{\int_{z_0}^{z}} (E + P\,dz) \right| = e^{\int_{z_0}^{z} \operatorname{tr} P\,dz} \tag{65}$$

From this formula it follows that the multiplicative integral is always a non-singular matrix provided only that the path of integration lies entirely in a domain in which $P(z)$ is regular.

If L is an arbitrary closed path in G and G is not a simply-connected domain, then (63) cannot hold. Moreover, the value of the integral

$$\overset{\frown}{\oint} (E + P\,dz)$$

is not determined by specification of the integrand and the closed path of integration L but also depends on the choice of the initial point of integration z_0 on L. For let us take on the closed curve L two points z_0 and z_1 and let us denote the portions of the path from z_0 to z_1 and from z_1 to z_0 (in the direction of integration) by L_1 and L_2, respectively. Then, by the formula IV′,[37]

$$\overset{\frown}{\oint}_{z_0} = \overset{\frown}{\int}_{L_2} \cdot \overset{\frown}{\int}_{L_1}, \quad \overset{\frown}{\oint}_{z_1} = \overset{\frown}{\int}_{L_1} \cdot \overset{\frown}{\int}_{L_2}$$

and therefore

$$\overset{\frown}{\oint}_{z_1} = \overset{\frown}{\int}_{L_1} \cdot \overset{\frown}{\oint}_{z_0} \cdot \overset{\frown}{\int}_{L_1}^{-1}. \tag{66}$$

The formula (66) shows that the symbol $\overset{\frown}{\oint} (E + P dz)$ determines a certain matrix to within a similarity transformation, i.e., determines only the elementary divisors of that matrix.

We consider an element $X(z)$ of the solution (64) in a neighborhood of z_0. Let L be an arbitrary closed path in G beginning and ending at z_0. After analytic continuation along L the element $X(z)$ goes over into an element $\tilde{X}(z)$. But the new element $\tilde{X}(z)$ satisfies the same differential equation (55), since $P(z)$ is a single-valued function in G. Therefore

$$\tilde{X} = XV,$$

where V is a non-singular constant matrix. From (64) it follows that

$$\tilde{X}(z_0) = \overset{\frown}{\oint}_{z_0} (E + P\,dz)\, X_0.$$

Comparing this equation with the preceding one, we find:

$$V = X_0^{-1} \overset{\frown}{\oint}_{z_0} (E + P\,dz)\, X_0. \tag{67}$$

In particular, for the matricant $X = \Omega_{z_0}^{z}$, we have $X_0 = E$, and then

$$V = \overset{\frown}{\oint}_{z_0} (E + P\,dz). \tag{68}$$

[37] To simplify the notation we have omitted the expression to be integrated, $E + Pdz$, which is the same for all the integrals.

§ 9. Isolated Singular Points

1. We shall now deal with the behavior of a solution (an integral matrix) in a neighborhood of an isolated singular point a.

Let the matrix function $P(z)$ be regular for the values of z satisfying the inequality

$$0 < |z - a| < R.$$

The set of these values forms a doubly-connected domain G. The matrix function $P(z)$ has in G an expansion in a Laurent series

$$P(z) = \sum_{n=-\infty}^{+\infty} P_n (z-a)^n . \tag{69}$$

An element $X(z)$ of the integral matrix, after going once around a in the positive direction along a path L, goes over into an element

$$X^+(z) = X(z)\, V ,$$

where V is a constant non-singular matrix.

Let U be the constant matrix that is connected with V by the relation

$$V = e^{2\pi i U}. \tag{70}$$

Then the matrix function $(z-a)^U$ after going around a along L goes over into $(z-a)^U V$. Therefore the matrix function

$$F(z) = X(z)\,(z-a)^{-U} , \tag{71}$$

which is analytic in G, goes over into itself (remains unchanged) by analytic continuation along L.[38] Therefore the matrix function $F(z)$ is regular in G and can be expanded in G in a Laurent series

$$F(z) = \sum_{n=-\infty}^{+\infty} F_n (z-a)^n . \tag{72}$$

From (71) it follows that:

$$X(z) = F(z)\,(z-a)^U . \tag{73}$$

Thus every integral matrix $X(z)$ can be represented in the form (73), where the single-valued function $F(z)$ and the constant matrix U depend on

[38] Hence it follows that when z traverses any other closed path in G, the function $F(z)$ returns to its original value.

the coefficient matrix $P(z)$. However, the algorithmic determination of U and of the coefficients F_n in (72) from the coefficients P_n in (69) is, in general, a complicated task.

A special case of the problem, where

$$P(z) = \sum_{n=-1}^{\infty} P_n (z-a)^n$$

will be analyzed completely in § 10. In this case, the point a is called a *regular singularity* of the system (55).

If the expansion (69) has the form

$$P(z) = \sum_{n=-q}^{\infty} P_n (z-a)^n \qquad (q>1;\ P_{-q} \neq O)$$

then a is called an *irregular singularity of the type of a pole.* Finally, if there is an infinity of non-zero matrix coefficients P_n with negative powers of $z-a$ in (69), then a is called an *essential singularity* of the given differential system.

From (73) it follows that under an arbitrary single circuit in the positive direction (along some closed path L) an integral matrix $X(z)$ is multiplied on the right by one and the same matrix

$$V = e^{2\pi i U}$$

If this circuit begins (and ends) at z_0, then by (67)

$$V = X(z_0)^{-1} \overset{\frown}{\oint_{z_0}} (E + P\,dz)\, X(z_0). \tag{74}$$

If instead of $X(z)$ we consider any other integral matrix $\hat{X}(z) = X(z)C$ (C is a constant matrix; $|C| \neq 0$), then, as is clear from (74), V is replaced by the similar matrix

$$\hat{V} = C^{-1}VC$$

Thus, the 'integral substitutions' V of the given system form a class of similar matrices.

From (74) it also follows that the integral

$$\overset{\frown}{\oint_{z_0}} (E + P\,dz) \tag{75}$$

is determined by the initial point z_0 and does not depend on the form of the curved path.[39] If we change the point z_0, then the various values of the integral that are so obtained are similar.[40]

These properties of the integral (75) can also be confirmed directly. For let L and L' be two closed paths in G around $z=a$ with the initial points z_0 and z_0' (see Fig. 6).

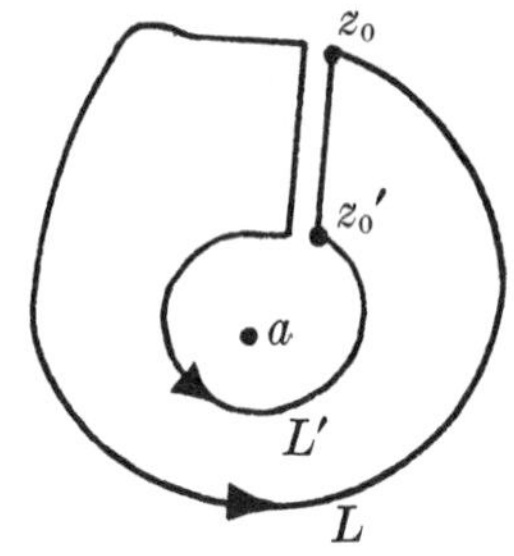

Fig. 6

The doubly-connected domain between L and L' can be made simply-connected by introducing the cut from z_0 to z_0'. The integral along the cut will be denoted by

$$T=\overset{\frown}{\int}_{z_0}^{z_0'}(E+P\,dz).$$

Since the multiplicative integral along a closed contour of a simply-connected domain is E, we have

$$\overset{\frown}{\int}_{L'} T \overset{\frown}{\int}_{L}^{-1} T^{-1}=E;$$

hence

$$\overset{\frown}{\int}_{L'}=T\overset{\frown}{\int}_{L} T^{-1}.$$

Thus, the integral $\overset{\frown}{\oint}(E+P\,dz)$, like V, is determined to within similarity, and we shall occasionally write (74) in the form

$$V\sim\overset{\frown}{\oint}(E+P\,dz);$$

meaning that the elementary divisors of the matrices on the left-hand and right-hand sides of the equation coincide.

2. As an example, we consider a system with a regular singularity

$$\frac{dX}{dz}=P(z)\,X$$

where

$$P(z)=\frac{P_{-1}}{z-a}+\sum_{n=0}^{+\infty}P_n(z-a)^n.$$

Let

$$Q(z)=\frac{P_{-1}}{z-a}.$$

[39] Under the condition, of course, that the path of integration goes around a once in the positive direction.

[40] This follows from (74), or from (66).

Using the formula VIII′ of the preceding section, we estimate the modulus of the difference

$$D = \widehat{\oint} (E + P\,dz) - \widehat{\oint} (E + Q\,dz)\,, \qquad (76)$$

taking as path of integration a circle of radius r $(r < R)$ in the positive direction. Then with

$$\operatorname{mod} P_{-1} \leqq p_{-1} I\,, \qquad \operatorname*{mod}_{|z-a|=r} \sum_{n=0}^{\infty} P_n (z-a)^n \leqq d(r)\, I\,, \qquad I = \|1\|;$$

we set in VIII′:

$$q = \frac{p_{-1}}{r}, \quad d = d(r)\,, \quad l = 2\pi r$$

and then obtain

$$\operatorname{mod} D \leq \frac{1}{n} e^{2\pi p_{-1}} (e^{2\pi n r d(r)} - 1)\, I\,.$$

Hence it is clear that[41]

$$\lim_{r \to 0} D = O\,. \qquad (77)$$

On the other hand, the system

$$\frac{dY}{dz} = QY$$

is a Cauchy system, and in that case we have for an arbitrary choice of the initial point z_0 and for every $r < R$

$$\widehat{\oint_{z_0}} (E + Q\,dz) = e^{2\pi i P_{-1}}\,.$$

Therefore it follows from (76) and (77) that:

$$\lim_{r \to 0} \widehat{\oint_{z_0}} (E + P\,dz) = e^{2\pi i P_{-1}}\,. \qquad (78)$$

But the elementary divisors of the integral $\widehat{\oint_{z_0}} (E + P dz)$ do not depend on z_0 and r and coincide with those of the integral substitution V.

From this Volterra in his well-known memoir (see [374]) and his book [63] (pp. 117-120) deduces that the matrices V and $e^{2\pi i P_{-1}}$ are similar, so that the integral substitution V is determined to within similarity by the 'residue' matrix P_{-1}.

But this assertion of Volterra is incorrect.

---

[41] Here we have used the fact that for a suitable choice of $d(r)$

$$\lim_{r \to 0} d(r) = d_0,$$

where d_0 is the greatest of the moduli of the elements of P_0.

From (74) and (78) we can only deduce that *the characteristic values of the integral substitution V coincide with those of the matrix* $e^{2\pi i P_{-1}}$. However, the elementary divisors of these matrices may be distinct. For example, for every $r \neq 0$ the matrix

$$\begin{Vmatrix} \alpha & r \\ 0 & \alpha \end{Vmatrix}$$

has one elementary divisor $(\lambda - \alpha)^2$, but the limit of the matrix for $r \to 0$, i.e., the matrix $\begin{Vmatrix} \alpha & 0 \\ 0 & \alpha \end{Vmatrix}$, has two elementary divisors $\lambda - \alpha$, $\lambda - \alpha$.

Thus, Volterra's assertion does not follow from (74) and (78). It is not even true in general, as the following example shows.

Let

$$P(z) = \begin{Vmatrix} 0 & 0 \\ 0 & -1 \end{Vmatrix} \frac{1}{z} + \begin{Vmatrix} 0 & 1 \\ 0 & 0 \end{Vmatrix}.$$

The corresponding system of differential equations has the form:

$$\frac{dx_1}{dz} = x_2, \qquad \frac{dx_2}{dz} = -\frac{x_2}{z}.$$

Integrating the system we find:

$$x_1 = c \ln z + d, \qquad x_2 = \frac{c}{z}.$$

The integral matrix

$$X(z) = \begin{Vmatrix} \ln z & 1 \\ z^{-1} & 0 \end{Vmatrix}$$

when the singular point $z = 0$ is encircled once in the positive direction, is multiplied on the right by the matrix

$$V = \begin{Vmatrix} 1 & 0 \\ 2\pi i & 1 \end{Vmatrix}.$$

This matrix has one elementary divisor $(\lambda - 1)^2$. At the same time the matrix

$$e^{2\pi i P_{-1}} = e^{2\pi i \begin{Vmatrix} 0 & 0 \\ 0 & -1 \end{Vmatrix}} = \begin{Vmatrix} 1 & 0 \\ 0 & 1 \end{Vmatrix} = E$$

has two elementary divisors $\lambda - 1$, $\lambda - 1$.

3. We now consider the case where the matrix $P(z)$ has a finite number of negative powers of $z - a$ (a is a regular or irregular singularity of the type of a pole):

$$P(z) = \frac{P_{-q}}{(z-a)^q} + \cdots + \frac{P_{-1}}{z-a} + \sum_{n=0}^{\infty} P_n (z-a)^n \qquad (q \geqq 1;\ P_{-q} \neq O).$$

We transform the given system

$$\frac{dX}{dz} = PX \tag{79}$$

by setting

$$X = A(z)\, Y, \tag{80}$$

where $A(z)$ is a matrix function that is regular at $z=0$ and assumes there the value E:

$$A(z) = E + A_1(z-a) + A_2(z-a)^2 + \cdots;$$

the power series on the right-hand side converges for $|z-a| < r_1$.

The well-known American mathematician G. D. Birkhoff has published a theorem in 1913 (see [117]) according to which the transformation (80) can always be chosen such that the coefficient matrix of the transformed system

$$\frac{dY}{dz} = P^*(z)\, Y \tag{79'}$$

contains only negative powers of $z-a$:

$$P^*(z) = \frac{P^*_{-q}}{(z-a)^q} + \cdots + \frac{P^*_{-1}}{z-a}.$$

Birkhoff's theorem with its complete proof is reproduced in the book *Ordinary Differential Equations*, by E. L. Ince.[42] Moreover, on the basis of these 'canonical' systems (79′) he investigates the behavior of the solution of an arbitrary system in the neighborhood of a singular point.

Nevertheless, *Birkhoff's proof contains an error, and the theorem is not true.* As a counter-example we can take the same example by which we have above refuted Volterra's claim.[43]

In this example $q=1$, $a=0$ and

$$P_{-1} = \begin{Vmatrix} 0 & 0 \\ 0 & -1 \end{Vmatrix}, \quad P_0 = \begin{Vmatrix} 0 & 1 \\ 0 & 0 \end{Vmatrix}, \quad P_n = O \quad \text{for } n = 1, 2, \ldots.$$

---

[42] See [20], pp. 632-41. Birkhoff and Ince formulate the theorem for the singular point $z=\infty$. This is no restriction, because every singular point $z=a$ can be carried by the transformation $z' = 1/(z-a)$ into $z' = \infty$.

[43] In the case $q=1$ the erroneous statement of Birkhoff coincides in essence with Volterra's mistake (see p. 145).

Applying Birkhoff's theorem and substituting in (79) the product AY for X in (79), we obtain after replacing $\frac{dY}{dz}$ by $\frac{P^*_{-1}}{z}$ and cancelling Y:

$$A\frac{P^*_{-1}}{z}+\frac{dA}{dz}=PA.$$

Equating the coefficients of $1/z$ and of the free terms we find:

$$P^*_{-1}=P_{-1},\quad A_1P_{-1}-P_{-1}A_1+A_1=P_0.$$

Setting

$$A_1=\begin{Vmatrix} a & b \\ c & d \end{Vmatrix},$$

we obtain:

$$\begin{Vmatrix} a & 0 \\ c & 0 \end{Vmatrix}-\begin{Vmatrix} 0 & 0 \\ -c & -d \end{Vmatrix}=\begin{Vmatrix} 0 & 1 \\ 0 & 0 \end{Vmatrix}.$$

This is a contradictory equation.

In the following section we shall examine, for the case of a regular singularity, what canonical form the system (79) can be transformed into by means of a transformation (80).

§ 10. Regular Singularities

In studying the behavior of a solution in a neighborhood of a singular point we can assume without loss of generality that the singular point is $z=0$.[44]

1. Let the given system be

$$\frac{dX}{dz}=P(z)X, \tag{81}$$

where

$$P(z)=\frac{P_{-1}}{z}+\sum_{m=0}^{\infty}P_mz^m \tag{82}$$

and the series $\sum\limits_{m=0}^{\infty}P_m z^m$ converges in the circle $|z|<r$.

We set

$$X=A(z)Y, \tag{83}$$

where

$$A(z)=E+A_1z+A_2z^2+\cdots. \tag{84}$$

[44] By the transformation $z'=z-a$ or $z'=1/z$ every finite point $z=a$ or $z=\infty$ can be carried into $z'=0$.

Leaving aside for the time being the problem of convergence of the series (84), let us try to determine the matrix coefficients A_m such that the transformed system

$$\frac{dY}{dz} = P^*(z)\,Y, \tag{85}$$

where

$$P^*(z) = \frac{P^*_{-1}}{z} + \sum_{m=0}^{\infty} P^*_m z^m \tag{86}$$

is of the simplest possible ('canonical') form.[45]

When we substitute the product AY for X in (81) and use (85), we obtain:

$$A(z)\,P^*(z)\,Y + \frac{dA}{dz}\,Y = P(z)\,A(z)\,Y.$$

Multiplying both sides of the equation by Y^{-1} on the right we find:

$$P(z)\,A(z) - A(z)\,P^*(z) = \frac{dA}{dz}.$$

When we replace here $P(z)$, $A(z)$, and $P^*(z)$ by the series (82), (84), and (86) and equate the coefficients of equal powers of z on the two sides, we obtain an infinite system of matrix equations for the unknown coefficients $A_1, A_2, \ldots$:[46]

$$\left.\begin{array}{rl} 1. & P_{-1} = P^*_{-1}, \\ 2. & P_{-1}A_1 - A_1(P_{-1}+E) + P_0 = P^*_0, \\ 3. & P_{-1}A_2 - A_2(P_{-1}+2E) + P_0A_1 - A_1P^*_0 + P_1 = P^*_1, \\ & \cdots\cdots\cdots\cdots\cdots\cdots\cdots\cdots \\ (m+2). & P_{-1}A_{m+1} - A_{m+1}[P_{-1}+(m+1)E] + \\ & + P_0A_m - A_mP^*_0 + P_1A_{m-1} - A_{m-1}P^*_1 + \cdots + P_m = P^*_m. \end{array}\right\} \tag{87}$$

2. We consider several cases separately:

1. *The matrix P_{-1} does not have distinct characteristic values that differ from each other by an integer.*

---

[45] We shall aim at having only a finite number (and indeed the smallest possible number) of non-zero coefficients P^*_m in (86).

[46] In all the equations beginning with the second we replace P^*_{-1} by P_{-1} in accordance with the first equation.

In this case the matrices P_{-1} and $P_{-1}+kE$ do not have characteristic values in common for any $k=1, 2, 3, \ldots$, and therefore (see Vol. I, Chapter VIII, § 3)[47] the matrix equation

$$P_{-1}U-U(P_{-1}+kE)=T$$

has one and only one solution for an arbitrary right-hand side T.

We shall denote this solution by

$$\Phi_k(P_{-1}, T).$$

We can therefore set all the matrices P_m^* $(m=0, 1, 2, \ldots)$ in (87) equal to zero and determine $A_1, A_2, \ldots$ successively by means of the equation

$$A_1=\Phi_1(P_{-1}, -P_0), \quad A_2=\Phi_2(P_{-1}, -P_1-P_0A_1), \ \ldots .$$

The transformed system is then a Cauchy system

$$\frac{dY}{dz}=\frac{P_{-1}}{z}Y,$$

and so the solution X of the original system (81) is of the form[48]

$$X=A(z)\,z^{P_{-1}}. \tag{88}$$

2. *Among the distinct characteristic values of P_{-1} there are some whose difference is an integer; furthermore, the matrix P_{-1} is of simple structure.*

We denote the characteristic values of P_{-1} by $\lambda_1, \lambda_2, \ldots, \lambda_n$ and order them in such a way that the inequalities

$$\operatorname{Re}(\lambda_1) \geqq \operatorname{Re}(\lambda_2) \geqq \ldots \geqq \operatorname{Re}(\lambda_n) \tag{89}$$

hold.

---

[47] However, we can also prove this without referring to Chapter VIII. The proposition in which we are interested is equivalent to the statement that the matrix equation

$$P_{-1}U=U(P_{-1}+kE) \tag{*}$$

has only the solution $U=O$. Since the matrices P_{-1} and $P_{-1}+kE$ have no characteristic values in common, there exists a polynomial $f(\lambda)$ for which

$$f(P_{-1})=O, \; f(P_{-1}+kE)=E.$$

But from (*) it follows that

$$f(P_{-1})U=Uf(P_{-1}+kE).$$

Hence $U=O$.

[48] The formula (88) defines one integral matrix of the system (81). Every integral matrix is obtained from (88) by multiplication on the right by an arbitrary constant non-singular matrix C.

Without loss of generality we can replace P_{-1} by a similar matrix. This follows from the fact that when we multiply both sides of (81) on the left by a non-singular matrix T and on the right by T^{-1}, we in fact replace all the P_m by TP_mT^{-1} ($m=-1, 0, 1, 2, \dots$); moreover, X is replaced by TXT^{-1}. Therefore we may assume in this case that P_{-1} is a diagonal matrix:

$$P_{-1}=\| \lambda_i \delta_{ik} \|_1^n. \tag{90}$$

We introduce a notation for the elements of P_m, P_m^* and A_m:

$$P_m=\| p_{ik}^{(m)} \|_1^n, \quad P_m^*=\| p_{ik}^{(m)*} \|_1^n, \quad A_m=\| x_{ik}^{(m)} \|_1^n. \tag{91}$$

In order to determine A_1, we use the second equation in (87). This matrix equation can be replaced by the scalar equations

$$(\lambda_i-\lambda_k-1)\, x_{ik}^{(1)}+p_{ik}^{(0)}=p_{ik}^{(0)*} \qquad (i, k=1, 2, \dots, n) \tag{92}$$

If none of the differences $\lambda_i-\lambda_k$ is 1, we can set $P_0^*=O$. We then have from (87_2) that $A_1=\Phi_1(P_{-1}-P_0)$.[49]

In that case the elements of A_1 are uniquely determined from (92):

$$x_{ik}^{(1)}=-\frac{p_{ik}^{(0)}}{\lambda_i-\lambda_k-1} \qquad (i, k=1, 2, \dots, n). \tag{93}$$

But if for some[50] i, k

$$\lambda_i-\lambda_k=1,$$

then the corresponding $p_{ik}^{(0)*}$ is determined from (92):

$$p_{ik}^{(0)*}=p_{ik}^{(0)},$$

and the corresponding $x_{ik}^{(1)}$ can be chosen quite arbitrarily.

For those i and k for which $\lambda_i-\lambda_k \neq 1$ we set:

$$p_{ik}^{(0)*}=0,$$

and find the corresponding $x_{ik}^{(1)}$ from (93).

Having determined A_1, we next determine A_2 from the third equation of (87). We replace this matrix equation by a system of n^2 scalar equations:

$$(\lambda_i-\lambda_k-2)\, x_{ik}^{(2)}=p_{ik}^{(1)*}-p_{ik}^{(1)}-(P_0A_1-A_1P_0^*)_{ik} \tag{94}$$
$$(i, k=1, 2, \dots, n).$$

Here we proceed exactly as in the determination of A_1.

---

49 We use the rotation introduced in dealing with the case 1.

50 By (89) this is only possible for $i<k$.

If $\lambda_i - \lambda_k \neq 2$, then we set:

$$p_{ik}^{(1)*} = 0;$$

and find from (94):

$$x_{ik}^{(2)} = -\frac{1}{\lambda_1 - \lambda_2 - 2}\,[p_{ik}^{(1)} - (P_0 A_1 - A_1 P_0^*)_{ik}].$$

But if $\lambda_i - \lambda_k = 2$, then it follows from (94) for these i and k that:

$$p_{ik}^{(1)*} = p_{ik}^{(1)} + (P_0 A_1 - A_1 P_0^*)_{ik}.$$

In this case $x_{ik}^{(2)}$ is chosen arbitrarily.

Continuing this process we determine all the matrices $P_{-1}^*, P_0^*, P_1^*, \ldots$ and $A_1, A_2, \ldots$ in succession.

Furthermore, only a finite number of the matrices P_m^* is different from zero and, as is easy to see, $P^*(z)$ is of the form[51]

$$P^*(z) = \left\| \begin{matrix} \frac{\lambda_1}{z} & a_{12} z^{\lambda_1 - \lambda_2 - 1} & \ldots & a_{1n} z^{\lambda_1 - \lambda_n - 1} \\ 0 & \frac{\lambda_2}{z} & \ldots & a_{2n} z^{\lambda_2 - \lambda_n - 1} \\ \cdot & \cdot & \cdot & \cdot \\ 0 & 0 & \ldots & \frac{\lambda_n}{z} \end{matrix} \right\|, \tag{95}$$

where $a_{ik} = 0$, when $\lambda_i - \lambda_k$ is not a positive integer, and $a_{ik} = p_{ik}^{(\lambda_i - \lambda_k - 1)*}$, when $\lambda_i - \lambda_k$ is a positive integer.

We denote by m_i the integral part of the numbers Re λ_i:[52]

$$m_i = [\mathrm{Re}\,(\lambda_i)] \qquad (i = 1, 2, \ldots, n). \tag{96}$$

Then, by (89),

$$m_1 \geqq m_2 \geqq \cdots \geqq m_n.$$

If $\lambda_i - \lambda_k$ is an integer, then

$$\lambda_i - \lambda_k = m_i - m_k.$$

---

[51] P_m^* $(m \geqq 0)$ can be different from zero only when there exist characteristic values λ_i and λ_k of P_{-1} such that $\lambda_i - \lambda_k - 1 = m$ (and, by (89), $i < k$). For a given m there corresponds to each such equation an element $p_{ik}^{(m)*} = a_{ik}$ of the matrix P_m^*; this element may be different from zero. All the remaining elements of P_m^* are zero.

[52] I.e., m_i is the largest integer not exceeding Re λ_i $(i = 1, 2, \ldots, n)$.

Therefore in the expression (95) for the canonical matrix $P^*(z)$ we can replace all the differences $\lambda_i - \lambda_k$ by $m_i - m_k$. Furthermore, we set:

$$\tilde{\lambda}_i = \lambda_i - m_i \qquad (i = 1, 2, \ldots, n), \tag{91'}$$

$$M = \| m_i \delta_{ik} \|_1^n, \quad U = \begin{Vmatrix} \tilde{\lambda}_1 & a_{12} & \ldots & a_{1n} \\ 0 & \tilde{\lambda}_2 & \ldots & a_{2n} \\ \cdot & \cdot & \cdot & \cdot \\ 0 & 0 & \ldots & \tilde{\lambda}_n \end{Vmatrix}. \tag{97}$$

Then it follows from (95) (see formula I on p. 134):

$$P^*(z) = z^M \frac{U}{z} z^{-M} + \frac{M}{z} = D_z (z^M z^U).$$

Hence $Y = z^M z^U$ is a solution of (85) and

$$X = A(z)\, z^M z^U \tag{98}$$

is a solution of (81).[53]

3. *The general case.* As we have explained above, we may replace P_{-1} without loss of generality by an arbitrary similar matrix. We shall assume that P_{-1} has the Jordan normal form[54]

$$P_{-1} = \{ \lambda_1 E_1 + H_1, \lambda_2 E_2 + H_2, \ldots, \lambda_u E_u + H_u \}, \tag{99}$$

with

$$\operatorname{Re}(\lambda_1) \geqq \operatorname{Re}(\lambda_2) \geqq \cdots \geqq \operatorname{Re}(\lambda_u). \tag{100}$$

Here E denotes the unit matrix and H the matrix in which the elements of the first superdiagonal are 1 and all the remaining elements zero. The orders of the matrices E_i and H_i in distinct diagonal blocks are, in general, different; their orders coincide with the degrees of the corresponding elementary divisors of P_{-1}.[55]

In accordance with the representation (99) of P_{-1} we split all the matrices P_m, P_m^*, A_m into blocks:

---

[53] The special form of the matrices (97) corresponds to the canonical form of P_{-1}. If P_{-1} does not have the canonical form, then the matrices M and U in (98) are similar to the matrices (97).

[54] See Vol. I, Chapter VI, § 6.

[55] To simplify the notation, the index that indicates the order of the matrices is omitted from E_i and H_i.

$$P_m = (P_{ik}^{(m)})_1^u, \quad P_m^* = (P_{ik}^{(m)*})_1^u, \quad A_m = (X_{ik}^{(m)})_1^u.$$

Then the second of the equations (87) may be replaced by a system of equations

$$(\lambda_i E_i + H_i) X_{ik}^{(1)} - X_{ik}^{(1)} [(\lambda_k + 1) E_k + H_k] + P_{ik}^{(0)} = P_{ik}^{(0)*}, \tag{101}$$

which can also be written as follows:

$$(\lambda_i - \lambda_k - 1) X_{ik}^{(1)} + H_i X_{ik}^{(1)} - X_{ik}^{(1)} H_k + P_{ik}^{(0)} = P_{ik}^{(0)*} \quad (i, k = 1, 2, \ldots, u). \tag{102}$$

Suppose that[56]

$$X_{ik}^{(1)} = \left\| \begin{matrix} x_{11} & x_{12} & \cdots \\ x_{21} & x_{22} & \cdots \\ \cdot & \cdot & \cdot \\ \cdot & \cdot & \cdot \end{matrix} \right\| = \| x_{st} \|, \quad P_{ik}^{(0)} = \| p_{st}^{(0)} \|, \quad P_{ik}^{(0)*} = \| p_{st}^{(0)*} \|.$$

Then the matrix equation (102) (for fixed i and k) can be replaced by a system of scalar equations of the form[57]

$$(\lambda_i - \lambda_k - 1) x_{st} + x_{s+1,t} - x_{s,t-1} + p_{st}^{(0)} = p_{st}^{(0)*} \tag{103}$$
$$(s = 1, 2, \ldots, v;\ t = 1, 2, \ldots, w;\ x_{v+1,t} = x_{s,0} = 0),$$

where v and w are the orders of the matrices $\lambda_i E_i + H_i$ and $\lambda_k E_k + H_k$ in (99).

If $\lambda_i - \lambda_k \neq 1$, then in (103) we can set all the $p_{st}^{(0)*}$ equal to zero and determine all the x_{st} uniquely from the recurrence relations (103). This means that in the matrix equations (102) we set

$$P_{ik}^{(0)*} = O$$

and determine $X_{ik}^{(1)}$ uniquely.

If $\lambda_i - \lambda_k = 1$, then the relations (103) assume the form

$$x_{s+1,t} - x_{s,t-1} + p_{st}^{(0)} = p_{st}^{(0)*} \tag{104}$$
$$(x_{v+1,t} = x_{s,0} = 0;\ s = 1, 2, \ldots, v;\ t = 1, 2, \ldots, w).$$

---

[56] To simplify the notation, we omit the indices i, k in the elements of the matrices X_{ik}, $P_{ik}^{(0)}$, $P_{ik}^{(0)*}$.

[57] The reader should bear in mind the properties of the matrix H that were developed on pp. 13-15 of Vol. I.

It is not difficult to show that the elements x_{st} of $X_{ik}^{(1)}$ can be determined from (104) so that the matrix $P_{ik}^{(0)*}$ has, depending on its dimensions $(v \times w)$, one of the forms

$$
\underset{(v=w)}{\begin{Vmatrix} a_0 & 0 & \cdots & 0 \\ a_1 & a_0 & \cdots & 0 \\ \vdots & \ddots & \ddots & \vdots \\ a_{v-1} & a_{v-2} \cdots & a_1 & a_0 \end{Vmatrix}}, \quad
\underset{(v<w)}{\begin{Vmatrix} a_0 & 0 & \cdots & & 0 & 0 & \cdots & 0 \\ a_1 & a_0 & \cdots & & 0 & 0 & \cdots & 0 \\ \vdots & \ddots & \ddots & & & & & \vdots \\ a_{v-1} & & & a_1 & a_0 & 0 & \cdots & 0 \end{Vmatrix}},
$$

$$
\underset{(v>w)}{\begin{Vmatrix} 0 & 0 & \cdots & 0 \\ \cdot & \cdot & \cdots & \cdot \\ 0 & 0 & \cdots & 0 \\ a_0 & 0 & \cdots & 0 \\ a_1 & a_0 & \cdots & 0 \\ \vdots & \ddots & \ddots & \vdots \\ a_{w-1} & \cdots & a_1 & a_0 \end{Vmatrix}}. \tag{105}
$$

We shall say of the matrices (105) that they have the *regular lower triangular form.*[58]

From the third of the equations (87) we can determine A_2. This equation can be replaced by the system

$$
(\lambda_i - \lambda_k - 2) X_{ik}^{(2)} + H_i X_{ik}^{(2)} - X_{ik}^{(2)} H_k + \{ P_0 A_1 - A_1 P_0 \}_{ik} + P_{ik}^{(1)} = P_{ik}^{(1)*} \tag{106}
$$
$$
(i, k = 1, 2, \ldots, u).
$$

In the same way that we determine A_1, we determine $X_{ik}^{(2)}$ uniquely with $P_{ik}^{(1)*} = O$ from (106) provided $\lambda_i - \lambda_k \neq 2$. But if $\lambda_i - \lambda_k = 2$, then $X_{ik}^{(2)}$ can be determined so that $P_{ik}^{(1)*}$ is of regular lower triangular form.

---

[58] Regular upper triangular matrices are defined similarly. The elements of $X_{ik}^{(1)}$ are not all uniquely determined from (104); there is a certain degree of arbitrariness in the choice of the elements x_{st}. This is immediately clear from (102): for $\lambda_i - \lambda_k = 1$ we may add to $X_{ik}^{(1)}$ an arbitrary matrix permutable with H, i.e., an arbitrary regular upper triangular matrix.

Continuing this process, we determine all the coefficient matrices A_1, $A_2, \ldots$ and P^*_{-1}, P^*_0, $P^*_1, \ldots$ in succession. Only a finite number of the coefficients P^*_m is different from zero, and the matrix $P^*(z)$ has the following block form:[59]

$$P^*(z) = \begin{pmatrix} \frac{\lambda_1 E_1 + H_1}{z} & B_{12} z^{\lambda_1 - \lambda_2 - 1} & \ldots & B_{1u} z^{\lambda_1 - \lambda_u - 1} \\ 0 & \frac{\lambda_2 E_2 + H_2}{z} & \ldots & B_{2u} z^{\lambda_2 - \lambda_u - 1} \\ \cdot & \cdot & \cdot & \cdot \\ 0 & 0 & \ldots & \frac{\lambda_u E_u + H_u}{z} \end{pmatrix}, \tag{107}$$

where

$$B_{ik} = \begin{cases} 0 & \text{if } \lambda_i - \lambda_k \text{ is not a positive integer,} \\ P_{ik}^{(\lambda_i - \lambda_k - 1)*} & \text{if } \lambda_i - \lambda_k \text{ is a positive integer.} \end{cases}$$

All the matrices B_{ik} $(i, k = 1, 2, \ldots, u;\ i < k)$ are of regular lower triangular form.

As in the preceding case, we denote by m_i the integral part of Re λ_i

$$m_i = [\mathrm{Re}\,(\lambda_i)] \qquad (i = 1, 2, \ldots, u) \tag{108}$$

and we set

$$\lambda_i = m_i + \tilde{\lambda}_i \qquad (i = 1, 2, \ldots, u). \tag{108'}$$

Then in the expression (107) for $P^*(z)$ we may again replace the difference $\lambda_i - \lambda_k$ everywhere by $m_i - m_k$. If we introduce the diagonal matrix M with integer elements and the upper triangular matrix U by means of the equations[60]

$$M = (m_i E_i \delta_{ik})_1^u, \quad U = \begin{pmatrix} \tilde{\lambda}_1 E_1 + H_1 & B_{12} & \ldots & B_{1u} \\ 0 & \tilde{\lambda}_2 E_2 + H_2 & \ldots & B_{2u} \\ \cdot & \cdot & \cdot & \cdot \\ 0 & 0 & \ldots & \tilde{\lambda}_u E_u + H_u \end{pmatrix}, \tag{109}$$

then we easily obtain, starting from (107), the following representation of $P^*(z)$:

$$P^*(z) = z^M \frac{U}{z} \cdot z^{-M} + \frac{M}{z} = D_z(z^M z^U).$$

---

[59] The dimensions of the square matrices E_i, H_i and the rectangular matrices B_{ik} are determined by the dimensions of the diagonal blocks in the Jordan matrix P_{-1}, i.e., by the degrees of the elementary divisors of P_{-1}.

[60] Here the splitting into blocks corresponds to that of P_{-1} and $P^*(z)$.

Hence it follows that the solution (85) can be given in the form

$$Y = z^M z^U$$

and the solution of (81) can be represented as follows:

$$X = A(z)\, z^M z^U. \tag{110}$$

Here $A(z)$ is the matrix series (84), M is a constant diagonal matrix whose elements are integers, and U is a constant triangular matrix. The matrices M and U are defined by (108), (108′), and (109).[61]

3. We now proceed to prove the convergence of the series

$$A(z) = E + A_1 z + A_2 z^2 + \cdots.$$

We shall use a lemma which is of independent interest.

LEMMA: *If the series*[62]

$$x = a_0 + a_1 z + a_2 z^2 + \cdots \tag{111}$$

formally satisfies the system

$$\frac{dx}{dz} = P(z)\, x \tag{112}$$

for which $z = 0$ *is a regular singularity, then* (111) *converges in every neighborhood of* $z = 0$ *in which the expansion of the coefficient matrix* $P(z)$ *in the series* (82) *converges.*

Proof. Let us suppose that

$$P(z) = \frac{P_{-1}}{z} + \sum_{q=0}^{\infty} P_q z^q,$$

where the series $\sum_{m=0}^{\infty} P^m z_m$ converges for $|z| < r$. Then there exist positive constants p_{-1} and p such that[63]

$$\operatorname{mod} P_{-1} \leqq p_{-1} I, \ \operatorname{mod} P_m \leqq \frac{p}{r^m} I, \quad I = \| 1 \| \qquad (m = 0, 1, 2, \ldots). \tag{113}$$

Substituting the series (111) for x in (112) and comparing the coefficients of like powers on both sides of (112), we obtain an infinite system of (column) vector equations

---

[61] See footnote 53.

[62] Here $x = (x_1, x_2, \ldots, x_n)$ is a column of unknown functions; $a_0, a_1, a_2, \ldots$ are constant columns; $P(z)$ is a square coefficient matrix.

[63] For the definition of the modulus of a matrix, see p. 128.

$$\left.\begin{aligned} P_{-1}a_0 &= o, \\ (E-P_{-1})\,a_1 &= P_0 a_0, \\ (2E-P_{-1})\,a_2 &= P_0 a_1 + P_1 a_0, \\ &\cdots\cdots\cdots\cdots \\ (mE-P_{-1})\,a_m &= P_0 a_{m-1} + P_1 a_{m-2} + \cdots + P_{m-1} a_0, \\ &\cdots\cdots\cdots\cdots \end{aligned}\right\} \tag{114}$$

It is sufficient to prove that every remainder of the series (111)

$$x^{(k)} = a_k z^k + a_{k+1} z^{k+1} + \cdots \tag{115}$$

converges in a neighborhood of $z=0$. The number k is subject to the inequality

$$k > np_{-1}.$$

Then k exceeds the moduli of all the characteristic values of P_{-1},[64] so that for $m \geqq k$ we have $|\, mE - P_{-1} \,| \neq 0$ and

$$(mE-P_{-1})^{-1} = \frac{1}{m}\left(E - \frac{P_{-1}}{m}\right)^{-1} = \frac{1}{m}E + \frac{1}{m^2}P_{-1} + \frac{1}{m^3}P_{-1}^2 + \cdots \tag{116}$$
$$(m = k, k+1, \ldots).$$

In the last part of this equation there is a convergent matrix series. With the help of this series and by using (114), we can express all the coefficients of (115) in terms of $a_0, a_1, \ldots, a_{k-1}$ by means of the recurrence relations

$$a_m = \left(\frac{1}{m}E + \frac{1}{m^2}P_{-1} + \frac{1}{m^3}P_{-1}^2 + \cdots\right)(f_{m-1} + P_0 a_{m-1} + \cdots + P_{m-k-1} a_k),$$
$$(m = k, k+1, \ldots) \tag{117}$$

where

$$f_{m-1} = P_{m-k} a_{k-1} + \cdots + P_{m-1} a_0 \qquad (m = k, k+1, \ldots). \tag{118}$$

Note that this series (115) formally satisfies the differential equation

---

[64] If λ_0 is a characteristic value of $A = \|a_{ik}\|_n^1$, then $|\,\lambda_0\,| \leqq n \cdot \max_{1 \leqq i, k \leqq n} |\,a_{ik}\,|$. For let $Ax = \lambda_0 x$, where $x = (x_1, x_2, \ldots, x_n) \neq 0$. Then

$$\lambda_0 x_i = \sum_{k=1}^{n} a_{ik} x_k \qquad (i = 1, 2, \ldots n).$$

Let $|\,x_j\,| = \max\{|\,x_1\,|, |\,x_2\,|, \ldots, |\,x_n\,|\}$. Then

$$|\,\lambda_0\,|\,|\,x_j\,| \leqq \sum_{k=1}^{n} |\,a_{jk}\,|\,|\,x_k\,| \leqq |\,x_j\,|\, n \max_{1 \leqq i, k \leqq n} |\,a_{ik}\,|.$$

Dividing through $|\,x_j\,|$, we obtain the required inequality.

$$\frac{dx^{(k)}}{dz} = P(z)\, x^{(k)} + f(z), \tag{119}$$

where

$$f(z) = \sum_{m=k-1}^{\infty} f_m z^m = P(z)\,(a_0 + a_1 z + \cdots + a_{k-1} z^{k-1}) - \\ - a_1 - 2\,a_2 z - \cdots - (k-1)\,a_{k-1} z^{k-2}. \tag{120}$$

From (120) it follows that the series

$$\sum_{m=k-1}^{\infty} f_m z^m$$

converges for $|z| < r$; hence there exists an integer $N > 0$ such that[65]

$$\operatorname{mod} f_m \leqq \left\| \frac{N}{r^m} \right\| \qquad (m = k-1, k, \ldots). \tag{121}$$

From the form of the recurrence relations (117) it follows that when the matrices P_{-1}, P_q, f_{m-1} in these relations are replaced by the majorant matrices $p_{-1}I$, $pr^{-q}I$, $\left\| \frac{N}{r^{m-1}} \right\|$ and the column a_m by $\| \alpha_m \|$ $(m = k, k+1, \ldots;\ q = 0, 1, 2, \ldots)$,[66] then we obtain relations that determine upper bounds $\| \alpha_m \|$ for mod a_m:

$$\operatorname{mod} a_m \leqq \| \alpha_m \|. \tag{122}$$

Therefore the series

$$\xi^{(k)} = \alpha_k z^k + \alpha_{k+1} z^{k+1} + \cdots \tag{123}$$

after term-by-term multiplication with the column $\| 1 \|$ becomes a majorant series for (115).

By replacing in (119) the matrix coefficients P_{-1}, P_q, f_m of the series

$$P(z) = \frac{P_{-1}}{z} + \sum_{q=0}^{\infty} P_q z^q, \qquad f(z) = \sum_{m=k-1}^{\infty} f_m z^m$$

by the corresponding majorant matrices $p_{-1}\, I$, $\frac{p}{r^q} I$, $\left\| \frac{N}{r^m} \right\|$, $\| \xi^{(k)} \|$, we obtain a differential equation for $\xi^{(k)}$:

$$\frac{d\xi^{(k)}}{dz} = n \left(\frac{p_{-1}}{z} + \frac{p}{1 - zr^{-1}} \right) \xi^{(k)} + \frac{N \dfrac{z^{k-1}}{r^{k-1}}}{1 - \dfrac{z}{r}}. \tag{124}$$

---

[65] Here $\| N/r^m \|$ denotes the column in which all the elements are equal to one and the same number, N/r^m.

[66] Here $\| \alpha_m \|$ denotes the column $(\alpha_m, \alpha_m, \ldots, \alpha_m)$ (α_m is a constant, $m = k, k+1, \ldots$).

This linear differential equation has the particular solution

$$\xi^{(k)} = \frac{N}{rk-1} \frac{z^{np_{-1}}}{\left(1 - zr^{-1}\right)^{npr}} \int_0^z z^{k-np_{-1}-1}\left(1 - \frac{z}{r}\right)^{npr-1} dz, \tag{125}$$

which is regular for $z = 0$ and can be expanded in a neighborhood of this point in the power series (123) which is convergent for $|z| < r$.

From the convergence of the majorant series (123) it follows that the series (115) is convergent for $|z| < r$, and the lemma is proved.

Note 1. This proof enables us to determine all the solutions of the differential system (112) that are regular at the singular point, provided such solutions exist.

For the existence of regular solutions (not identically zero) it is necessary and sufficient that the residue matrix P_{-1} have a non-negative integral characteristic value. If s is the greatest integral characteristic value, then columns $a_0, a_1, \ldots, a_s$ that do not all vanish can be determined from the first $s + 1$ of the equations (114); for the determinant of the corresponding linear homogeneous equation is zero:

$$\Delta = |P_{-1}|\,|E - P_{-1}| \cdots |sE - P_{-1}| = 0.$$

From the remaining equations of (114) the columns $a_{s+1}, a_{s+2}, \ldots$ can be expressed uniquely in terms of $a_0, a_1, \ldots, a_s$. The series (111) so obtained converges, by the lemma. Thus, the linearly independent solutions of the first $s + 1$ equations (114) determine all the linearly independent solutions of the system (112) that are regular at the singular point $z = 0$.

If $z = 0$ is a singular point, then a regular solution (111) at that point (if such a solution exists) is not uniquely determined when the initial value a_0 is given. However, a solution that is regular at a regular singularity is uniquely determined when $a_0, a_1, \ldots, a_s$ are given, i.e., when the initial values at $z = 0$ of this solution and the initial values of its first s derivatives are given (s is the largest non-negative integral characteristic value of the residue matrix P_{-1}).

Note 2. The proof of the lemma remains valid for $P_{-1} = O$. In this case an arbitrary positive number can be chosen for p_{-1} in the proof of the lemma. For $P_{-1} = O$ the lemma states the well-known proposition on the existence of a regular solution in a neighborhood of a regular point of the system. In this case the solution is uniquely determined when the initial value a_0 is given.

4. Suppose given the system

$$\frac{dX}{dz} = P(z)\,X, \tag{126}$$

where

$$P(z) = \frac{P_{-1}}{z} + \sum_{m=0}^{\infty} P_m z^m$$

and the series on the right-hand side converges for $|z| < r$.

Suppose, further, that by setting

$$X = A(z)\,Y \tag{127}$$

and substituting for $A(z)$ the series

$$A(z) = A_0 + A_1 z + A_2 z^2 + \cdots, \tag{128}$$

we obtain after formal transformations:

$$\frac{dY}{dz} = P^*(z)\,Y, \tag{129}$$

where

$$P^*(z) = \frac{P^*_{-1}}{z} + \sum_{m=0}^{\infty} P^*_m z^m,$$

and that here, as in the expression for $P(z)$, the series on the right-hand side converges for $|z| < r$.

We shall show that the series (128) also converges in the neighborhood $|z| < r$ of $z = 0$.

Indeed, it follows from (126), (127), and (129) that the series (128) formally satisfies the following matrix differential equation

$$\frac{dA}{dz} = P(z)\,A - AP^*(z). \tag{130}$$

We shall regard A as a vector (column) in the space of all matrices of order n, i.e., a space of dimension n^2. If in this space a linear operator $\widehat{P}(z)$ on A, depending analytically on a parameter z, is defined by the equation

$$\widehat{P}(z)\,[A] = P(z)\,A - AP^*(z), \tag{131}$$

then the differential equation (130) can be written in the form

$$\frac{dA}{dz} = \widehat{P}(z)\,[A]. \tag{132}$$

The right-hand side of this equation can be considered as the product of the matrix $r\,(z)$ of order n^2 and the column A of n^2 elements. From (131) it is clear that $z = 0$ is a regular singularity of the system (132). The series (128) formally satisfies this system. Therefore, by applying the lemma, we conclude that (128) converges in the neighborhood $|z| < r$ of $z = 0$.

In particular, the series for $A(z)$ in (110) also converges.

Thus, we have proved the following theorem:

THEOREM 2. *Every system*

$$\frac{dX}{dz} = P(z)\,X, \tag{133}$$

with a regular singularity at $z=0$

$$P(z) = \frac{P_{-1}}{z} + \sum_{m=0}^{\infty} P_m z^m,$$

has a solution of the form

$$X = A(z)\, z^M z^U, \tag{134}$$

where $A(z)$ *is a matrix function that is regular for* $z=0$ *and becomes the unit matrix* E *at that point, and where* M *and* U *are constant matrices,* M *being of simple structure and having integral characteristic values, whereas the difference between any two distinct characteristic values of* U *is not an integer.*

If the matrix P_{-1} *is reduced to the Jordan form by means of a non-singular matrix* T

$$P_{-1} = T\,\{\lambda_1 E_1 + H_1,\ \lambda_2 E_2 + H_2,\ \ldots,\ \lambda_s E_s + H_s\}\,T^{-1} \tag{135}$$
$$(\mathrm{Re}\,(\lambda_1) \geqq \mathrm{Re}\,(\lambda_2) \geqq \cdots \geqq \mathrm{Re}\,(\lambda_s)),$$

then M *and* U *can be chosen in the form*

$$M = T\,\{m_1 E_1,\ m_2 E_2,\ \ldots,\ m_s E_s\}\,T^{-1}, \tag{136}$$

$$U = T \begin{pmatrix} \tilde{\lambda}_1 E_1 + H_1 & B_{12} & \ldots & B_{1s} \\ O & \tilde{\lambda}_2 E_2 + H_2 & \ldots & B_{2s} \\ \cdot & \cdot & \cdot & \cdot \\ O & O & \ldots & \tilde{\lambda}_s E_s + H_s \end{pmatrix} T^{-1}, \tag{137}$$

where

$$m_i = [\lambda_i], \quad \tilde{\lambda}_i = \lambda_i - m_i \qquad (i = 1, 2, \ldots, s). \tag{138}$$

The B_{ik} *are regular lower triangular matrices* ($i, k = 1, 2, \ldots, s$) *and* $B_{ik} = O$ *if* $\lambda_i - \lambda_k$ *is not a positive integer* ($i, k = 1, 2, \ldots, s$).

In the particular case where none of the differences $\lambda_i - \lambda_k$ ($i, k = 1, 2, 3, \ldots, s$) *is a positive integer, we can set in* (134) $M = O$ *and* $U = P_{-1}$; *i.e., in this case the solution can be represented in the form*

$$X = A(z)\, z^{P_{-1}}. \tag{139}$$

Note 1. We wish to point out that in this section we have developed an algorithm to determine the coefficients of the series $A(z)=\sum_{m=0}^{\infty} A_m z^m$ $(A_0=E)$ in terms of the coefficients P_m of the series for $P(z)$. Moreover, the theorem also determines the integral substitution V by which the solution (134) is multiplied when a circuit is made once in the positive direction around the singular point $z=0$:

$$V=e^{2\pi i U}.$$

Note 2. From the enunciation of the theorem it follows that

$$B_{ik}=O \quad \text{for} \quad \tilde{\lambda}_i \neq \tilde{\lambda}_k \qquad (i,\, k=1,\, 2,\, \ldots,\, s).$$

Therefore the matrices

$$\tilde{\Lambda}=T\{\tilde{\lambda}_1 E_1,\, \tilde{\lambda}_2 E_2,\, \ldots,\, \tilde{\lambda}_s E_s\}\, T^{-1} \quad \text{and} \quad \tilde{U}=T\begin{pmatrix} O & B_{12} & \ldots & B_{1s} \\ O & O & \ldots & B_{2s} \\ \cdot & \cdot & \cdot & \cdot \\ O & O & \ldots & O \end{pmatrix} T^{-1} \tag{140}$$

are permutable:

$$\tilde{\Lambda}\tilde{U}=\tilde{U}\tilde{\Lambda}.$$

Hence

$$z^M z^U = z^M z^{\tilde{\Lambda}+\tilde{U}} = z^M z^{\tilde{\Lambda}} z^{\tilde{U}} = z^{\Lambda} z^{\tilde{U}}, \tag{141}$$

where

$$\Lambda = M + \tilde{\Lambda} = T\{\lambda_1,\, \lambda_2,\, \ldots,\, \lambda_n\}\, T^{-1} \tag{142}$$

and where $\lambda_1, \lambda_2, \ldots, \lambda_n$ are all the characteristic values of P_{-1} arranged in the order $\operatorname{Re}\lambda_1 \geqq \operatorname{Re}\lambda_2 \geqq \ldots \geqq \operatorname{Re}\lambda_n$.

On the other hand,

$$z^{\tilde{U}} = h(\tilde{U}),$$

where $h(\lambda)$ is the Lagrange-Sylvester interpolation polynomial for $f(\lambda)=z^{\lambda}$.

Since all the characteristic values of $\tilde{U}$ are zero, $h(\lambda)$ depends linearly on $f(0), f'(0), \ldots, f^{(g-1)}(0)$, i.e., on $1, \ln z, \ldots, (\ln z)^{g-1}$ (g is the least exponent for which $\tilde{U}^g=O$). Therefore

$$h(\lambda)=\sum_{j=0}^{g-1} h_j(\lambda)\,(\ln z)^j$$

and

$$z^{\tilde{U}}=h(\tilde{U})=\sum_{j=0}^{g-1} h_j(\tilde{U})\,(\ln z)^j = T\begin{pmatrix} 1 & q_{12} & \ldots & q_{1n} \\ 0 & 1 & \ldots & q_{2n} \\ \cdot & \cdot & \cdot & \cdot \\ 0 & 0 & \ldots & 1 \end{pmatrix} T^{-1}, \tag{143}$$

where q_{ij} $(i, j = 1, 2, \ldots, n;\ i < j)$ are polynomials in $\ln z$ of degree less than g.

By (134), (141), (142), and (143) a particular solution of (126) can be chosen in the form

$$X = A(z) \begin{Vmatrix} z^{\lambda_1} & 0 & \ldots & 0 \\ 0 & z^{\lambda_2} & \ldots & 0 \\ \cdot & \cdot & \cdot & \cdot \\ 0 & 0 & \ldots & z^{\lambda_n} \end{Vmatrix} \begin{Vmatrix} 1 & q_{12} & \ldots & q_{1n} \\ 0 & 1 & \ldots & q_{2n} \\ \cdot & \cdot & \cdot & \cdot \\ 0 & 0 & \ldots & 1 \end{Vmatrix}. \tag{144}$$

Here $\lambda_1, \lambda_2, \ldots, \lambda_n$ are the characteristic values of P_{-1} arranged in the order $\operatorname{Re}\lambda_1 \geqq \operatorname{Re}\lambda_2 \geqq \ldots \geqq \operatorname{Re}\lambda_n$, and q_{ij} $(i, j = 1, 2, \ldots, n;\ i < j)$ are polynomials in $\ln z$ of degree not higher than $g - 1$, where g is the maximal number of characteristic values λ_i that differ from each other by an integer; $A(z)$ is a matrix function, regular at $z = 0$, and $A(0) = T$ $(|T| \neq 0)$. If P_{-1} has the Jordan form, then $T = E$.

§ 11. Reducible Analytic Systems

1. As an application of the theorem of the preceding section we shall investigate in what cases the system

$$\frac{dX}{dt} = Q(t)X, \tag{145}$$

where

$$Q(t) = \sum_{m=1}^{\infty} \frac{Q_m}{t^m} \tag{146}$$

is a convergent series for $t > t_0$, is reducible (in the sense of Lyapunov), i.e., in what cases the system has a solution of the form

$$X = L(t)e^{Bt}, \tag{147}$$

where $L(t)$ is a Lyapunov matrix (i.e., $L(t)$ satisfies the conditions 1.-3. on p. 117) and B is a constant matrix.[67] Here X and Q are matrices with complex elements and t is a real variable.

We make the transformation

$$z = \frac{1}{t}.$$

[67] If the equation (147) holds, then the Lyapunov transformation $X = L(t)Y$ carries the system (145) into the system $\frac{dY}{dt} = BY$.

Then the system (145) assumes the form

$$\frac{dX}{dz} = P(z)\,X, \tag{148}$$

where

$$P(z) = -z^{-2}Q\left(\frac{1}{z}\right) = -\frac{Q_1}{z} - \sum_{m=0}^{\infty} Q_{m+2} z^m. \tag{149}$$

The series on the right-hand side of the expression for $P(z)$ converges for $|z| < 1/t_0$. Two cases can arise:

1) $Q_1 = O$. In that case $z = 0$ is not a singular point of the system (148). The system has a solution that is regular and normalized at $z = 0$. This solution is given by a convergent power series

$$X(z) = E + X_1 z + X_2 z^2 + \cdots \qquad \left(|z| < \frac{1}{t_0}\right).$$

Setting

$$L(t) = X\left(\frac{1}{t}\right), \qquad B = O,$$

we obtain the required representation (147). The system is reducible.

2) $Q_1 \neq O$. In that case the system (148) has a regular singularity at $z = 0$.

Without loss of generality we may assume that the residue matrix $P_{-1} = -Q_1$ is reduced to the Jordan form in which the diagonal elements $\lambda_1, \lambda_2, \ldots, \lambda_n$ are arranged in the order $\operatorname{Re}\lambda_1 \geqq \operatorname{Re}\lambda_2 \geqq \ldots \geqq \operatorname{Re}\lambda_n$.

Then in (144) $T = E$, and therefore the system (148) has the solution

$$X = A(z) \left\| \begin{matrix} z^{\lambda_1} & 0 & \ldots & 0 \\ 0 & z^{\lambda_2} & \ldots & 0 \\ \cdot & \cdot & \cdot & \cdot \\ 0 & 0 & \ldots & z^{\lambda_n} \end{matrix} \right\| \left\| \begin{matrix} 1 & q_{12} & \ldots & q_{1n} \\ 0 & 1 & \ldots & q_{2n} \\ \cdot & \cdot & \cdot & \cdot \\ 0 & 0 & \ldots & 1 \end{matrix} \right\|,$$

where the function $A(z)$ is regular for $z = 0$ and assumes at this point the value E, and where q_{ik} $(i, k = 1, 2, \ldots, n;\ i < k)$ are polynomials in $\ln z$. When we replace z by $1/t$, we have:

$$X = A\left(\frac{1}{t}\right) \left\| \begin{matrix} \left(\frac{1}{t}\right)^{\lambda_1} & 0 & \ldots & 0 \\ 0 & \left(\frac{1}{t}\right)^{\lambda_2} & \ldots & 0 \\ \cdot & \cdot & \cdot & \cdot \\ 0 & 0 & \ldots & \left(\frac{1}{t}\right)^{\lambda_n} \end{matrix} \right\| \left\| \begin{matrix} 1 & q_{12}\left(\ln\frac{1}{t}\right) & \ldots & q_{1n}\left(\ln\frac{1}{t}\right) \\ 0 & 1 & \ldots & q_{2n}\left(\ln\frac{1}{t}\right) \\ \cdot & \cdot & \cdot & \cdot \\ 0 & 0 & \ldots & 1 \end{matrix} \right\|. \tag{150}$$

Since $X = A(1/t)Y$ is a Lyapunov transformation, the system (145) is reducible to a system with constant coefficients if and only if the product

$$L_1(t) = \left\| \begin{matrix} t^{-\lambda_1} & 0 & \dots & 0 \\ 0 & t^{-\lambda_2} & \dots & 0 \\ \cdot & \cdot & \cdot & \cdot \\ 0 & 0 & \dots & t^{-\lambda_n} \end{matrix} \right\| \left\| \begin{matrix} 1 & q_{12}\left(\ln\frac{1}{t}\right) & \dots & q_{1n}\left(\ln\frac{1}{t}\right) \\ 0 & 1 & \dots & q_{2n}\left(\ln\frac{1}{t}\right) \\ \cdot & \cdot & \cdot & \cdot \\ 0 & 0 & \dots & 1 \end{matrix} \right\| e^{-Bt}, \tag{151}$$

where B is a constant matrix, is a Lyapunov matrix, i.e., when the matrices $L_1(t), \frac{dL_1}{dt}$, and $L_1^{-1}(t)$ are bounded. It follows from the theorem of Erugin (§ 4) that the matrix B can be assumed here to have real characteristic values.

Since $L_1(t)$ and $L_1^{-1}(t)$ are bounded for $t > t_0$, all the characteristic values of B must be zero. This follows from the expression for e^{Bt} and e^{-Bt} obtained from (151). Moreover, all the numbers $\lambda_1, \lambda_2, \dots, \lambda_n$ must be pure imaginary, because by (151) the fact that the elements of the last row of $L_1(t)$ and of the first column of $L_1^{-1}(t)$ are bounded implies that $\operatorname{Re} \lambda_n \geqq 0$ and $\operatorname{Re} \lambda_1 \leqq 0$.

But if all the characteristic values of P_{-1} are pure imaginary, then the difference between any two distinct characteristic values of P_{-1} cannot be an integer. Therefore the formula (139) holds

$$X = A(z)\, z^{P_{-1}} = A\left(\frac{1}{t}\right) t^{Q_1},$$

and for the reducibility of the system it is necessary and sufficient that the matrix

$$L_2(t) = t^{Q_1} e^{-Bt} \tag{152}$$

together with its inverse be bounded for $t > t_0$.

Since all the characteristic values of B must be zero, the minimal polynomial of B is of the form λ^d. We denote by

$$\psi(\lambda) = (\lambda - \mu_1)^{c_1} (\lambda - \mu_2)^{c_2} \cdots (\lambda - \mu_u)^{c_u} \qquad (\mu_i \neq \mu_k \text{ for } i \neq k)$$

the minimal polynomial of Q_1. As $Q_1 = -P_{-1}$, the numbers $\mu_1, \mu_2, \dots, \mu_u$ differ only in sign from the corresponding numbers λ_i and are therefore all pure imaginary. Then (see the formulas (12), (13) on p. 116)

$$t^{Q_1} = \sum_{k=1}^{u} [U_{k0} + U_{k1} \ln t + \cdots + U_{k, c_k-1} (\ln t)^{c_k-1}] \, t^{\mu_k}, \tag{153}$$

$$e^{Bt} = V_0 + V_1 t + \cdots + V_{d-1} t^{d-1}. \tag{154}$$

Substituting these expressions in the equation

$$L_2(t) \, e^{Bt} = t^{Q_1},$$

we obtain

$$[L_2(t) \, V_{d-1} + (*)] \, t^{d-1} = Z_0(t) \, (\ln t)^{c-1}, \tag{155}$$

where c is the greatest of the numbers $c_1, c_2, \ldots, c_n$, $(*)$ denotes a matrix that tends to zero for $t \to \infty$, and $Z_0(t)$ is a bounded matrix for $t > t_0$.

Since the matrices on both sides of (155) must be of equal order of magnitude for $t \to \infty$, we have

$$d = c = 1,$$

i.e.,

$$B = O,$$

and the matrix Q_1 has simple elementary divisors.

Conversely, if Q_1 has simple elementary divisors and pure imaginary characteristic values $\mu_1, \mu_2, \ldots, \mu_n$, then

$$X = A(z) \, z^{-Q_1} = A(z) \, \| z^{-\mu_i} \delta_{ik} \|_1^n$$

is a solution of (149). Setting $z = 1/t$, we find:

$$X = A\left(\frac{1}{t}\right) \| t^{\mu_i} \delta_{ik} \|_1^n .$$

The function $X(t)$ as well as $\frac{dX(t)}{dt}$ and the inverse matrix $X^{-1}(t)$ are bounded for $t > t_0$. Therefore the system is reducible ($B = O$). Thus we have proved the following theorem:[68]

Theorem 3: *The system*

$$\frac{dX}{dt} = Q(t) \, X,$$

where the matrix $Q(t)$ can be represented in a series convergent for $t > t_0$

$$Q(t) = \frac{Q_1}{t} + \frac{Q_2}{t^2} + \cdots,$$

is reducible if and only if all the elementary divisors of the residue matrix Q_1 are simple and all its characteristic values pure imaginary.

---

[68] See Erugin [13]. The theorem is proved for the case where Q_1 does not have distinct characteristic values that differ from each other by an integer.

§ 12. Analytic Functions of Several Matrices and their Application to the Investigation of Differential Systems. The Papers of Lappo-Danilevskiĭ

1. An analytic function of m matrices $X_1, X_2, \ldots, X_m$ of order n can be given by a series

$$F(X_1, X_2, \ldots, X_m) = \alpha_0 + \sum_{\nu=1}^{\infty} \sum_{j_1, j_2, \ldots, j_\nu}^{(1, \ldots, m)} \alpha_{j_1 j_2 \ldots j_\nu} X_{j_1} X_{j_2} \cdots X_{j_\nu} \tag{156}$$

convergent for all matrices X_j of order n that satisfy the inequality

$$\operatorname{mod} X_j < R_j \qquad (j = 1, 2, \ldots, m). \tag{157}$$

Here the coefficients

$$\alpha_0, \ \alpha_{j_1 j_2 \ldots j_\nu} \qquad (j_1, j_2, \ldots, j_\nu = 1, 2, \ldots, m; \ \nu = 1, 2, 3, \ldots)$$

are complex numbers, R_j $(j = 1, 2, \ldots, m)$ are constant matrices of order n with positive elements, and X_j $(j = 1, 2, \ldots, m)$ are permutable matrices of the same order with complex elements.

The theory of analytic functions of several matrices was developed by I. A. Lappo-Danilevskiĭ. He used this theory as a basis for fundamental investigations on systems of linear differential equations with rational coefficients.

A system with rational coefficients can always be reduced to the form

$$\frac{dX}{dz} = \sum_{j=1}^{m} \left\{ \frac{U_{j0}}{(z - a_j)^{s_j}} + \frac{U_{j1}}{(z - a_j)^{s_j - 1}} + \cdots + \frac{U_{j, s_j - 1}}{z - a_j} \right\} X \tag{158}$$

after a suitable transformation of the independent variable, where U_{jk} are constant matrices of order n, a_j are complex numbers, and s_j are positive integers $(k = 0, 1, \ldots, s_{j-1};\ j = 1, 2, \ldots, m)$.[69]

We shall illustrate some of Lappo-Danilevskiĭ's results in the special case of the so-called *regular* systems. The latter are characterized by the condition $s_1 = s_2 = \ldots = s_m = 1$ and can be written in the form

$$\frac{dX}{dz} = \sum_{j=1}^{m} \frac{U_j}{z - a_j} X. \tag{159}$$

[69] In the system (158) all the coefficients are regular rational fractions in z. Arbitrary rational coefficients can be reduced to this form by carrying a finite point $z = c$ that is regular (for all coefficients) by means of a fractional linear transformation on z into $z = \infty$.

Following Lappo-Danilevskiĭ, we introduce special analytic functions, namely hyperlogarithms, which are defined by the following recurrence relations:

$$l_b(z;\, a_{j_1}) = \int_b^z \frac{dz}{z - a_{j_1}},$$

$$l_b(z;\, a_{j_1}, a_{j_2}, \ldots, a_{j_\nu}) = \int_b^z \frac{l_b(z;\, a_{j_2}, a_{j_3}, \ldots, a_{j_\nu})}{z - a_{j_1}}\, dz.$$

Regarding $a_1, a_2, \ldots, a_m, \infty$ as branch points of logarithmic type, we construct the corresponding Riemann surface $S(a_1, a_2, \ldots, a_m;\, \infty)$. Every hyperlogarithm is a single-valued function on this surface. On the other hand, the matricant Ω_b^z of the system (159) (i.e., the solution normalized at $z = b$) after analytic continuation can also be regarded as a single-valued function on $S(a_1, a_2, \ldots, a_m;\, \infty)$; here b can be chosen as an arbitrary finite point on S other than $a_1, a_2, \ldots, a_m$.

For the normalized solution Ω_b^z Lappo-Danilevskiĭ gives an explicit expression in terms of the defining matrices $U_1, U_2, \ldots, U_m$ of (159) in the form of a series

$$\Omega_b^z = E + \sum_{\nu=1}^{\infty} \sum_{j_1, \ldots, j_\nu}^{(1, \ldots, m)} l_b(z;\, a_{j_1}, a_{j_2}, \ldots, a_{j_\nu})\, U_{j_1} U_{j_2} \cdots U_{j_\nu}. \tag{160}$$

This expansion converges uniformly in z for arbitrary $U_1, U_2, \ldots, U_m$ and represents Ω_b^z in any finite domain on $S(a_1, a_2, \ldots, a_m;\, \infty)$ provided only that the domain does not contain $a_1, a_2, \ldots, a_m$ in the interior or on the boundary.

If the series (156) converges for arbitrary matrices $X_1, X_2, \ldots, X_m$, then the corresponding function $F(X_1, X_2, \ldots, X_m)$ is called *entire*. Ω_b^z is an entire function of the matrices $U_1, U_2, \ldots, U_m$.

If in (160) we let the argument z go around the point a_j once in the positive direction along a contour that does not enclose other points a_i (for $i \neq j$), then we obtain the expression for the *integral substitution* V_j corresponding to the point $z = a_j$:

$$V_j = E + \sum_{\nu=1}^{\infty} \sum_{j_1, \ldots, j_\nu}^{(1, \ldots, m)} p_j(b;\, a_{j_1}, a_{j_2}, \ldots, a_{j_\nu})\, U_{j_1} U_{j_2} \ldots U_{j_\nu} \tag{161}$$

$$(j = 1, 2, \ldots, m),$$

where in a readily understandable notation

$$p_j(b;a_{j_1})=\int\limits_{(a_j)}\frac{dz}{z-a_{j_1}},$$

$$p_j(b;a_{j_1},a_{j_2},\ldots,a_{j_\nu})=\int\limits_{(a_j)}\frac{l_b(z;a_{j_2},a_{j_3},\ldots,a_{j_\nu})}{z-a_{j_1}}\,dz$$

$$\begin{pmatrix} j_1, j_2, \ldots, j_\nu, j=1,2,\ldots,m \\ \nu=1,2,3,\ldots \end{pmatrix}.$$

The series (161), like (160), is an entire function of $U_1, U_2, \ldots, U_m$.

2. Generalizing the theory of analytic functions to the case[70] of a countably infinite set of matrix arguments $X_1, X_2, X_3, \ldots$, Lappo-Danilevskiĭ has used it to study the behavior of a solution of a system in a neighborhood of an irregular singularity.[71] We quote the basic result.

The normalized solution Ω_b^z of the system

$$\frac{dX}{dz}=\sum_{j=-q}^{+\infty}P_j z^j X,$$

where the power series on the right-hand side converges for $|z|<r$ $(r>1)$,[72] can be represented by a series

$$\begin{aligned}\Omega_b^z=E+\sum_{\nu=1}^{\infty}\sum_{j_1,j_2,\ldots,j_\nu=-q}^{\infty}P_{j_1}\cdots P_{j_\nu}\times\\ \times\sum_{\mu=0}^{\nu}b^{j_{\mu+1}+\cdots+j_\nu+\nu-\mu}z^{j_1+\cdots+j_\mu+\mu}\sum_{\lambda=0}^{n-\mu}\alpha^{*(\lambda)}_{j_{\mu+1},\ldots,j_\nu}\ln^\lambda b\sum_{\varkappa=0}^{\mu}\alpha^{(\varkappa)}_{j_1,\ldots,j_\mu}\ln^\varkappa z.\end{aligned}\tag{162}$$

Here $\alpha^{*(\lambda)}_{j_{\mu+1},\ldots,j_\nu}$ and $\alpha^{(\varkappa)}_{j_1,\ldots,j_\mu}$ are scalar coefficients that are defined by special formulas. The series (162) converges for arbitrary matrices $P_1, P_2, \ldots$ in an annulus

$$\varrho<|z|<r$$

(ϱ is any positive number less than r). The point b must also lie in this annulus ($\varrho<|b|<r$).

[70] See [29], Vol. I, Memoir 1.

[71] See [29], Vol. I, Memoir 3. See also [252], [253], [254], [146], and [147].

[72] The restriction $r>1$ is not essential, since this condition can always be obtained by replacing z by αz, where α is a suitably chosen positive number.

Since in this book we cannot possibly describe the contents of the papers of Lappo-Danilevskiĭ in sufficient detail, we have had to restrict ourselves to giving above statements of a few basic results and we must refer the reader to the appropriate literature.

All the papers of Lappo-Danilevskiĭ that deal with differential equations have been published posthumously in three volumes ([29] : *Mémoires sur la théorie des systèmes des équations différentielles linéaires* (1934-36)). Moreover, his fundamental results are expounded in the papers [252], [253], [254] and the small book [28]. A concise exposition of some of the results can also be found in the book by V. I. Smirnov [56], Vol. III.

CHAPTER XV

THE PROBLEM OF ROUTH-HURWITZ AND RELATED QUESTIONS

§ 1. Introduction

In Chapter XIV, § 3 we explained that according to Lyapunov's theorem the zero solution of the system of differential equations

$$\frac{dx_i}{dt} = \sum_{k=1}^{n} a_{ik}x_k + (**) \tag{1}$$

(a_{ik} $(i, k = 1, 2, \dots, n)$ are constant coefficients) with arbitrary terms (**) of the second and higher orders in $x_1, x_2, \dots, x_n$ is stable if all the characteristic values of the matrix $A = \| a_{ik} \|_1^n$, i.e., all the roots of the secular equation $\Delta(\lambda) \equiv | \lambda E - A | = 0$, have negative real parts.

Therefore the task of establishing necessary and sufficient conditions under which all the roots of a given algebraic equation lie in the left half-plane is of great significance in a number of applied fields in which the stability of mechanical and electrical systems is investigated.

The importance of this algebraic task was clear to the founders of the theory of governors, the British physicist J. C. Maxwell and the Russian scientific research engineer I. A. Vyshnegradskiĭ who, in their papers on governors,[1] established and extensively applied the above-mentioned algebraic conditions for equations of a degree not exceeding three.

In 1868 Maxwell proposed the mathematical problem of discovering corresponding conditions for algebraic equations of arbitrary degree. Actually this problem had already been solved in essence by the French mathematician Hermite in a paper [187] published in 1856. In this paper he had established a close connection between the number of roots of a complex polynomial $f(z)$ in an arbitrary half-plane (and even inside an arbitrary triangle) and the signature of a certain quadratic form. But Hermite's

[1] J. C. Maxwell, 'On governors' Proc. Roy. Soc. London, vol. 10 (1868); I. A. Vyshnegradskiĭ, 'On governors with direct action' (1876). These papers were reprinted in the survey 'Theory of automatic governors' (Izd. Akad. Nauk SSSR, 1949). See also the paper by A. A. Andronov and I. N. Voznesenskiĭ, 'On the work of J. C. Maxwell, I. A. Vyshnegradskiĭ, and A. Stodol in the theory of governors of machines.'

results had not been carried to a stage at which they could be used by specialists working in applied fields and therefore his paper did not receive due recognition.

In 1875 the British applied mathematician Routh [47], [48], using Sturm's theorem and the theory of Cauchy indices, set up an algorithm to determine the number k of roots of a real polynomial in the right half-plane ($\operatorname{Re} z > 0$). In the particular case $k = 0$ this algorithm then gives a criterion for stability.

At the end of the 19th century, the Austrian research engineer A. Stodola, the founder of the theory of steam and gas turbines, unaware of Routh's paper, again proposed the problem of finding conditions under which all the roots of an algebraic equation have negative real parts, and in 1895 A. Hurwitz [204] on the basis of Hermite's paper gave another solution (independent of Routh's). The determinantal inequalities obtained by Hurwitz are known nowadays as the inequalities of Routh-Hurwitz.

However, even before Hurwitz' paper appeared, the founder of the modern theory of stability, A. M. Lyapunov, had proved in his celebrated dissertation ('The general problem of stability of motion,' Kharkov, 1892)[2] a theorem which yields necessary and sufficient conditions for all the roots of the characteristic equation of a real matrix $A = \| a_{ik} \|_1^n$ to have negative real parts. These conditions are made use of in a number of papers on the theory of governors.[3]

A new criterion of stability was set up in 1914 by the French mathematicians Liénard and Chipart [259]. Using special quadratic forms, these authors obtained a criterion of stability which has a definite advantage over the Routh-Hurwitz criterion (the number of determinantal inequalities in the Liénard-Chipart criterion is roughly half of that in the Routh-Hurwitz criterion).

The famous Russian mathematicians P. L. Chebyshev and A. A. Markov have proved two remarkable theorems on continued-fraction expansions of a special type. These theorems, as will be shown in § 16, have an immediate bearing on the Routh-Hurwitz problem.

The reader will see that in the sphere of problems we have outlined, the theory of quadratic forms (Vol. I, Chapter X) and, in particular, the theory of Hankel forms (Vol. I, Chapter X, § 10) forms an essential tool.

§ 2. Cauchy Indices

1. We begin with a discussion of the so-called Cauchy indices.

---

[2] See [32], § 20.

[3] See, for example, [102].

DEFINITION 1: *The Cauchy index of a real rational function* $R(x)$ *between the limits* a *and* b (notation: $I_a^b R(x)$; *a and b are real numbers or* $\pm\infty$) *is the difference between the numbers of jumps of* $R(x)$ *from* $-\infty$ *to* $+\infty$ *and that of jumps from* $+\infty$ *to* $-\infty$ *as the argument changes from* a *to* b.[5]

According to this definition, if

$$R(x)=\sum_{i=1}^{p}\frac{A_i}{x-\alpha_i}+R_1(x),$$

where A_i, α_i $(i=1, 2, \ldots, p)$ are real numbers and $R_1(x)$ is a rational function[6] without real poles, then[7]

$$I_{-\infty}^{+\infty} R(x)=\sum_{i=1}^{p}\operatorname{sgn} A_i \tag{2}$$

and, in general,

$$I_a^b R(x)=\sum_{a<\alpha_i<b}\operatorname{sgn} A_i \quad (a<b). \tag{2'}$$

In particular, if $f(x)=a_0(x-\alpha_1)^{n_1}\cdots(x-\alpha_m)^{n_m}$ is a real polynomial $(\alpha_i \neq \alpha_k$ for $i\neq k$; $i, k=1, 2, \ldots, m)$ and if among its roots $\alpha_1, \alpha_2, \ldots, \alpha_m$ only the first p are real, then

$$\frac{f'(x)}{f(x)}=\sum_{j=1}^{m}\frac{n_j}{x-\alpha_j}=\sum_{i=1}^{p}\frac{n_i}{x-\alpha_i}+R_1(x), \tag{2''}$$

where $R_1(x)$ is a real rational function without real poles.

Therefore, by (2′): *The index*

$$I_a^b\frac{f'(x)}{f(x)} \quad (a<b)$$

is equal to the number of distinct real roots of $f(x)$ *in the interval* (a, b).

An arbitrary real rational function $R(x)$ can always be represented in the form

$$R(x)=\sum_{i=1}^{p}\left\{\frac{A_1^{(i)}}{x-\alpha_i}+\cdots+\frac{A_{n_i}^{(i)}}{(x-\alpha_i)^{n_i}}\right\}+R_1(x),$$

where all the α and A are real numbers ($A^{(i)}\neq 0$; $i=1, 2, \ldots, p$) and $R_1(x)$ has no real poles.

Then

[5] In counting the number of jumps, the extreme values of x—the limits a and b—are not included.

[6] The poles of a rational function are those values of the argument for which the function becomes infinite.

[7] By sign a (a is a real number) we mean $+1$, -1, or 0 according as $a>0$, $a<0$, or $a=0$.

$$I_{-\infty}^{+\infty} R(x) = \sum_{n_i \text{ odd}} \operatorname{sgn} A_{n_i}^{(i)} \tag{3}$$

and, in general,[8]

$$I_a^b R(x) = \sum_{a < \alpha_i < b,\ n_i \text{ odd}} \operatorname{sgn} A_{n_i}^{(i)} \quad (a < b). \tag{3'}$$

2. One of the methods of computing the index $I_a^b R(x)$ is based on the classical *theorem of Sturm.*

We consider a sequence of real polynomials

$$f_1(x),\ f_2(x),\ \ldots,\ f_m(x) \tag{4}$$

that has the two following properties with respect to the interval (a, b):[9]

1. For every value x $(a < x < b)$, if any $f_k(x)$ vanishes, the two adjacent functions $f_{k-1}(x)$ and $f_{k+1}(x)$ have values different from zero and of opposite signs; i.e., for $a < x < b$ it follows from $f_k(x) = 0$ that

$$f_{k-1}(x) f_{k+1}(x) < 0.$$

2. The last function $f_m(x)$ in (4) does not vanish in the interval (a, b); i.e., $f_m(x) \neq 0$ for $a < x < b$.

Such a sequence (4) of polynomials is called a *Sturm chain in the interval* (a, b).

We denote by $V(x)$ the number variations of sign in (4) for a fixed value x.[10] Then the value of $V(x)$, as x varies from a to b, can only change when one of the functions in (4) passes through zero. But by 1., when the functions $f_k(x)$ $(k = 2, \ldots, m-1)$ pass through zero, the value of $V(x)$ does not change. When $f_1(x)$ passes through zero, then one variation of sign in (4) is lost or gained according as the ratio $f_2(x)/f_1(x)$ goes from $-\infty$ to $+\infty$ or vice versa. Hence we have:

Theorem 1 (Sturm): *If $f_1(x), f_2(x), \ldots, f_m(x)$ is a Sturm chain in (a, b) and $V(x)$ is the number of variations of sign in the chain, then*

$$I_a^b \frac{f_2(x)}{f_1(x)} = V(a) - V(b). \tag{5}$$

[8] In (3) the sum is extended over all the values i for which the corresponding n_i is odd. In (3′) the sum is extended over all the i for which n_i is odd and $a < \alpha_i < b$.

[9] Here a may be $-\infty$ and b may be $+\infty$.

[10] If $a < x < b$ and $f_1(x) \neq 0$, then by 1. in the determination of $V(x)$ a zero value in (4) may be omitted or an arbitrary sign may be attributed to this value. If a is finite, then $V(a)$ must be interpreted as $V(a + \varepsilon)$, where ε is a positive number sufficiently small that in the half-closed interval $(a, a + \varepsilon]$ none of the functions $f_i(x)$ vanishes. In exactly the same way, if b is finite, $V(b)$ is to be interpreted as $V(b - \varepsilon)$, where the number ε is defined similarly.

Note. Let us multiply all the terms of a Sturm chain by one and the same arbitrary polynomial $d(x)$. The chain of polynomials so obtained is called a *generalized Sturm chain.* Since the multiplication of all the terms of (4) by one and the same polynomial alters neither the left-hand nor the right-hand side of (5), Sturm's theorem remains valid for generalized Sturm chains.

Note that if $f(x)$ and $g(x)$ are any two polynomials (where the degree of $f(x)$ is not less than that of $g(x)$), then we can always construct a generalized Sturm chain (4) beginning with $f_1(x) \equiv f(x)$, $f_2(x) \equiv g(x)$ by means of the Euclidean algorithm.

For if we denote by $-f_3(x)$ the remainder on dividing $f_1(x)$ by $f_2(x)$, by $-f_4(x)$ the remainder on dividing $f_2(x)$ by $f_3(x)$, etc., then we have the chain of identities

$$\begin{gathered} f_1(x) = q_1(x) f_2(z) - f_3(x), \\ \cdots\cdots\cdots\cdots\cdots\cdots \\ f_{k-1}(x) = q_{k-1}(x) f_k(x) - f_{k+1}(x), \\ \cdots\cdots\cdots\cdots\cdots\cdots \\ f_{m-1}(x) = q_{m-1}(x) f_m(x), \end{gathered} \tag{6}$$

where the last remainder $f_m(x)$ that is not identically zero is the greatest common divisor of $f(x)$ and $g(x)$ and also of all the functions of the sequence (4) so constructed. If $f_m(x) \neq 0$ $(a < x < b)$ then this sequence (4) satisfies the conditions 1., 2. by (6) and is a Sturm chain. If the polynomial $f_m(x)$ has roots in the interval (a, b), then (4) is a generalized Sturm chain, because it becomes a Sturm chain when all the terms are divided by $f_m(x)$.

From what we have shown it follows that the index of every rational function $R(x)$ can be determined by Sturm's theorem. For this purpose it is sufficient to represent $R(x)$ in the form $Q(x) + \frac{g(x)}{f(x)}$, where $Q(x)$, $f(x)$, $g(x)$ are polynomials and the degree of $g(x)$ does not exceed that of $f(x)$. If we then construct the generalized Sturm chain for $f(x)$, $g(x)$, we have

$$I_a^b R(x) = I_a^b \frac{g(x)}{f(x)} = V(a) - V(b).$$

By means of Sturm's theorem we can determine the number of distinct real roots of a polynomial $f(x)$ in the interval (a, b), since this number, as we have seen, is $I_a^b \frac{f'(x)}{f(x)}$.

§ 3. Routh's Algorithm

1. Routh's problem consists in determining the number k of roots of a real polynomial $f(x)$ in the right half-plane ($\operatorname{Re} z > 0$).

To begin with, we treat the case where $f(z)$ *has no roots on the imaginary axis.* In the right half-plane we construct the semicircle of radius R with its center at the origin and we consider the domain bounded by this semicircle and the segment of the imaginary axis (Fig. 7). For sufficiently large R all the zeros of $f(z)$ with positive real parts lie inside this domain. Therefore $\arg f(z)$ increases by $2k\pi$ on going in the positive direction along the contour of the domain.[11] On the other hand, the increase of $\arg f(z)$ along the semicircle of radius R for $R \to \infty$ is determined by the increase of the argument of the highest term $a_0 z^n$ and is therefore $n\pi$. Hence the increase of $\arg f(z)$ along the imaginary axis ($R \to \infty$) is given by the expression

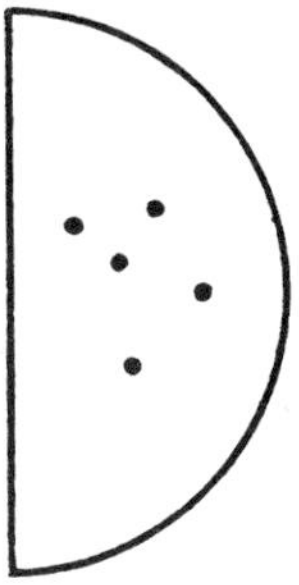

Fig. 7

$$\Delta_{-\infty}^{+\infty} \arg f(i\omega) = (n - 2k)\pi. \tag{7}$$

We introduce a somewhat unusual notation for the coefficients of $f(z)$; namely, we set

$$f(z) = a_0 z^n + b_0 z^{n-1} + a_1 z^{n-2} + b_1 z^{n-3} + \cdots \qquad (a_0 \neq 0).$$

Then

$$f(i\omega) = U(\omega) + iV(\omega), \tag{8}$$

where for even n

$$\left.\begin{aligned} U(\omega) &= (-1)^{\frac{n}{2}} (a_0\omega^n - a_1\omega^{n-2} + a_2\omega^{n-4} - \cdots), \\ V(\omega) &= (-1)^{\frac{n}{2}-1} (b_0\omega^{n-1} - b_1\omega^{n-3} + b_2\omega^{n-5} - \cdots) \end{aligned}\right\} \tag{8'}$$

and for odd n

$$\left.\begin{aligned} U(\omega) &= (-1)^{\frac{n-1}{2}} (b_0\omega^{n-1} - b_1\omega^{n-3} + b_2\omega^{n-5} - \cdots), \\ V(\omega) &= (-1)^{\frac{n-1}{2}} (a_0\omega^n - a_1\omega^{n-2} + a_2\omega^{n-4} - \cdots). \end{aligned}\right\} \tag{8''}$$

---

[11] For if $f(z) = a_0 \prod_{i=1}^{n} (z - z_i)$, then $\Delta \arg f(z) = \sum_{i=1}^{n} \Delta \arg (z - z_i)$. If the point z_i lies inside the domain in question, then $\Delta \arg (z - z_i) = 2\pi$; if z_i lies outside the domain, then $\Delta \arg (z - z_i) = 0$.

Following Routh, we make use of the Cauchy index. Then[12]

$$\frac{1}{\pi}\Delta_{-\infty}^{+\infty}\arg f(i\omega)=\begin{cases} I_{-\infty}^{+\infty}\dfrac{U(\omega)}{V(\omega)} & \text{for } \lim\limits_{\omega\to\infty}\dfrac{U(\omega)}{V(\omega)}=0, \\ -I_{-\infty}^{+\infty}\dfrac{V(\omega)}{U(\omega)} & \text{for } \lim\limits_{\omega\to\infty}\dfrac{V(\omega)}{U(\omega)}=0. \end{cases} \tag{9}$$

The equations (8′) and (8″) show that for even n the lower formula in (9) must be taken and for odd n, the upper. Then we easily obtain from (7), (8′), (8″), and (9) that for every n (even or odd)[13]

$$I_{-\infty}^{+\infty}\frac{b_0\omega^{n-1}-b_1\omega^{n-3}+\cdots}{a_0\omega^n-a_1\omega^{n-2}+\cdots}=n-2k. \tag{10}$$

2. In order to determine the index on the left-hand side of (10) we use Sturm's theorem (see the preceding section). We set

$$f_1(\omega)=a_0\omega^n-a_1\omega^{n-2}+\cdots,\qquad f_2(\omega)=b_0\omega^{n-1}-b_1\omega^{n-3}+\cdots \tag{11}$$

and, following Routh, construct a generalized Sturm chain (see p. 176)

$$f_1(\omega),\, f_2(\omega),\, f_3(\omega),\, \ldots,\, f_m(\omega). \tag{12}$$

by the Euclidean algorithm.

First we consider the *regular case*: $m=n+1$. In this case the degree of each function in (12) is one less than that of the preceding, and the last function $f_m(\omega)$ is of degree zero.[14]

From Euclid's algorithm (see (6)) it follows that

$$f_3(\omega)=\frac{a_0}{b_0}\omega f_2(\omega)-f_1(\omega)=c_0\omega^{n-2}-c_1\omega^{n-4}+c_2\omega^{n-6}-\cdots,$$

where

$$c_0=a_1-\frac{a_0}{b_0}b_1=\frac{b_0a_1-a_0b_1}{b_0},\quad c_1=a_2-\frac{a_0}{b_0}b_2=\frac{b_0a_2-a_0b_2}{b_0},\ \cdots. \tag{13}$$

Similarly

$$f_4(\omega)=\frac{b_0}{c_0}\omega f_3(\omega)-f_2(\omega)=d_0\omega^{n-3}-d_1\omega^{n-5}+\cdots,$$

where

$$d_0=b_1-\frac{b_0}{c_0}c_1=\frac{c_0b_1-b_0c_1}{c_0},\quad d_1=b_2-\frac{b_0}{c_0}c_2=\frac{c_0b_2-b_0c_2}{c_0},\ \cdots. \tag{13'}$$

The coefficients of the remaining polynomials $f_5(\omega), \ldots, f_{n+1}(\omega)$ are similarly determined.

---

[12] Since $\arg f(i\omega)=\operatorname{arccot}\frac{U(\omega)}{V(\omega)}=\arctan\frac{V(\omega)}{U(\omega)}$.

[13] We recall that the formula (10) was derived under the assumption that $f(z)$ has no roots on the imaginary axis.

[14] In the regular case (12) is the ordinary (not generalized) Sturm chain.

Each polynomial

$$f_1(\omega),\ f_2(\omega),\ \ldots,\ f_{n+1}(\omega) \tag{14}$$

is an even or an odd function and two adjacent polynomials always have opposite parity.

We form the *Routh scheme*

$$\left.\begin{matrix} a_0, & a_1, & a_2, & \ldots, \\ b_0, & b_1, & b_2, & \ldots, \\ c_0, & c_1, & c_2, & \ldots, \\ d_0, & d_1, & d_2, & \ldots, \\ \cdot & \cdot & \cdot & \cdot\ \cdot\ \cdot \end{matrix}\right\}. \tag{15}$$

The formulas (13), (13′) show that every row in this scheme is determined by the two preceding rows according to the following rule:

From the numbers of the upper row we subtract the corresponding numbers of the lower row multiplied by the number that makes the first difference zero. Omitting this zero difference, we obtain the required row.

The regular case is obviously characterized by the fact that the repeated application of this rule never yields a zero in the sequence

$$b_0,\ c_0,\ d_0,\ \ldots .$$

Figs. 8 and 9 show the skeleton of Routh's scheme for an even n ($n=6$) and an odd n ($n=7$). Here the elements of the scheme are indicated by dots.

In the regular case, the polynomials $f_1(\omega)$ and $f_2(\omega)$ have the greatest common divisor $f_{n+1}(\omega) = \text{const.} \neq 0$. Therefore these polynomials, and hence $U(\omega)$ and $V(\omega)$ (see (8′), (8″), and (11)), do not vanish simultaneously; i.e., $f(i\omega) = U(\omega) + iV(\omega) \neq 0$ for real ω. Therefore: *In the regular case the formula* (10) *holds.*

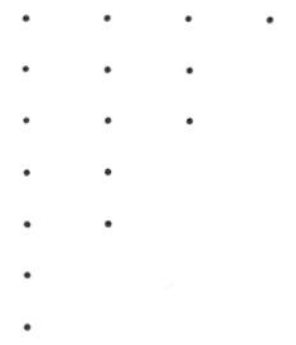

Fig. 8

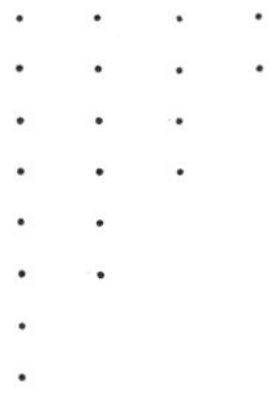

Fig. 9

When we apply Sturm's theorem in the interval $(-\infty, +\infty)$ to the left-hand side of this formula and make use of (14), we obtain by (10):

$$V(-\infty) - V(+\infty) = n - 2k. \tag{16}$$

In our case[15]

$$V(+\infty) = V(a_0, b_0, c_0, d_0, \ldots)$$

and

[15] The sign of $f_k(\omega)$ for $\omega = +\infty$ coincides with the sign of the highest coefficient and for $\omega = -\infty$ differs from it by the factor $(-1)^{n-k+1}$ ($k = 1, 2, \ldots, n+1$).

$$V(-\infty) = V(a_0, -b_0, c_0, -d_0, \ldots).$$

Hence

$$V(-\infty) = n - V(+\infty). \tag{17}$$

From (16) and (17) we find:

$$k = V(a_0, b_0, c_0, d_0, \ldots). \tag{18}$$

Thus we have proved the following theorem:

THEOREM 2 (Routh): *The number of roots of the real polynomial $f(z)$ in the right half-plane* Re $z > 0$ *is equal to the number of variations of sign in the first column of Routh's scheme.*

3. We consider the important special case where all the roots of $f(z)$ have negative real parts ('case of stability'). If in this case we construct for the polynomials (11) the generalized Sturm chain (14), then, since $k = 0$, the formula (16) can be written as follows:

$$V(-\infty) - V(+\infty) = n. \tag{19}$$

But $0 \leqq V(-\infty) \leqq m - 1 \leqq n$ and $0 \leqq V(+\infty) \leqq m - 1 \leqq n$. Therefore (19) is possible only when $m = n + 1$ (regular case!) and $V(+\infty) = 0$, $V(-\infty) = m - 1 = n$. The formula (18) then implies:

ROUTH'S CRITERION. *All the roots of the real polynomial $f(z)$ have negative real parts if and only if in the carrying out of Routh's algorithm all the elements of the first column of Routh's scheme are different from zero and of like sign.*

4. In deriving Routh's theorem we have made use of the formula (10). In what follows we shall have to generalize this formula. The formula (10) was deduced under the assumption that $f(z)$ has no roots on the imaginary axis. We shall now show that in the general case, where the polynomial $f(z) = a_0 z^n + b_0 z^{n-1} + a_1 z^{n-2} + \cdots$ $(a_0 \neq 0)$ has k roots in the right half-plane and s roots on the imaginary axis, the formula (10) is replaced by

$$I_{-\infty}^{+\infty} \frac{b_0\omega^{n-1} - b_1\omega^{n-3} + b_2\omega^{n-5} - \cdots}{a_0\omega^n - a_1\omega^{n-2} + a_2\omega^{n-4} - \cdots} = n - 2k - s. \tag{20}$$

For

$$f(z) = d(z) f^*(z),$$

where the real polynomial $d(z) = z^s + \ldots$ has s roots on the imaginary axis and the polynomial $f^*(z)$ of degree $n^* = n - s$ has no such roots.

For the sake of definiteness, we consider the case where s is even (the case where s is odd is analyzed similarly).

Let

$$f(i\omega) = U(\omega) + iV(\omega) = d(i\omega)\,[U^*(\omega) + iV^*(\omega)].$$

Since in our case $d(i\omega)$ is a real polynomial in ω, we have

$$\frac{U(\omega)}{V(\omega)} = \frac{U^*(\omega)}{V^*(\omega)}.$$

Since n and n^* have equal parity, we find by using (8′), (8″), and the notation (11):

$$\frac{f_2(\omega)}{f_1(\omega)} = \frac{f_2^*(\omega)}{f_1^*(\omega)}.$$

We apply formula (10) to $f^*(z)$. Therefore

$$I_{-\infty}^{+\infty}\frac{f_2(\omega)}{f_1(\omega)} = I_{-\infty}^{+\infty}\frac{f_2^*(\omega)}{f_1^*(\omega)} = n^* - 2k = n - 2k - s,$$

and this is what we had to prove.

§ 4. The Singular Case. Examples

1. In the preceding section we have examined the regular case where in Routh's scheme none of the numbers $b_0, c_0, d_0, \ldots$ vanish.

We now proceed to deal with the singular cases, where among the numbers $b_0, c_0, \ldots$ there occurs a zero, say, $h_0 = 0$. Routh's algorithm stops with the row in which h_0 occurs, because to obtain the numbers of the following row we would have to divide by h_0.

The singular cases can be of two types:

1) *In the row in which h_0 occurs there are numbers different from zero.* This means that at some place of (12) the degree drops by more than one.

2) *All the numbers of the row in which h_0 occurs vanish simultaneously.* Then this row is the $(m+1)$-th, where m is the number of terms in the generalized Sturm chain (12). In that case, the degrees of the functions in (12) decrease by unity from one function to the next, but the degree of the last function $f_m(\omega)$ is greater than zero. In both cases the number of functions in (12) is $m < n+1$.

Since the ordinary Routh's algorithm comes to an end in both cases, Routh gives a special rule for continuing the scheme in the cases 1), 2).

2. In case 1), according to Routh, we have to substitute for $h_0 = 0$ a 'small' value ε of definite (but arbitrary) sign and continue to fill in the scheme. Then the subsequent elements of the first column of the scheme are rational functions of ε. The signs of these elements are determined by the 'smallness' and the sign of ε. If any one of these elements vanishes identically in ε, then we replace this element by another small value η and continue the algorithm.

Example:

$$f(z) = z^4 + z^3 + 2z^2 + 2z + 1.$$

Routh's scheme (with a small parameter ε):

$$\begin{array}{lll} 1, & 2, & 1 \\ 1, & 2 & \\ \varepsilon, & 1 & \\ 2-\frac{1}{\varepsilon} & & \\ 1 & & \end{array} \qquad k = V\left(1, 1, \varepsilon, 2 - \frac{1}{\varepsilon}, 1\right) = 2.$$

This special method of varying the elements of the scheme is based on the following observation:

Since we assume that there is no singularity of the second type, the functions $f_1(\omega)$ and $f_2(\omega)$ are relatively prime. Hence it follows that the polynomial $f(z)$ has no roots on the imaginary axis.

In Routh's scheme all the elements are expressed rationally in terms of the elements of the first two rows, i.e., the coefficients of the given polynomial. But it is not difficult to observe in the formulas (13), (13′) and the analogous formulas for the subsequent rows that, once we have given arbitrary values to the elements of any two adjacent rows of Routh's scheme and to the first element of the preceding row, we can express all the elements in the first two rows, i.e., the coefficients of the original polynomial, in integral rational form in terms of these elements. Thus, for example, all the numbers a, b can be represented as integral rational functions of

$$a_0, b_0, c_0, \ldots, g_0, g_1, g_2, \ldots, h_0, h_1, h_2, \ldots$$

Therefore, in replacing $g_0 = 0$ by ε we in fact modify our original polynomial. Instead of the scheme for $f(z)$ we have the Routh scheme for a polynomial $F(z, \varepsilon)$, where $F(z, \varepsilon)$ is an integral rational function of z and ε which reduces to $f(z)$ for $\varepsilon = 0$. Since the roots of $F(z, \varepsilon)$ change continuously with a change of the parameter ε and since there are no roots on the imaginary axis for $\varepsilon = 0$, the number k of roots in the right half-plane is the same for $F(z, \varepsilon)$ and $F(z, 0) = f(z)$ for values of ε of small modulus.

3.. Let us now proceed to a singularity of the second type. Suppose that in Routh's scheme

$$a_0 \neq 0,\ b_0 \neq 0,\ \ldots,\ e_0 \neq 0,\ g_0 = 0,\ g_1 = 0,\ g_2 = 0,\ \ldots$$

In this case, the last polynomial in the generalized Sturm chain (16) is of the form:

$$f_m(\omega) = e_0\omega^{n-m+1} - e_1\omega^{n-m-1} + \cdots.$$

Routh proposes to replace $f_{m+1}(\omega)$, which is zero, by $f'_m(\omega)$; i.e., he proposes to write instead of $g_0, g_1, \ldots$ the corresponding coefficients

$$(n-m+1)\, e_0, \quad (n-m-1)\, e_1, \ \ldots$$

and to continue the algorithm.

The logical basis for this rule is as follows:

By formula (20)

$$I_{-\infty}^{+\infty} \frac{f_2(\omega)}{f_1(\omega)} = n - 2k - s$$

(the s roots of $f(z)$ on the imaginary axis coincide with the real roots of $f_m(\omega)$). Therefore, if these real roots are simple, then (see p. 174)

$$I_{-\infty}^{+\infty} \frac{f'_m(\omega)}{f_m(\omega)} = s$$

and therefore

$$I_{-\infty}^{+\infty} \frac{f_2(\omega)}{f_1(\omega)} + I_{-\infty}^{+\infty} \frac{f'_m(\omega)}{f_m(\omega)} = n - 2k.$$

This formula shows that the missing part of Routh's scheme must be filled by the Routh scheme for the polynomials $f_m(\omega)$ and $f'_m(\omega)$. The coefficients of $f'_m(\omega)$ are used to replace the elements of the zero row in Routh's scheme.

But if the roots of $f_m(\omega)$ are not simple, then we denote by $d(\omega)$ the greatest common divisor of $f_m(\omega)$ and $f'_m(\omega)$, by $e(\omega)$ the greatest common divisor of $d(\omega)$ and $d'(\omega)$, etc., and we have:

$$I_{-\infty}^{+\infty} \frac{f'_m(\omega)}{f_m(\omega)} + I_{-\infty}^{+\infty} \frac{d'(\omega)}{d(\omega)} + I_{-\infty}^{+\infty} \frac{e'(\omega)}{e(\omega)} + \cdots = s.$$

Thus the required number k can be found if the missing part of Routh's scheme is filled by the Routh scheme for $f_m(\omega)$ and $f'_m(\omega)$, then the scheme for $d(\omega)$ and $d'(\omega)$, then that for $e(\omega)$ and $e'(\omega)$, etc., i.e., Routh's rule has to be applied several times to dispose of a singularity of the second type.

Example. $f(z) = z^{10} + z^9 - z^8 - 2z^7 + z^6 + 3z^5 + z^4 - 2z^3 - z^2 + z + 1$.

Scheme

$$
\begin{array}{llrrrrrr}
\omega^{10} & & 1 & -1 & 1 & 1 & -1 & 1 \\
\omega^{9} & & 1 & -2 & 3 & -2 & 1 & \\
\omega^{8} & & 1 & -2 & 3 & -2 & 1 & \\
\omega^{7} & \left\{ \begin{array}{l} \\ \end{array} \right. & \begin{array}{r} 8 \\ 2 \end{array} & \begin{array}{r} -12 \\ -3 \end{array} & \begin{array}{r} 12 \\ 3 \end{array} & \begin{array}{r} -4 \\ -1 \end{array} & & \\
\omega^{6} & & -1 & 3 & -3 & 2 & & \\
\omega^{5} & \left\{ \begin{array}{l} \\ \end{array} \right. & \begin{array}{r} 3 \\ 1 \end{array} & \begin{array}{r} -3 \\ -1 \end{array} & \begin{array}{r} 3 \\ 1 \end{array} & & & \\
\omega^{4} & \left\{ \begin{array}{l} \\ \end{array} \right. & \begin{array}{r} 2 \\ 1 \end{array} & \begin{array}{r} -2 \\ -1 \end{array} & \begin{array}{r} 2 \\ 1 \end{array} & & & \\
\omega^{3} & \left\{ \begin{array}{l} \\ \end{array} \right. & \begin{array}{r} 4 \\ 2 \end{array} & \begin{array}{r} -2 \\ -1 \end{array} & & & & \\
\omega^{2} & & -1 & 2 & & & & \\
\omega & & 1 & & & & & \\
\omega^{0} & \left\{ \begin{array}{l} \\ \end{array} \right. & \begin{array}{r} 2 \\ 1 \end{array} & & & & &
\end{array}
$$

$$k = V(1, 1, 1, 2, -1, 1, 1, 2, -1, 1, 1) = 4.$$

Note. All the elements of any one row may be multiplied by one and the same number without changing the signs of the elements of the first column. This remark has been used in constructing the scheme.

4. However, the application of both rules of Routh does not enable us to determine the number k in all the cases. The application of the first rule (introduction of small parameters $\varepsilon, \ldots$) is justified only when $f(z)$ has no roots on the imaginary axis.

If $f(z)$ has roots on the imaginary axis, then by varying the parameter ε some of these roots may pass over into the right half-plane and change k.

Example. $f(z) = z^6 + z^5 + 3z^4 + 3z^3 + 3z^2 + 2z + 1$.

Scheme

$$
\begin{array}{llll}
\omega^{6} & 1 & 3 & 3 \quad 1 \\
\omega^{5} & 1 & 3 & 2 \\
\omega^{4} & \varepsilon & 1 & 1 \\
\omega^{3} & 3 - \dfrac{1}{\varepsilon} & 2 - \dfrac{1}{\varepsilon} & \\
\omega^{2} & 1 - \dfrac{2\varepsilon - 1}{3 - \dfrac{1}{\varepsilon}} & 1 & \\
\omega & u & & \\
\omega^{0} & 1 & &
\end{array}
$$

$$\left(u = 2 - \frac{1}{\varepsilon} - \frac{3 - \dfrac{1}{\varepsilon}}{1 - \dfrac{2\varepsilon - 1}{3 - \dfrac{1}{\varepsilon}}} = -\varepsilon + \cdots \right)$$

$$V\left(1, 1, \varepsilon, 3 - \frac{1}{\varepsilon}, 1, -\varepsilon, 1\right) = \begin{cases} 4 & \text{for } \varepsilon > 0, \\ 2 & \text{for } \varepsilon < 0. \end{cases}$$

The question of the value of k remains open.

In the general case, where $f(z)$ has roots on the imaginary axis, we have to proceed as follows:

Setting $f(z) = F_1(z) + F_2(z)$, where

$$F_1(z) = a_0 z^n + a_1 z^{n-2} + \dots, \qquad F_2(z) = b_0 z^{n-1} + b_1 z^{n-2} + \dots,$$

we must find the greatest common divisor $d(z)$ of $F_1(z)$ and $F_2(z)$. Then $f(z) = d(z) f^*(z)$.

If $f(z)$ has a root z for which $-z$ is also a root (all the roots on the imaginary axis have this property), then it follows from $f(z) = 0$ and $f(-z) = 0$ that $F_1(z) = 0$ and $F_2(z) = 0$, i.e., z is a root of $d(z)$. Therefore $f^*(z)$ has no roots z for which $-z$ is also a root of $f^*(z)$.

Then

$$k = k_1 + k_2,$$

where k_1 and k_2 are the respective numbers of roots of $f^*(z)$ and $d(z)$ in the right half-plane; k_1 is determined by Routh's algorithm and $k_2 = (q - s)/2$, where q is the degree of $d(z)$ and s the number of real roots of $d(i\omega)$.[16]

In the last example,

$$d(z) = z^2 + 1, \quad f^*(z) = z^4 + z^3 + 2z^2 + 2z + 1.$$

Therefore (see example on p. 182), we have $k_2 = 0$, $k_1 = 2$, and hence

$$k = 2.$$

§ 5. Lyapunov's Theorem

1. From the investigations of A. M. Lyapunov published in 1892 in his monograph 'The General Problem of Stability of Motion' there follows a theorem[17] that gives necessary and sufficient conditions for all the roots of the characteristic equation $|\lambda E - A| = 0$ of a real matrix $A = \| a_{ik} \|_1^n$ to have negative real parts. Since every polynomial

$$f(\lambda) = a_0 \lambda^n + a_1 \lambda^{n-1} + \dots + a_n \qquad (a_0 \neq 0)$$

[16] $d(i\omega)$ is a real polynomial or becomes one after cancelling i. The number of its real roots can be determined by Sturm's theorem.

[17] See [32], § 20.

can be represented as a characteristic determinant $|\lambda E - A|$,[18] Lyapunov's theorem is of general character and is applicable to an arbitrary algebraic equation $f(\lambda)=0$.

Suppose given a real matrix $A=\|a_{ik}\|_1^n$ and a homogeneous polynomial of dimension m in the variables $x_1, x_2, \ldots, x_n$:

$$V(\underbrace{x, x, \ldots, x}_{m}) \qquad (x=(x_1, x_2, \ldots, x_n)).$$

Let us find the total derivative with respect to t of $V(x, x, \ldots, x)$ under the assumption that x is a solution of the differential system

$$\frac{dx}{dt}=Ax.$$

Then

$$\begin{aligned}\frac{d}{dt} V(x, x, \ldots, x) &= V(Ax, x, \ldots, x)\\ &\quad + V(x, Ax, \ldots, x)+\cdots+V(x, x, \ldots, Ax)\\ &= W(x, x, \ldots, x),\end{aligned} \tag{21}$$

where $W(x, x, \ldots, x)$ is again a homogeneous polynomial of dimension m in $x_1, x_2, \ldots, x_n$. The equation (21) defines a linear operator $\widehat{A}$ which associates with every homogeneous polynomial of dimension m $V(x, x, \ldots, x)$ a certain homogeneous polynomial $W(x, x, \ldots, x)$ of the same dimension m

$$W=\widehat{A}(V).$$

We restrict ourselves to the case $m=2$.[19] Then $V(x,x)$ and $W(x,x)$ are quadratic forms in the variables $x_1, x_2, \ldots, x_n$ connected by the equation

$$\frac{d}{dt} V(x, x)=V(Ax, x)+V(x, Ax)=W(x, x); \tag{22}$$

hence[20]

$$W=\widehat{A}(V)=A^{\mathsf{T}}V+VA. \tag{23}$$

---

[18] For this purpose it is sufficient to set, for example:

$$A=\left\|\begin{matrix} 0 & 0 & \ldots & 0 & -\frac{a_n}{a_0}\\ 1 & 0 & \ldots & 0 & -\frac{a_{n-1}}{a_0}\\ \cdot & & & & \\ & \cdot & & & \\ \cdot & \cdot & \cdot & \cdot & \cdot \\ & & \cdot & & \\ & & & \cdot & \\ 0 & 0 & \ldots & 1 & -\frac{a_1}{a_0}\end{matrix}\right\|.$$

[19] A. M. Lyapunov has proved his theorem for every positive integer m.

[20] Because $V(x, y)=x^{\mathsf{T}}Vy$.

Here $V = \| v_{ik} \|_1^n$ and $W = \| w_{ik} \|_1^n$ are symmetric matrices formed, respectively, from the coefficients of the forms $V(x, x)$ and $W(x, x)$. The linear operator $\widehat{\boldsymbol{A}}$ in the space of matrices of order n is completely determined by specification of the matrix $A = \| a_{ik} \|_1^n$.

If $\lambda_1, \lambda_2, \ldots, \lambda_n$ are the characteristic values of the matrix A, then every characteristic value of the operator $\widehat{\boldsymbol{A}}$ can be represented in the form $\lambda_i + \lambda_k$ $(1 \leqq i, k \leqq n)$.[21]

Therefore, if the matrix $A = \| a_{ik} \|_1^n$ has no zero characteristic value and no two that are opposites, then the operator $\widehat{\boldsymbol{A}}$ is non-singular. In this case the matrix W in (23) determines the matrix V uniquely.

If V is symmetric, then the matrix W defined by (23) is also symmetric. If $\widehat{\boldsymbol{A}}$ is a non-singular operator, then the converse statement also holds: Every symmetric matrix W corresponds by (23) to a symmetric matrix V. For in this case we find, by going over to the transposed matrices on both sides of (23), that the matrix V^{T}, as well as V, satisfies (23). By the uniqueness of the solution, $V^{\mathsf{T}} = V$.

Thus: *If the matrix $A = \| a_{ik} \|_1^n$ has no zero and no two opposite characteristic values, then every quadratic form $W(x, x)$ corresponds to one and only one quadratic form $V(x, x)$ connected with $W(x, x)$ by* (22).

Now we can formulate Lyapunov's theorem.

THEOREM 3 (Lyapunov): *If all the characteristic values of the real matrix $A = \| a_{ik} \|_1^n$ have negative real parts, then to every negative-definite quadratic form $W(x, x)$ there corresponds a positive-definite quadratic form $V(x, x)$ connected with $W(x, x)$—taking*

$$\frac{dx}{dt} = Ax \tag{24}$$

into account—by the equation

$$\frac{d}{dt} V(x, x) = W(x, x). \tag{25}$$

Conversely, if for every negative-definite form $W(x, x)$ there exists a positive-definite form $V(x, x)$ connected with $W(x, x)$ by the equation (25) *—taking* (24) *into account—then all the characteristic values of the matrix $A = \| a_{ik} \|_1^n$ have negative real parts.*

Proof. 1. Suppose that all the characteristic values of A have negative real parts. Then for every solution $x = e^{At} x_0$ of (24) we have $\lim_{t \to +\infty} x = o$.[22] Suppose that the forms $V(x, x)$ and $W(x, x)$ are connected by (25) and that

[21] See footnote 18.

[22] See Vol. I, Chapter V, § 6.

$W(x, x) < 0$ $(x \neq o)$.[23]

Let us assume that for some $x_0 \neq o$

$$V_0 = V(x_0, x_0) \leq 0.$$

But $\frac{d}{dt} V(x, x) = W(x, x) < 0$ $(x = e^{At} x_0)$. Therefore for $t > 0$ the value of $V(x, x)$ is negative and decreases for $t \to \infty$, which results in a contradiction to the equation $\lim_{t \to \infty} V(x, x) = \lim_{x \to o} V(x, x) = 0$. Therefore $V(x, x) > 0$ for $x \neq o$, i.e., $V(x, x)$ is a positive-definite quadratic form.

2. Suppose, conversely, that in (25)

$$W(x, x) < 0, \quad V(x, x) > 0 \qquad (x \neq o).$$

From (25) it follows that

$$V(x, x) = V(x_0, x_0) + \int_0^t W(x, x)\, dt \qquad (x = e^{At} x_0). \tag{25'}$$

We shall show that for every $x_0 \neq o$ the column $x = e^{At} x_0$ comes arbitrarily near to zero for arbitrarily large values of $t > 0$. Assume the contrary. Then there exists a number $\nu > 0$ such that

$$W(x, x) < -\nu < 0 \qquad (x = e^{At} x_0, \quad x_0 \neq o, \quad t > 0).$$

But then from (25′)

$$V(x, x) < V(x_0, x_0) - \nu t,$$

and so for sufficiently large values of t we have $V(x, x) < 0$, which contradicts our assumption.

From what we have shown, it follows that for certain sufficiently large values of t the value of $V(x, x)$ $(x = e^{At} x_0, x_0 \neq 0)$ will be arbitrarily near to zero. But $V(x, x)$ decreases monotonically for $t > 0$, since $\frac{d}{dt} V(x, x) = W(x, x) < 0$. Therefore $\lim_{t \to \infty} V(x, x) = 0$.

Hence it follows that for every $x_0 \neq o$, $\lim_{t \to \infty} e^{At} x_0 = o$, i.e., $\lim_{t \to \infty} e^{At} = O$. This is only possible if all the characteristic values of A have negative real parts (see Vol. I, Chapter V, § 6).

The theorem is now completely proved.

For the form $W(x, x)$ in Lyapunov's theorem we can take any negative-definite form, in particular, the form $-\sum_{i=1}^{n} x_i^2$. In this case the theorem admits of the following matrix formulation:

---

[23] The form $W(x, x)$ is given arbitrarily. The form $V(x, x)$ is uniquely determined by (25), because A has in this case neither the characteristic value zero nor pairs of opposite characteristic values.

THEOREM 3′: *All the characteristic values of the real matrix* $A = \| a_{ik} \|_1^n$ *have negative real parts if and only if the matrix equation*

$$A'V + VA = -E \tag{26}$$

has as its solution V *the coefficient matrix of some positive-definite quadratic form* $V(x, x) > 0$.

2. From this theorem we derive a criterion for determining the stability of a non-linear system from its linear approximation.[24]

Suppose that it is required to prove the asymptotic stability of the zero solution of the non-linear system of differential equations (1) (p. 172) in the case where the coefficients a_{ik} $(i, k = 1, 2, \ldots, n)$ in the linear terms on the right-hand side form a matrix $A = \| a_{ik} \|_1^n$ having only characteristic values with negative real parts. Then, if we determine a positive-definite form $V(x, x)$ by the matrix equation (26) and calculate its total derivative with respect to time under the assumption that $x = (x_1, x_2, \ldots, x_n)$ is a solution of the given system (1), we have:

$$\frac{d}{dt} V(x, x) = -\sum_{i=1}^{n} x_i^2 + R(x_1, x_2, \ldots, x_n),$$

where $R(x_1, x_2, \ldots, x_n)$ is a series containing terms of the third and higher total degree in $x_1, x_2, \ldots, x_n$. Therefore, in some sufficiently small neighborhood of $(0, 0, \ldots, 0)$ we have simultaneously for every $x \neq o$

$$V(x, x) > 0, \quad \frac{d}{dt} V(x, x) < 0.$$

By Lyapunov's general criterion of stability[25] this also indicates the asymptotic stability of the zero solution of the system of differential equations.

If we express the elements of V from the matrix equation (26) in terms of the elements of A and substitute these expressions in the inequalities

$$v_{11} > 0, \quad \begin{vmatrix} v_{11} & v_{12} \\ v_{21} & v_{22} \end{vmatrix} > 0, \quad \ldots, \quad \begin{vmatrix} v_{11} & v_{12} & \cdots & v_{1n} \\ v_{21} & v_{22} & \cdots & v_{2n} \\ \cdot & \cdot & \cdot & \cdot \\ v_{n1} & v_{n2} & \cdots & v_{nn} \end{vmatrix} > 0,$$

then we obtain the inequalities that the elements of a matrix $A = \| a_{ik} \|_1^n$ must satisfy in order that all the characteristic values of the matrix should

[24] See [32], § 26; [9], pp. 113 ff.; [36], pp. 66 ff.

[25] See [32], § 16; [9], pp. 19-21 and 31-33; [36], pp. 32-34.

have negative real parts. However, these inequalities can be obtained in a considerably simpler form from the criterion of Routh-Hurwitz, which will be discussed in the following section.

Note. Lyapunov's theorem (3) or (3′) can be generalized immediately to the case of an arbitrary *complex* matrix $A = \| a_{ik} \|_1^n$. The quadratic forms $V(x, x)$ and $W(x, x)$ are then replaced by Hermitian forms

$$V(x, x) = \sum_{i,k=1}^{n} v_{ik} \bar{x}_i x_k, \quad W(x, x) = \sum_{i,k=1}^{n} w_{ik} \bar{x}_i x_k.$$

Correspondingly, the matrix equation (26) is replaced by the equation

$$A^* V + VA = -E \qquad (A^* = \overline{A}').$$

§ 6. The Theorem of Routh-Hurwitz

1. In the preceding sections we have explained the method of Routh, unsurpassed in its simplicity, of determining the number k of roots in the right half-plane of a real polynomial whose coefficients are given as explicit numbers. If the coefficients of the polynomial depend on parameters and it is required to determine for what values of the parameters the number k has one value or another—in particular, the value 0 ('domain of stability')[26] —then it is desirable to have explicit expressions for the values of $c_0, d_0, \ldots$ in terms of the coefficients of the given polynomial. In solving this problem, we obtain a method of determining k and, in particular, a stability criterion in a form in which it was established by Hurwitz [204].

We again consider the polynomial

$$f(z) = a_0 z^n + b_0 z^{n-1} + a_1 z^{n-2} + b_1 z^{n-3} + \cdots \quad (a_0 \neq 0).$$

By the *Hurwitz matrix* we mean the square matrix of order n

$$H = \left\| \begin{matrix} b_0 & b_1 & b_2 & \ldots & b_{n-1} \\ a_0 & a_1 & a_2 & \ldots & a_{n-1} \\ 0 & b_0 & b_1 & \ldots & b_{n-2} \\ 0 & a_0 & a_1 & \ldots & a_{n-2} \\ 0 & 0 & b_0 & \ldots & b_{n-3} \\ \cdot & \cdot & \cdot & \cdot & \cdot \\ \cdot & \cdot & \cdot & \cdot & \cdot \end{matrix} \right\| \quad \left(\begin{matrix} a_k = 0 \text{ for } k > \left[\frac{n}{2}\right], \\ b_k = 0 \text{ for } k > \left[\frac{n-1}{2}\right] \end{matrix} \right). \tag{27}$$

---

[26] For this is precisely the situation in planning new mechanical or electrical systems of governors.

We transform the matrix by subtracting from the second, fourth, ... rows the first, third, ..., row, multiplied by a_0/b_0.[27] We obtain the matrix

$$\begin{Vmatrix} b_0 & b_1 & b_2 & \dots & b_{n-1} \\ 0 & c_0 & c_1 & \dots & c_{n-2} \\ 0 & b_0 & b_1 & \dots & b_{n-2} \\ 0 & 0 & c_0 & \dots & c_{n-3} \\ 0 & 0 & b_0 & \dots & b_{n-3} \\ \cdot & \cdot & \cdot & \cdot & \cdot \\ \cdot & \cdot & \cdot & \cdot & \cdot \end{Vmatrix}.$$

In this matrix $c_0, c_1, \dots$ is the third row of Routh's scheme supplemented by zeros ($c_k = 0$ for $k > [n/2] - 1$).

We transform this matrix again by subtracting from the third, fifth, ... rows the second, fourth, ... row, multiplied by b_0/c_0:

$$\begin{Vmatrix} b_0 & b_1 & b_2 & b_3 & \dots \\ 0 & c_0 & c_1 & c_2 & \dots \\ 0 & 0 & d_0 & d_1 & \dots \\ 0 & 0 & c_0 & c_1 & \dots \\ 0 & 0 & 0 & d_0 & \dots \\ 0 & 0 & 0 & c_0 & \dots \\ \cdot & \cdot & \cdot & \cdot & \cdot \\ \cdot & \cdot & \cdot & \cdot & \cdot \end{Vmatrix},$$

Continuing this process, we ultimately arrive at a triangular matrix of order n

$$R = \begin{Vmatrix} b_0 & b_1 & b_2 & \dots \\ 0 & c_0 & c_1 & \dots \\ 0 & 0 & d_0 & \dots \\ \cdot & \cdot & \cdot & \cdot \\ \cdot & \cdot & \cdot & \cdot \end{Vmatrix}. \tag{28}$$

which we call the *Routh matrix*. It is obtained from Routh's scheme (see (15)) by: 1) deleting the first row; 2) shifting the rows to the right so that their first elements come to lie on the main diagonal; and 3) completing it by zeros to a square matrix of order n.

---

[27] We begin by dealing with the regular case where $b_0 \neq 0$, $c_0 \neq 0$, $d_0 \neq 0$, ...

DEFINITION 2: *Two matrices $A = \| a_{ik} \|_1^n$ and $B = \| b_{ik} \|_1^n$ will be called equivalent if and only if for every $p \leqq n$ the corresponding minors of order p in the first p rows are equal:*

$$A\begin{pmatrix}1 & 2 & \ldots & p \\ i_1 & i_2 & \ldots & i_p\end{pmatrix} = B\begin{pmatrix}1 & 2 & \ldots & p \\ i_1 & i_2 & \ldots & i_p\end{pmatrix} \quad \begin{pmatrix}i_1, i_2, \ldots, i_p = 1, 2, \ldots, n \\ p = 1, 2, \ldots, n\end{pmatrix}.$$

Since we do not change the values of the minors of order p in the first p rows when we subtract from any row of the matrix an arbitrary multiple of any preceding row, the Hurwitz and Routh matrices H and R are equivalent in the sense of Definition 2:

$$H\begin{pmatrix}1 & 2 & \ldots & p \\ i_1 & i_2 & \ldots & i_p\end{pmatrix} = R\begin{pmatrix}1 & 2 & \ldots & p \\ i_1 & i_2 & \ldots & i_p\end{pmatrix} \quad \begin{pmatrix}i_1, i_2, \ldots, i_p = 1, 2, \ldots, n, \\ p = 1, 2, \ldots, n\end{pmatrix}. \tag{29}$$

The equivalence of the matrices H and R enables us to express all the elements of R, i.e. of the Routh scheme, in terms of the minors of the Hurwitz matrix H and, therefore, in terms of the coefficients of the given polynomial. For when we give to p in (29) the values 1, 2, 3, ... in succession, we obtain

$$\left.\begin{array}{lll}
H\begin{pmatrix}1\\1\end{pmatrix} = b_0, & H\begin{pmatrix}1\\2\end{pmatrix} = b_1, & H\begin{pmatrix}1\\3\end{pmatrix} = b_2, \ldots, \\
H\begin{pmatrix}1 & 2\\1 & 2\end{pmatrix} = b_0c_0, & H\begin{pmatrix}1 & 2\\1 & 3\end{pmatrix} = b_0c_1, & H\begin{pmatrix}1 & 2\\1 & 4\end{pmatrix} = b_0c_2, \ldots, \\
H\begin{pmatrix}1 & 2 & 3\\1 & 2 & 3\end{pmatrix} = b_0c_0d_0, & H\begin{pmatrix}1 & 2 & 3\\1 & 2 & 4\end{pmatrix} = b_0c_0d_1, & H\begin{pmatrix}1 & 2 & 3\\1 & 2 & 5\end{pmatrix} = b_0c_0d_2, \ldots \\
\ldots\ldots\ldots & \ldots\ldots\ldots & \ldots\ldots\ldots
\end{array}\right\}, \tag{30}$$

etc.

Hence we find the following expressions for the elements of Routh's scheme:

$$\left.\begin{array}{lll}
b_0 = H\begin{pmatrix}1\\1\end{pmatrix}, & b_1 = H\begin{pmatrix}1\\2\end{pmatrix}, & b_2 = H\begin{pmatrix}1\\3\end{pmatrix}, \ldots, \\
c_0 = \dfrac{H\begin{pmatrix}1 & 2\\1 & 2\end{pmatrix}}{H\begin{pmatrix}1\\1\end{pmatrix}}, & c_1 = \dfrac{H\begin{pmatrix}1 & 2\\1 & 3\end{pmatrix}}{H\begin{pmatrix}1\\1\end{pmatrix}}, & c_2 = \dfrac{H\begin{pmatrix}1 & 2\\1 & 4\end{pmatrix}}{H\begin{pmatrix}1\\1\end{pmatrix}}, \ldots, \\
d_0 = \dfrac{H\begin{pmatrix}1 & 2 & 3\\1 & 2 & 3\end{pmatrix}}{H\begin{pmatrix}1 & 2\\1 & 2\end{pmatrix}}, & d_1 = \dfrac{H\begin{pmatrix}1 & 2 & 3\\1 & 2 & 4\end{pmatrix}}{H\begin{pmatrix}1 & 2\\1 & 2\end{pmatrix}}, & d_2 = \dfrac{H\begin{pmatrix}1 & 2 & 3\\1 & 2 & 5\end{pmatrix}}{H\begin{pmatrix}1 & 2\\1 & 2\end{pmatrix}}, \ldots, \\
\ldots\ldots\ldots & \ldots\ldots\ldots & \ldots\ldots\ldots
\end{array}\right\} \tag{31}$$

The successive principal minors of H are usually called the *Hurwitz determinants.* We shall denote them by

$$\Delta_1 = H\begin{pmatrix}1\\1\end{pmatrix} = b_0, \quad \Delta_2 = H\begin{pmatrix}1&2\\1&2\end{pmatrix} = \begin{vmatrix} b_0 & b_1 \\ a_0 & a_1 \end{vmatrix}, \ldots$$

$$\ldots, \Delta_n = H\begin{pmatrix}1 & 2 & \ldots & n\\1 & 2 & \ldots & n\end{pmatrix} = \begin{vmatrix} b_0 & b_1 & \ldots & b_{n-1} \\ a_0 & a_1 & \ldots & a_{n-1} \\ 0 & b_0 & \ldots & b_{n-2} \\ 0 & a_0 & \ldots & a_{n-2} \\ \cdot & \cdot & \cdot & \cdot \end{vmatrix}. \tag{32}$$

Note 1. By the formulas (30),[28]

$$\Delta_1 = b_0, \quad \Delta_2 = b_0 c_0, \quad \Delta_3 = b_0 c_0 d_0, \quad \ldots . \tag{33}$$

From $\Delta_1 \neq 0, \ldots, \Delta_p \neq 0$ it follows that the first p of the numbers $b_0, c_0, \ldots$ are different from zero, and vice versa; in this case the p successive rows of Routh's scheme beginning with the third are completely determined and the formulas (31) hold for them.

Note 2. The regular case (all the $b_0, c_0, \ldots$ have a meaning and are different from zero) is characterized by the inequalities

$$\Delta_1 \neq 0, \quad \Delta_2 \neq 0, \quad \ldots, \quad \Delta_n \neq 0.$$

Note 3. The definition of the elements of Routh's scheme by means of the formulas (31) is more general than that by means of Routh's algorithm. Thus, for example, if $b_0 = H\begin{pmatrix}1\\1\end{pmatrix} = 0$, then Routh's algorithm does not give us anything except the first two rows formed from the coefficients of the given polynomial. However if for $\Delta_1 = 0$ the remaining determinants $\Delta_2, \Delta_3, \ldots$ are different from zero, then by omitting the row of c's we can determine by means of the formulas (31) all the remaining rows of Routh's scheme.

By the formulas (33),

$$b_0 = \Delta_1, \quad c_0 = \frac{\Delta_2}{\Delta_1}, \quad d_0 = \frac{\Delta_3}{\Delta_2}, \ldots$$

and therefore

[28] If the coefficients of $f(z)$ are given numerically, then the formulas (33)—reducing this computation, as they do, to the formation of the Routh scheme—give by far the simplest method for computing the Hurwitz determinants.

$$V(a_0, b_0, c_0, \ldots) = V\left(a_0, \Delta_1, \frac{\Delta_2}{\Delta_1}, \ldots, \frac{\Delta_n}{\Delta_1}\right) = V(a_0, \Delta_1, \Delta_3, \ldots) + V(1, \Delta_2, \Delta_4, \ldots).$$

Hence Routh's theorem can be restated as follows:

THEOREM 4 (Routh-Hurwitz): *The number of real roots of the polynomial* $f(z) = a_0 z^n + \ldots$ *in the right half-plane is determined by the formula*

$$k = V\left(a_0, \Delta_1, \frac{\Delta_2}{\Delta_1}, \frac{\Delta_3}{\Delta_2}, \ldots, \frac{\Delta_n}{\Delta_{n-1}}\right) \tag{34}$$

or (what is the same) *by*

$$k = V(a_0, \Delta_1, \Delta_3, \ldots) + V(1, \Delta_2, \Delta_4, \ldots). \tag{34'}$$

Note. This statement of the Routh-Hurwitz theorem assumes that we have the regular case

$$\Delta_1 \neq 0, \quad \Delta_2 \neq 0, \quad \ldots, \quad \Delta_n \neq 0.$$

In the following section we shall show how this formula can be used in the singular cases where some of the Hurwitz determinants Δ_i are zero.

2. We now consider the special case where all the roots of $f(z)$ are in the left half-plane $\operatorname{Re} z < 0$. By Routh's criterion, all the $a_0, b_0, c_0, d_0, \ldots$ must then be different from zero and of like sign. Since we are concerned here with the regular case, we obtain from (34) for $k = 0$ the following criterion:

CRITERION OF ROUTH-HURWITZ: *All the roots of the real polynomial* $f(z) = a_0 z^n + \ldots$ $(a_0 \neq 0)$ *have negative real parts if and only if the inequalities*

$$\left.\begin{array}{r} a_0\Delta_1 > 0, \ \ \Delta_2 > 0, \ \ a_0\Delta_3 > 0, \ \ \Delta_4 > 0, \ \ldots, \ a_0\Delta_n > 0 \ (\textit{for odd } n), \\ \Delta_n > 0 \ (\textit{for even } n) \end{array}\right\} \tag{35}$$

hold.

Note. If $a_0 > 0$, these conditions can be written as follows:

$$\Delta_1 > 0, \quad \Delta_2 > 0, \ \ldots, \ \Delta_n > 0. \tag{36}$$

If we use the usual notation for the coefficients of the polynomial

$$f(z) = a_0 z^n + a_1 z^{n-1} + a_2 z^{n-2} + \cdots + a_{n-1} z + a_n,$$

then for $a_0 > 0$ the Routh-Hurwitz conditions (36) can be written in the form of the following determinantal inequalities:

$$|a_1|>0,\quad \begin{vmatrix} a_1 & a_3 \\ a_0 & a_2 \end{vmatrix}>0,\quad \begin{vmatrix} a_1 & a_3 & a_5 \\ a_0 & a_2 & a_4 \\ 0 & a_1 & a_3 \end{vmatrix}>0,\ \ldots,\ \begin{vmatrix} a_1 & a_3 & a_5 & \ldots & 0 \\ a_0 & a_2 & a_4 & \ldots & 0 \\ 0 & a_1 & a_3 & \ldots & 0 \\ 0 & a_0 & a_2 & \ldots & 0 \\ \cdot & \cdot & \cdot & \cdot & \cdot \\ \cdot & \cdot & \cdot & \cdot & a_n \end{vmatrix}>0. \tag{36'}$$

A real polynomial $f(z)=a_0z^n+\ldots$ whose coefficients satisfy (35), i.e., whose roots have negative real parts, is often called a *Hurwitz polynomial.*

3. In conclusion, we mention a remarkable property of Routh's scheme.

Let $f_0, f_1, \ldots$ and $g_0, g_1, \ldots$ be the $(m+1)$-th and $(m+2)$-th rows of the scheme $(f_0=\Delta_m/\Delta_{m-1},\ g_0=\Delta_{m+1}/\Delta_m)$. Since these two rows together with the subsequent rows form a Routh scheme of their own, the elements of the $(m+p+1)$-th row (of the original scheme) can be expressed in terms of the elements of the $(m+1)$-th and $(m+2)$-th rows $f_0, f_1, \ldots$ and $g_0, g_1, \ldots$ by the same formulas as the $(p+1)$-th row can in terms of the elements of the first two rows $a_0, a_1, \ldots$ and $b_0, b_1, \ldots$; that is, if we set

$$\widetilde{H}=\begin{Vmatrix} g_0 & g_1 & g_2 & \ldots \\ f_0 & f_1 & f_2 & \ldots \\ 0 & g_0 & g_1 & \ldots \\ 0 & f_0 & f_1 & \ldots \\ \cdot & \cdot & \cdot & \cdot \\ \cdot & \cdot & \cdot & \cdot \end{Vmatrix},$$

then we have

$$\frac{H\begin{pmatrix} 1\ldots m+p-1 & m+p \\ 1\ldots m+p-1 & m+p+k-1 \end{pmatrix}}{H\begin{pmatrix} 1\ldots m+p-1 \\ 1\ldots m+p-1 \end{pmatrix}}=\frac{\widetilde{H}\begin{pmatrix} 1\ldots p-1 & p \\ 1\ldots p-1 & p+k-1 \end{pmatrix}}{\widetilde{H}\begin{pmatrix} 1\ldots p-1 \\ 1\ldots p-1 \end{pmatrix}}. \tag{37}$$

The Hurwitz determinant Δ_{m+p} is equal to the product of the first $m+p$ numbers in the sequence $b_0, c_0, \ldots$:

$$\Delta_{m+p}=b_0c_0\ \ldots\ f_0g_0\ \ldots\ l_0.$$

But

$$\Delta_m=b_0c_0\ \ldots\ f_0,\quad \widetilde{\Delta}_p=g_0\ \ldots\ l_0.$$

Therefore the following important relation[29] holds:

$$\Delta_{m+p}=\Delta_m\widetilde{\Delta}_p. \tag{38}$$

[29] Here $\widetilde{\Delta}_p$ is the minor of order p in the top left-hand corner of $\widetilde{H}$.

The formula (38) holds whenever the numbers $f_0, f_1, \ldots$ and $g_0, g_1, \ldots$ are well defined, i.e., under the conditions $\Delta_{m-1} \neq 0$, $\Delta_m \neq 0$.

The formula (37) has a meaning if in addition to the conditions $\Delta_{m-1} \neq 0$, $\Delta_m \neq 0$ we also have $\Delta_{m+p-1} \neq 0$. From this condition it follows that the denominator of the fraction on the right-hand side of (37) is also different from zero: $\tilde{\Delta}_{p-1} \neq 0$.

§ 7. Orlando's Formula

1. In the discussion of the cases where some of the Hurwitz determinants are zero we shall have to use the following formula of Orlando [294], which expresses the determinant Δ_{n-1} in terms of the highest coefficient a_0 and the roots $z_1, z_2, \ldots, z_n$ of $f(z)$:[30]

$$\Delta_{n-1} = (-1)^{\frac{n(n-1)}{2}} a_0^{n-1} \prod_{i<k}^{1,\ldots,n} (z_i + z_k). \tag{39}$$

For $n = 2$ this reduces to the well-known formula for the coefficient b_0 in the quadratic equation $a_0 z^2 + b_0 z + a_1 = 0$:

$$\Delta_1 = b_0 = -a_0 (z_1 + z_2).$$

Let us assume that the formula (39) is true for polynomials of degree n, $f(z) = a_0 z^n + b_0 z^{n-1} + \cdots$ and show that it is then true for polynomials of degree $n+1$

$$\begin{aligned} F(z) &= (z+h) f(z) \\ &= a_0 z^{n+1} + (b_0 + h a_0) z^n + (a_1 + h b_0) z^{n-1} + \cdots \qquad (h = -z_{n+1}). \end{aligned}$$

For this purpose we form the auxiliary determinant of order $n+1$

$$D = \begin{vmatrix} b_0 & b_1 & \ldots & b_{n-1} & h^n \\ a_0 & a_1 & \ldots & a_{n-1} & -h^{n-1} \\ 0 & b_0 & \ldots & b_{n-2} & h^{n-2} \\ 0 & a_0 & \ldots & a_{n-2} & -h^{n-3} \\ \cdot & \cdot & \cdot & \cdot & \cdot \\ \cdot & \cdot & \cdot & \cdot & \cdot \\ 0 & 0 & \ldots & \ldots & (-1)^n \end{vmatrix} \qquad \left(\begin{matrix} a_k = 0 & \text{for} & k > \left[\frac{n}{2}\right], \\ b_k = 0 & \text{for} & k > \left[\frac{n-1}{2}\right] \end{matrix} \right).$$

---

[30] The coefficients of $f(z)$ may be arbitrary complex numbers.

We multiply the first row of D by a_0 and add to it the second row multiplied by $-b_0$, the third multiplied by a_1, the fourth by $-b_1$, etc. Then in the first row all the elements except the last are zero, and the last element is $f(h)$. Hence we deduce that

$$D=(-1)^n \Delta_{n-1} f(h).$$

On the other hand, when we add to each row of D (except the last) the next multiplied by h we obtain, apart from a factor $(-1)^n$, the Hurwitz determinant Δ_n^* of order n for the polynomial $F(z)$:

$$D=(-1)^n \begin{vmatrix} b_0+ha_0 & b_1+ha_1 & \dots \\ a_0 & a_1+hb_0 & \dots \\ 0 & b_0+ha_0 & \dots \\ 0 & a_0 & \dots \\ \cdot & \cdot & \cdot \\ \cdot & \cdot & \cdot \end{vmatrix} =(-1)^n \Delta_n^*.$$

Thus

$$\Delta_n^* = \Delta_{n-1} f(h) = a_0 \Delta_{n-1} \prod_{i=1}^{n} (h-z_i).$$

When we replace Δ_{n-1} by its expression (39) and set $h=-z_{n+1}$, we obtain

$$\Delta_n^* = (-1)^{\frac{(n+1)n}{2}} a_0^n \prod_{i<k}^{1,\dots,n+1} (z_i+z_k).$$

Thus, by mathematical induction Orlando's formula is established for polynomials of every degree.

From Orlando's formula it follows that: $\Delta_{n-1}=0$ *if and only if the sum of two roots of* $f(z)$ *is zero.*[31]

Since $\Delta_n = c\Delta_{n-1}$, where c is the constant term of the polynomial $f(z)$ $(c=(-1)^n a_0 z_1 z_2 \dots z_n)$, it follows from (39) that:

$$\Delta_n = (-1)^{\frac{n(n+1)}{2}} a_0^n z_1 z_2 \cdots z_n \prod_{i<k}^{1,\dots,n} (z_i+z_k). \tag{40}$$

The last formula shows that: Δ_n *vanishes if and only if* $f(z)$ *has a pair of opposite roots* z *and* $-z$.

[31] In particular, $\Delta_{n-1}=0$ when $f(z)$ has at least one pair of conjugate pure imaginary roots or multiple zero roots.

§ 8. Singular Cases in the Routh-Hurwitz Theorem

In discussing the singular cases where some of the Hurwitz determinants are zero, we may assume that $\Delta_n \neq 0$ (and consequently $\Delta_{n-1} \neq 0$).

For if $\Delta_n = 0$, then, as we have seen at the end of the preceding section, the real polynomial $f(z)$ has a root z' for which $-z'$ is also a root. If we set $f(z) = F_1(z) + F_2(z)$, where

$$F_1(z) = a_0 z^n + a_1 z^{n-2} + \cdots, \quad F_2(z) = b_0 z^{n-1} + b_1 z^{n-3} + \cdots,$$

then we can deduce from $f(z') = f(-z') = 0$ that $F_1(z') = F_2(z') = 0$. Therefore z' is a root of the greatest common divisor $d(z)$ of the polynomials $F_1(z)$ and $F_2(z)$. Setting $f(z) = d(z) f^*(z)$, we reduce the Routh-Hurwitz problem for $f(z)$ to that for the polynomial $f^*(z)$ for which the last Hurwitz determinant is different from zero.

1. To begin with, we examine the case where

$$\Delta_1 = \ldots = \Delta_p = 0, \quad \Delta_{p+1} \neq 0, \quad \ldots, \quad \Delta_n \neq 0. \tag{41}$$

From $\Delta_1 = 0$ it follows that $b_0 = 0$; from $\Delta_2 = \begin{vmatrix} 0 & b_1 \\ a_0 & a_1 \end{vmatrix} = -a_0 b_1 = 0$ it follows that $b_1 = 0$. But then we have automatically

$$\Delta_3 = \begin{vmatrix} 0 & b_1 & b_2 \\ a_0 & a_1 & a_2 \\ 0 & 0 & b_1 \end{vmatrix} = -a_0 b_1^2 = 0.$$

From

$$\Delta_4 = \begin{vmatrix} 0 & 0 & b_2 & b_3 \\ a_0 & a_1 & a_2 & a_3 \\ 0 & 0 & 0 & b_2 \\ 0 & a_0 & a_1 & a_2 \end{vmatrix} = -a_0^2 b_2^2 = 0$$

it follows that $b_2 = 0$ and then $\Delta_5 = -a_0^2 b_2^3 = 0$, etc.

This argument shows that in (41) p is always an *odd* number $p = 2h - 1$. Then $b_0 = b_1 = b_2 = \ldots = b_{h-1} = 0$, $b_h \neq 0$, and[32]

$$\Delta_{p+1} = \Delta_{2h} = (-1)^{\frac{h(h+1)}{2}} a_0^h b_h^h, \quad \Delta_{p+2} = \Delta_{2h+1} = (-1)^{\frac{h(h+1)}{2}} a_0^h b_h^{h+1} = \Delta_{p+1} b_h. \tag{42}$$

Let us vary the coefficients $b_0, b_1, \ldots, b_{h-1}$ in such a way that for the new, slightly altered values $b_0^*, b_1^*, \ldots, b_{h-1}^*$ all the Hurwitz determinants $\Delta_1^*, \Delta_2^*, \ldots, \Delta_n^*$ become different from zero and $\Delta_{p+1}^*, \ldots, \Delta_n^*$ keep their previous signs. We shall take $b_0^*, b_1^*, \ldots, b_{h+1}^*$ as 'small' values of different orders of 'smallness'; indeed, we shall assume that every b_{j-1}^* is in abso-

[32] From (42) it follows that for odd h sign $\Delta_{p+2} = (-1)^{\frac{h+1}{2}}$ sign a_0, and for even h sign $\Delta_{p+1} = (-1)^{\frac{h}{2}}$.

lute value 'considerably' smaller than b_j^* ($j=1, 2, \ldots, h$; $b_h^*=b_h$). The latter means that in computing the sign of an integral algebraic expression in the b_i^* we can neglect terms in which some b_i^* have an index less than j in comparison with terms where all the b_i^* have an index at least j. We can then easily find the 'sign-determining' terms of $\Delta_1^*, \Delta_2^*, \ldots, \Delta_p^*$ $(p=2h-1)$:[33]

$$\Delta_1^*=b_0^*,\ \Delta_2^*=-a_0b_1^*+\cdots,\ \Delta_3^*=-a_0b_1^{*2}+\cdots,\ \Delta_4^*=-a_0^2b_2^{*2}+\cdots,$$
$$\Delta_5^*=-a_0^2b_2^{*3}+\cdots,\ \Delta_6^*=a_0^3b_3^{*3}+\cdots,$$

etc.; in general,

$$\left.\begin{aligned} \Delta_{2j}^*&=(-1)^{\frac{j(j+1)}{2}}a_0^jb_j^{*j}+\cdots \qquad (j=1, 2, \ldots, h-1),\\ \Delta_{2j+1}^*&=(-1)^{\frac{j(j+1)}{2}}a_0^jb_j^{*j+1}+\cdots \qquad (j=0, 1, \ldots, h-1).\end{aligned}\right\} \tag{43}$$

We choose $b_0^*, b_1^*, \ldots, b_{2h-1}$ as positive; then the sign of Δ_i^* is determined by the formula

$$\operatorname{sign}\Delta_i^*=(-1)^{\frac{j(j+1)}{2}}\operatorname{sign}a_0^j \qquad \left(j=\left[\frac{i}{2}\right],\ i=1, 2, \ldots, p\right). \tag{44}$$

In any small variation of the coefficients of the polynomial the number k remains unchanged, because $f(z)$ has no roots on the imaginary axis. Therefore, starting from (44) we determine the number of roots in the right half-plane by the formula

$$k=V\left(a_0, \Delta_1^*, \frac{\Delta_2^*}{\Delta_1^*}, \ldots, \frac{\Delta_{p+1}}{\Delta_p^*}, \frac{\Delta_{p+2}}{\Delta_{p+1}}\right)+V\left(\frac{\Delta_{p+2}}{\Delta_{p+1}}, \ldots, \frac{\Delta_n}{\Delta_{n-1}}\right). \tag{45}$$

An elementary calculation based on (42) and (44) shows that

$$V\left(a_0, \Delta_1^*, \frac{\Delta_2^*}{\Delta_1^*}, \ldots, \frac{\Delta_{p+1}}{\Delta_p^*}, \frac{\Delta_{p+2}}{\Delta_{p+1}}\right)=h+\frac{1-(-1)^h\varepsilon}{2}\left(\begin{gathered}p=2h-1\\ \varepsilon=\operatorname{sign}\left(a_0\frac{\Delta_{p+2}}{\Delta_{p+1}}\right)\end{gathered}\right) \tag{46}$$

Note that the value on the left-hand side of (46) does not depend on the method of varying the coefficients and retains one and the same sign for arbitrary small variations. This follows from (45), because k does not change its value under small variations of the coefficients.

[33] Essentially the same terms have already been computed above for $\Delta_1, \Delta_2, \ldots, \Delta_p$.

2. Suppose now that for $s > 0$

$$\Delta_{s+1} = \cdots = \Delta_{s+p} = 0 \tag{47}$$

and that all the remaining Hurwitz determinants are different from zero.

We denote by $\tilde{a}_0, \tilde{a}_1, \ldots$ and $\tilde{b}_0, \tilde{b}_1, \ldots$ the elements of the $(s+1)$-th rows in Routh's scheme ($\tilde{a}_0 = \Delta_s/\Delta_{s-1}$, $\tilde{b}_0 = \Delta_{s+1}/\Delta_s$). We denote the corresponding determinants by $\tilde{\Delta}_1, \tilde{\Delta}_2, \ldots, \tilde{\Delta}_{n-s}$. By formula (38) (p. 195),

$$\Delta_{s+1} = \Delta_s\tilde{\Delta}_1, \ \ldots, \ \Delta_{s+p} = \Delta_s\tilde{\Delta}_p, \ \Delta_{s+p+1} = \Delta_s\tilde{\Delta}_{p+1}, \ \Delta_{s+p+2} = \Delta_s\tilde{\Delta}_{p+2}. \tag{48}$$

Then by 1. it follows that p is odd, say $p = 2h - 1$.[34]

Let us vary the coefficients of $f(z)$ in such a way that all the Hurwitz determinants become different from zero and that those that were different from zero before the variation retain their sign. Since the formula (46) is applicable to the determinants $\tilde{\Delta}$, we then obtain, starting from (48):

$$V\left(\frac{\Delta_s}{\Delta_{s-1}}, \frac{\overset{*}{\Delta}_{s+1}}{\Delta_s}, \ldots, \frac{\Delta_{s+p+1}}{\overset{*}{\Delta}_{s+p}}, \frac{\Delta_{s+p+2}}{\Delta_{s+p+1}}\right)$$

$$= h + \frac{1-(-1)^h\varepsilon}{2} \quad \left(\begin{array}{c} p = 2h-1, \\ \varepsilon = \operatorname{sign}\left(\dfrac{\Delta_s}{\Delta_{s-1}}\dfrac{\Delta_{s+p+2}}{\Delta_{s+p+1}}\right)\end{array}\right), \tag{49}$$

$$k = V\left(a_0, \Delta_1, \ldots, \frac{\Delta_s}{\Delta_{s-1}}\right) + V\left(\frac{\Delta_s}{\Delta_{s-1}}, \frac{\overset{*}{\Delta}_{s+1}}{\Delta_s}, \ldots, \frac{\Delta_{s+p+2}}{\Delta_{s+p+1}}\right) + V\left(\frac{\Delta_{s+p+2}}{\Delta_{s+p+1}}, \ldots, \frac{\Delta_n}{\Delta_{n-1}}\right).$$

The value on the left-hand side of (49) again does not depend on the method of variation.

3. Finally, let us assume that among the Hurwitz determinants there are ν groups of zero determinants. We shall show that for every such group (47) the value on the left-hand side of (49) does not depend on the method of variation and is determined by that formula.[35] We have proved this statement for $\nu = 1$. Let us assume that it is true for $\nu - 1$ groups and then show that it is also true for ν groups. Suppose that (47) is the second of the ν groups; we determine $\tilde{\Delta}_1, \tilde{\Delta}_2$ in the same way as was done under 2.; then for this variation

---

[34] In accordance with footnote 32, for $p = 2h - 1$ and odd h,

$$\operatorname{sign}\Delta_{s+p+2} = (-1)^{\frac{h+1}{2}}\operatorname{sign}\Delta_{s-1};$$

and for even h,

$$\operatorname{sign}\Delta_{s+p+1} = (-1)^{\frac{h}{2}}\operatorname{sign}\Delta_s.$$

[35] From (47) and $\Delta_s \neq 0, \Delta_{s+p+1} \neq 0$ it follows by (48) and (42) that $\Delta_{s-1} \neq 0$, $\Delta_{s+p+2} \neq 0$.

$$V\left(\frac{\Delta_s^*}{\Delta_{s-1}^*}, \ldots, \frac{\Delta_n^*}{\Delta_{n-1}^*}\right) = V\left(\tilde{a}_0^*, \tilde{\Delta}_1^*, \ldots, \frac{\tilde{\Delta}_{n-s}^*}{\tilde{\Delta}_{n-s-1}^*}\right).$$

Since we have only $\nu - 1$ groups of zero determinants on the right-hand side of this equation, our statement holds for the right-hand side and hence for the left-hand side of the equation. In other words, the formula (49) holds for the second, ..., ν-th group of zero Hurwitz determinants. But then it follows from the formula

$$k = V\left(a_0^*, \Delta_1^*, \frac{\Delta_2^*}{\Delta_1^*}, \ldots, \frac{\Delta_n^*}{\Delta_{n-1}^*}\right)$$

that the value of $V\left(\frac{\Delta_s}{\Delta_{s-1}}, \frac{\overset{\bullet}{\Delta}_{s+1}}{\Delta_s}, \frac{\overset{\bullet}{\Delta}_{s+2}}{\Delta_{s+1}^*}, \ldots, \frac{\Delta_{s+p+2}}{\Delta_{s+p+1}}\right)$ does not depend on the method of variation for the first group of zero determinants, and therefore that (49) holds for this group as well.

Thus we have proved the following theorem:

THEOREM 5: *If some of the Hurwitz determinants are zero, but $\Delta_n \neq 0$, then the number of roots of the real polynomial $f(z)$ in the right half-plane is determined by the formula*

$$k = V\left(a_0, \Delta_1, \frac{\Delta_2}{\Delta_1}, \ldots, \frac{\Delta_n}{\Delta_{n-1}}\right)$$

in which for the calculation of the value of V for every group of p successive zero determinants (p is always odd!)

$$(\Delta_s \neq 0)\ \Delta_{s+1} = \cdots = \Delta_{s+p} = 0 \quad (\Delta_{s+p+1} \neq 0)$$

we have to set

$$V\left(\frac{\Delta_s}{\Delta_{s-1}}, \frac{\Delta_{s+1}}{\Delta_s}, \ldots, \frac{\Delta_{s+p+2}}{\Delta_{s+p+1}}\right) = h + \frac{1-(-1)^h\varepsilon}{2} \tag{50}$$

where[36]

$$p = 2h - 1 \quad \textit{and} \quad \varepsilon = \operatorname{sign}\left(\frac{\Delta_s}{\Delta_{s-1}} \frac{\Delta_{s+p+2}}{\Delta_{s+p+1}}\right).$$

§ 9. The Method of Quadratic Forms. Determination of the Number of Distinct Real Roots of a Polynomial

Routh obtained his algorithm by applying Sturm's theorem to the computation of the Cauchy index of a regular rational fraction of special type (see formula (10) on p. 178). Of the two polynomials in this fraction—numera-

[36] For $s = 1$ $\frac{\Delta_s}{\Delta_{s-1}}$ is to be replaced by Δ_1; and for $s = 0$, by a_0.

tor and denominator—one contains only even, the other only odd powers of the argument z.

In this and in the following sections we shall explain the deeper and more comprehensive method of quadratic forms, due to Hermite, in its application to the Routh-Hurwitz problem. By means of this method we shall obtain an expression for the index of an arbitrary rational fraction in terms of the coefficients of the numerator and denominator. The method of quadratic forms enables us to apply the results of Frobenius' subtle investigations in the theory of Hankel forms (Vol. I, Chapter X, § 10) to the Routh-Hurwitz problem and to establish a close connection of certain remarkable theorems of Chebyshev and Markov with the problem of stability.

1. We shall acquaint the reader with the method of quadratic forms first in the comparatively simple problem of determining the number of distinct real roots of a polynomial.

In the solution of this problem we may restrict ourselves to the case where $f(z)$ is a real polynomial. For suppose that $f(z) = u(z) + iv(z)$ is a complex polynomial ($u(z)$ and $v(z)$ being real polynomials). Each real root of $f(z)$ makes $u(z)$ and $v(z)$ vanish simultaneously. Therefore the complex polynomial $f(z)$ has the same real roots as the real polynomial $d(z)$, the greatest common divisor of $u(z)$ and $v(z)$.

Thus, let $f(z)$ be a real polynomial with the distinct roots $\alpha_1, \alpha_2, \ldots, \alpha_q$ of the respective multiplicities $n_1, n_2, \ldots, n_q$:

$$f(z) = a_0 (z - \alpha_1)^{n_1} (z - \alpha_2)^{n_2} \cdots (z - \alpha_q)^{n_q}$$

$$(a_0 \neq 0;\ \alpha_i \neq \alpha_k \quad \text{for} \quad i \neq k;\ i, k = 1, 2, \ldots, q).$$

We introduce Newton's sums

$$s_p = \sum_{j=1}^{q} n_j \alpha_j^p \quad (p = 0, 1, 2, \ldots).$$

With these sums we form the Hankel forms

$$S_n(x, x) = \sum_{i,k=0}^{n-1} s_{i+k} x_i x_k,$$

where n is an arbitrary integer, $n \geqq q$.

Then the following theorem holds:

THEOREM 6: *The number of all the distinct roots of $f(z)$ is equal to the rank, and the number of all the distinct real roots to the signature, of the form $S_n(x, x)$.*

Proof. From the definition of the form $S_n(x, x)$ we immediately obtain the following representation:

$$S_n(x, x) = \sum_{j=1}^{q} n_j (x_0 + \alpha_j x_1 + \alpha_j^2 x_2 + \cdots + \alpha_j^{n-1} x_{n-1})^2. \tag{51}$$

Here to each root α_j of $f(z)$ there corresponds the square of a linear form $Z_j = x_0 + \alpha_j x_1 + \ldots + \alpha_j^{n-1} x_{n-1}$ $(j = 1, 2, \ldots, q)$. The forms $Z_1, Z_2, \ldots, Z_q$ are linearly independent, since their coefficients form the Vandermonde matrix $\| \alpha_j^h \|$ whose rank is equal to the number of distinct α_j, i.e., to q. Therefore (see Vol. I, p. 297) the rank of the form $S_n(x, x)$ is q.

In the representation (51) to each real root α_j there corresponds a positive square. To each pair of conjugate complex roots α_j and $\bar{\alpha}_j$ there correspond two complex conjugate forms:

$$Z_j = P_j + iQ_j, \quad \bar{Z}_j = P_j - iQ_j;$$

the corresponding terms in (51) together give one positive and one negative square:

$$n_j Z_j^2 + n_j \bar{Z}_j^2 = 2 n_j P_j^2 - 2 n_j Q_j^2.$$

Hence it is easy to see[37] that the signature of $S_n(x, x)$, i.e., the difference between the number of positive and negative squares, is equal to the number of distinct real α_j.

This proves the theorem.

2. Using the rule for determining the signature of a quadratic form that we established in Chapter X (Vol. I, p. 303), we obtain from the theorem the following corollary:

COROLLARY: *The number of distinct real roots of the real polynomial $f(z)$ is equal to the excess of permanences of sign over variations of sign in the sequence*

$$1,\ s_0,\ \begin{vmatrix} s_0 & s_1 \\ s_1 & s_2 \end{vmatrix},\ \ldots,\ \begin{vmatrix} s_0 & s_1 & \ldots & s_{n-1} \\ s_1 & s_2 & \ldots & s_n \\ \cdot & \cdot & \cdot & \cdot \\ s_{n-1} & s_n & \ldots & s_{2n-2} \end{vmatrix}, \tag{52}$$

where the s_p $(p = 0, 1, \ldots)$ are Newton's sums for $f(z)$ and n is any integer not less than the number q of distinct roots of $f(z)$ (in particular, n can be chosen as the degree of $f(z)$).

---

[37] The quadratic form $S_n(x, x)$ is representable as an (algebraic) sum of q squares of the real forms Z_j (for real α_j) and P_j and Q_j (for complex α_j). These forms are linearly independent, since the rank of $S_n(x, x)$ is q.

This rule for determining the number of distinct real roots is directly applicable only when all the numbers in (52) are different from zero. However, since we deal here with the computation of the signature of a Hankel form, by the results of Vol. I, Chapter X, § 10, the rule with proper refinements remains valid in the general case (for further details see § 11 of that chapter).

From our theorem it follows that: *All the forms*

$$S_n(x, x) \quad (n = q, q+1, \ldots)$$

have the same rank and the same signature.

In applying Theorem 6 (or its corollary) to determine the number of distinct real roots, we may take n to be the degree of $f(z)$.

The number of distinct real roots of the real polynomial $f(z)$ is equal to the index $I_{-\infty}^{+\infty} \frac{f'(z)}{f(z)}$ (see p. 175). Therefore the corollary to Theorem 6 gives the formula

$$I_{-\infty}^{+\infty} \frac{f'(z)}{f(z)} = n - 2V\left(1, s_0, \begin{vmatrix} s_0 & s_1 \\ s_1 & s_2 \end{vmatrix}, \ldots, \begin{vmatrix} s_0 & s_1 & \ldots & s_{n-1} \\ s_1 & s_2 & \ldots & s_n \\ \cdot & \cdot & \cdot & \cdot \\ s_{n-1} & s_n & \ldots & s_{2n-2} \end{vmatrix}\right),$$

where $s_p = \sum_{j=1}^{q} n_j \alpha_j^p$ $(p = 0, 1, \ldots)$ are Newton's sums and n is the degree of $f(z)$.

In § 11 we shall establish a similar formula for the index of an arbitrary rational fraction. The information on infinite Hankel matrices that will be required for this purpose will be given in the next section.

§ 10. Infinite Hankel Matrices of Finite Rank

1. Let

$$s_0, s_1, s_2, \ldots$$

be a sequence of complex numbers. This determines an infinite symmetric matrix

$$S = \begin{Vmatrix} s_0 & s_1 & s_2 & \ldots \\ s_1 & s_2 & s_3 & \ldots \\ s_2 & s_3 & s_4 & \ldots \\ \cdot & \cdot & \cdot & \cdot \end{Vmatrix},$$

which is usually called a *Hankel matrix*. Together with the infinite Hankel matrices we shall consider[38] the finite Hankel matrices $S_n = \| s_{i+k} \|_0^{n-1}$ and their associated Hankel forms

$$S_n(x, x) = \sum_{i,k=0}^{n-1} s_{i+k} x_i x_k .$$

The successive principal minors of S will be denoted by $D_1, D_2, D_3, \ldots$

$$D_p = | s_{i+k} |_0^{p-1} \qquad (p = 1, 2, \ldots).$$

Infinite matrices may be of finite or of infinite rank. In the latter case, the matrices have non-zero minors of arbitrarily large order. The following theorem gives a necessary and sufficient condition for a sequence of numbers $s_0, s_1, s_2, \ldots$ to generate an infinite Hankel matrix $S = \| s_{i+k} \|_0^\infty$ of finite rank.

THEOREM 7: *The infinite matrix $S = \| s_{i+k} \|_0^\infty$ is of finite rank r if and only if there exist r numbers $\alpha_1, \alpha_2, \ldots, \alpha_r$ such that*

$$s_q = \sum_{g=1}^{r} \alpha_g s_{q-g} \qquad (q = r, r+1, \ldots) \tag{53}$$

and r is the least number having this property.

Proof. If the matrix $S = \| s_{i+k} \|_0^\infty$ has finite rank r, then its first $r+1$ rows $R_1, R_2, \ldots, R_{r+1}$ are linearly dependent. Therefore there exists a number $h \leqq r$ such that $R_1, R_2, \ldots, R_h$ are linearly independent and R_{h+1} is a linear combination of them:

$$R_{h+1} = \sum_{g=1}^{h} \alpha_g R_{h-g+1} .$$

We consider the rows $R_{q+1}, R_{q+2}, \ldots, R_{q+h+1}$, where q is any non-negative integer. From the structure of S it is immediately clear that the rows $R_{q+1}, R_{q+2}, \ldots, R_{q+h+1}$ are obtained from $R_1, R_2, \ldots, R_{h+1}$ by a 'shortening' process in which the elements in the first q columns are omitted. Therefore

$$R_{q+h+1} = \sum_{g=1}^{h} \alpha_g R_{q+h-g+1} \qquad (q = 0, 1, 2, \ldots).$$

Thus, every row of S beginning with the $(h+1)$-th can be expressed linearly in terms of the h preceding rows and therefore in terms of the linearly

---

[38] See Vol. I, Chapter X, § 10.

independent first h rows. Hence it follows that the rank of S is $r=h$.[39] The linear dependence

$$R_{q+h+1}=\sum_{g=1}^{h}\alpha_g R_{q+h-g+1}$$

after replacement of h by r and written in more convenient notation yields (53).

Conversely, if (53) holds, then every row (column) of S is a linear combination of the first r rows (columns). Therefore all the minors of S whose orders exceed r are zero and S is of rank at most r. But the rank cannot be less than r, since then, as we have already shown, there would be relations of the form (53) with a smaller value than r, and this contradicts the second condition of the theorem. The proof of the theorem is now complete.

COROLLARY: *If the infinite Hankel matrix $S=\| s_{i+k}\|_0^\infty$ is of finite rank r, then*

$$D_r=| s_{i+k}|_0^{r-1}\neq 0.$$

For it follows from the relations (53) that every row (column) of S is a linear combination of the first r rows (columns). Therefore every minor of S of order r can be represented in the form aD_r, where a is a constant. Hence it follows that $D_r\neq 0$.

Note. For finite Hankel matrices of rank r the inequality $D_r\neq 0$ need not hold. For example $S_2=\begin{Vmatrix} s_0 & s_1\\ s_1 & s_2\end{Vmatrix}$ for $s_0=s_1=0$, $s_2\neq 0$ is of rank 1, whereas $D_1=s_0=0$.

2. We shall now explain certain remarkable connections between infinite Hankel matrices and rational functions.

Let

$$R(z)=\frac{g(z)}{h(z)}$$

be a proper rational fractional function, where

$$h(z)=a_0z^m+\cdots+a_m \quad (a_0\neq 0),\qquad g(z)=b_1z^{m-1}+b_2z^{m-2}+\cdots+b_m.$$

We write the expansion of $R(z)$ in a power series of negative powers of z:

$$R(z)=\frac{g(z)}{h(z)}=\frac{s_0}{z}+\frac{s_1}{z^2}+\frac{s_2}{z^3}+\cdots.$$

---

[39] The statement 'The number of linearly independent rows in a rectangular matrix is equal to its rank' is true not only for finite rows but also for infinite rows.

If all the poles of $R(z)$, i.e., all the values of z for which $R(z)$ becomes infinite, lie in the circle $|z| \leqq a$, then the series on the right-hand side of the expansion converges for $|z| > a$. We multiply both sides by the denominator $h(z)$:

$$(a_0 z^m + a_1 z^{m-1} + \cdots + a_m)\left(\frac{s_0}{z} + \frac{s_1}{z^2} + \frac{s_2}{z^3} + \cdots\right) = b_1 z^{m-1} + b_2 z^{m-2} + \cdots + b_m.$$

Equating coefficients of equal powers of z on both sides of this identity, we obtain the following system of relations:

$$\left.\begin{array}{l} a_0 s_0 = b_1, \\ a_0 s_1 + a_1 s_0 = b_2, \\ \cdot\ \cdot\ \cdot\ \cdot\ \cdot\ \cdot\ \cdot \\ a_0 s_{m-1} + a_1 s_{m-2} + \cdots + a_{m-1} s_0 = b_m, \end{array}\right\} \tag{54}$$

$$a_0 s_q + a_1 s_{q-1} + \cdots + a_m s_{q-m} = 0 \qquad (q = m, m+1, \ldots). \tag{54'}$$

Setting

$$\alpha_g = -\frac{a_g}{a_0} \qquad (g = 1, 2, \ldots, m),$$

we can write the relations (54′) in the form (53) (for $r = m$). Therefore, by Theorem 7, the infinite Hankel matrix

$$S = \| s_{i+k} \|_0^\infty$$

formed from the coefficients $s_0, s_1, s_2, \ldots$ is of finite rank ($\leqq m$).

Conversely, if the matrix $S = \| s_{i+k} \|_0^\infty$ is of finite rank r, then the relations (53) hold, which can be written in the form (54′) (for $m = r$). Then, when we define the numbers $b_1, b_2, \ldots, b_m$ by the equations (54) we have the expansion

$$\frac{b_1 z^{m-1} + \cdots + b_m}{a_0 z^m + a_1 z^{m-1} + \cdots + a_m} = \frac{s_0}{z} + \frac{s_1}{z^2} + \cdots.$$

The least degree of the denominator m for which this expansion holds is the same as the least integer m for which the relations (53) hold. By Theorem 7, this least value of m is the rank of $S = \| s_{i+k} \|_0^\infty$.

Thus we have proved the following theorem:

THEOREM 8: *The matrix* $S = \| s_{i+k} \|_0^\infty$ *is of finite rank if and only if the sum of the series*

$$R(z) = \frac{s_0}{z} + \frac{s_1}{z^2} + \frac{s_2}{z^3} + \cdots$$

is a rational function of z. *In this case the rank of* S *is the same as the number of poles of* $R(z)$, *counting each pole with its proper multiplicity.*

§ 11. Determination of the Index of an Arbitrary Rational Fraction by the Coefficients of Numerator and Denominator

1. Suppose given a rational function. We write its expansion in a series of descending powers of z:[40]

$$R(z) = s_{-u-1}z^u + \cdots + s_{-2}z + s_{-1} + \frac{s_0}{z} + \frac{s_1}{z^2} + \cdots. \tag{55}$$

The sequence of coefficients of the negative powers of z

$$s_0, s_1, s_2, \ldots$$

determines an infinite Hankel matrix $S = \| s_{i+k} \|_0^\infty$.

We have thus established a correspondence

$$R(z) \sim S.$$

Obviously two rational functions whose difference is an integral function correspond to one and the same matrix S. However, not every matrix $S = \| s_{i+k} \|_0^\infty$ corresponds to some rational function. In the preceding section we have seen that an infinite matrix S corresponds to a rational function if and only if it is of finite rank. This rank is equal to the number of poles of $R(z)$ (multiplicities taken into account), i.e., to the degree of the denominator $f(z)$ in the reduced fraction $g(z)/f(z) = R(z)$. By means of the expansion (55) we have a one-to-one corespondence between proper rational functions $R(z)$ and Hankel matrices $S = \| s_{i+k} \|_0^\infty$ of finite rank.

We mention some properties of the correspondence:

1. If $R_1(z) \sim S_1$, $R_2(z) \sim S_2$, then for arbitrary numbers c_1, c_2

$$c_1R_1(z) + c_2R_2(z) \sim c_1S_1 + c_2S_2.$$

In what follows we shall have to deal with the case where the coefficients of the numerator and the denominator of $R(z)$ are integral rational functions of a parameter α; R is then a rational function of z and α. From the expansion (54) it follows that in this case the numbers $s_0, s_1, s_2, \ldots$, i.e., the elements of S, depend rationally on α. Differentiating (55) term by term with respect to α, we obtain:

2. If $R(z, \alpha) \sim S(\alpha)$, then $\frac{\partial R}{\partial \alpha} \sim \frac{\partial S}{\partial \alpha}$.[41]

---

[40] The series (55) converges outside every circle (with center at $z = 0$) containing all the poles of $R(z)$.

[41] If $S = \| s_{i+k} \|_0^\infty$, then $\frac{\partial S}{\partial \alpha} = \| \frac{\partial s_{i+k}}{\partial \alpha} \|_0^\infty$.

2. Let us write down the expansion of $R(z)$ in partial fractions:

$$R(z) = Q(z) + \sum_{j=1}^{q} \left\{ \frac{A_1^{(j)}}{z - \alpha_j} + \frac{A_2^{(j)}}{(z-\alpha_j)^2} + \cdots + \frac{A_{\nu_j}^{(j)}}{(z-\alpha_j)^{\nu_j}} \right\}, \tag{56}$$

where $Q(z)$ is a polynomial; we shall show how to construct the matrix S corresponding to $R(z)$ from the numbers α and A.

For this purpose we consider first the simple rational function

$$\frac{1}{z-\alpha} = \sum_{p=0}^{\infty} \frac{\alpha^p}{z^{p+1}}.$$

It corresponds to the matrix

$$S_\alpha = \| \alpha^{i+k} \|_0^\infty .$$

The form $S_{\alpha n}(x, x)$ associated with this matrix is

$$S_{\alpha n}(x, x) = \sum_{i,k=0}^{n-1} \alpha^{i+k} x_i x_k = (x_0 + \alpha x_1 + \cdots + \alpha^{n-1} x_{n-1})^2 .$$

If

$$R(z) = Q(z) + \sum_{j=1}^{q} \frac{A^{(j)}}{z - \alpha_j},$$

then by 1. the corresponding matrix S is determined by the formula

$$S = \sum_{j=1}^{q} A^{(j)} S_{\alpha_j} = \left\| \sum_{j=1}^{q} A^{(j)} \alpha_j^{i+k} \right\|_0^\infty$$

and the corresponding quadratic form is

$$S_n(x, x) = \sum_{j=1}^{q} A^{(j)} (x_0 + \alpha_j x_1 + \cdots + \alpha_j^{n-1} x_{n-1})^2 .$$

In order to proceed to the general case (56), we first differentiate the relation

$$\frac{1}{z - \alpha} \sim S_\alpha = \| \alpha^{i+k} \|_0^\infty$$

$h-1$ times term by term. By 1. and 2., we obtain:

$$\frac{1}{(z-\alpha)^h} \sim \frac{1}{(h-1)!} \frac{\partial^{h-1} S_\alpha}{\partial \alpha^{h-1}} = \left\| \binom{i+k}{h-1} \alpha^{i+k-h+1} \right\|_0^\infty, \ \binom{i+k}{h-1} = 0 \text{ for } i+k < h-1.$$

Therefore, by using rule 1. again we find in the general case, where $R(z)$ has the expansion (56):

$$R(z) \sim S = \sum_{j=1}^{q} \left(A_1^{(j)} + A_2^{(j)} \frac{\partial}{\partial \alpha_j} + \cdots + \frac{1}{(\nu_j - 1)!} A_{\nu_j}^{(j)} \frac{\partial^{\nu_j - 1}}{\partial \alpha_j^{\nu_j - 1}} \right) S_{\alpha_j}. \tag{57}$$

By carrying out the differentiation, we obtain:

$$S = \left\| \sum_{j=1}^{q} A_1^{(j)} \alpha_j^{i+k} + A_2^{(j)} \binom{i+k}{1} \alpha_j^{i+k-1} + \cdots + A_{\nu_j}^{(j)} \binom{i+k}{\nu_j - 1} \alpha_j^{i+k-\nu_j+1} \right\|_0^{\infty}.$$

The corresponding Hankel form $S_n(x, x) = \sum_{i,k=0}^{n-1} s_{i+k} x_i x_k$ is

$$S_n(x, x) = \sum_{j=1}^{q} \left(A_1^{(j)} + A_2^{(j)} \frac{\partial}{\partial \alpha_j} + \cdots + \frac{1}{(\nu_j - 1)!} A_{\nu_j}^{(j)} \frac{\partial^{\nu_j - 1}}{\partial \alpha_j^{\nu_j - 1}} \right) (x_0 + \alpha_j x_1 + \cdots + \alpha_j^{n-1} x_{n-1})^2. \tag{57'}$$

3. Now we are in a position to enunciate and prove the fundamental theorem:[42]

THEOREM 9: *If*

$$R(z) \sim S$$

and m is the rank of S,[43] then the Cauchy index $I_{-\infty}^{+\infty} R(z)$ is equal to the signature[44] of the form $S_n(x, x)$ for any $n \geqq m$:

$$I_{-\infty}^{+\infty} R(z) = \sigma\,[S_n(x, x)].$$

Proof. Suppose that the expansion (56) holds. Then, by (57),

$$S = \sum_{j=1}^{q} T_{\alpha_j},$$

where each term is of the form

$$T_\alpha = \left(A_1 + A_2 \frac{\partial}{\partial \alpha} + \cdots + \frac{1}{(\nu - 1)!} A_\nu \frac{\partial^{\nu-1}}{\partial \alpha^{\nu-1}} \right) S_\alpha, \quad S_\alpha = \left\| \alpha^{i+k} \right\|_0^{\infty} \tag{58}$$

and

$$S_n(x, x) = \sum_{j=1}^{q} T_{\alpha_j}(x, x) = \sum_{\alpha_j \text{ real}} T_{\alpha_j}(x, x) + \sum_{\alpha_j \text{ complex}} [T_{\alpha_j}(x, x) + T_{\bar{\alpha}_j}(x, x)]$$

---

[42] This theorem was proved by Hermite in 1856 for the simplest case where $R(z)$ has no multiple poles [187]. In the general case it was proved by Hurwitz [204] (see also [25], pp. 17-19). The proof in the text differs from Hurwitz' proof.

[43] As we have already mentioned, m is the degree of the denominator in the reduced representation of the rational fraction $R(z)$ (see Theorem 8 on p. 207).

[44] We denote the signature of $S_n(x, x)$ by $\sigma[S_n(x, x)]$.

By Theorem 8, the rank of the matrix T_{α_j}, and hence of the form $T_{\alpha_j}(x, x)$, is ν_j $(j=1, 2, \ldots, q)$ and the rank of $S_n(x, x)$ is $m=\sum_{j=1}^{q}\nu_j$. But if the rank of the sum of certain real quadratic forms is equal to the sum of the ranks of the constituent forms, then the same relation holds for the signatures:

$$\sigma\left[S_n(x, x)\right]=\sum_{\alpha_j \text{ real}}\sigma\left[T_{\alpha_j}(x, x)\right]+\sum_{\alpha_j \text{ complex}}\sigma\left[T_{\alpha_j}(x, x)+T_{\overline{\alpha}_j}(x, x)\right]. \tag{59}$$

We consider two cases separately:

1) α is real. Under any variation of the parameters $A_1, A_2, \ldots, A_{\nu-1}$ and α in

$$\frac{A_1}{z-\alpha}+\frac{A_2}{(z-\alpha)^2}+\cdots+\frac{A_\nu}{(z-\alpha)^\nu} \tag{60}$$

the rank of the corresponding matrix T_α remains unchanged $(=\nu)$; therefore the signature of $T_\alpha(x, x)$ also remains unchanged (see Vol. I, p. 309). Therefore $\sigma[T_\alpha(x, x)]$ does not change if we set in (59) and (60): $A_1=\ldots=A_{\nu-1}=0$ and $\alpha=0$, i.e., if for T_α we take the matrix

$$\frac{1}{(\nu-1)!}\frac{\partial^{\nu-1}S_\alpha}{\partial\alpha^{\nu-1}}=\begin{Vmatrix} \overbrace{0\ \ 0\ \ldots\ 0}^{\nu-1} & A_\nu & 0 & 0 & \ldots \\ 0 & & & \cdot^{\cdot^{\cdot}} & \\ \vdots & & \cdot^{\cdot^{\cdot}} & & \\ 0 & \cdot^{\cdot^{\cdot}} & & & \\ A_\nu & & & & \\ 0 & & & & \\ 0 & & & & \\ \vdots & & & & \end{Vmatrix}.$$

The corresponding quadratic form is equal to

$$\left.\begin{aligned} &2A_\nu\,(x_0x_{\nu-1}+x_1x_{\nu-2}+\cdots+x_{s-1}x_s) \text{ for } \nu=2s, \\ &A_\nu\,[2\,(x_0x_{\nu-1}+\cdots+x_{s-2}x_s)+x_{s-1}^2] \text{ for } \nu=2s-1, \end{aligned}\right\}\quad (s=1, 2, 3, \ldots).$$

But the signature of the upper form is always zero and that of the lower form is sign A_ν. Thus, if α is real, then

$$\sigma\left[T_\alpha(x,x)\right]=\begin{cases} 0, & \text{for even } \nu \\ \operatorname{sign} A_\nu, & \text{for odd } \nu \end{cases} \tag{61}$$

2) α is complex.

$$T_\alpha(x,x)=\sum_{k=1}^{\nu}(P_k+iQ_k)^2,\quad T_{\bar\alpha_j}(x,x)=\sum_{k=1}^{\nu}(P_k-iQ_k)^2,$$

where P_k, Q_k $(k=1,2,\ldots,\nu)$ are real linear forms in the variables x_0, x_1, $x_2,\ldots,x_{n-1}$. Then

$$T_\alpha(x,x)+T_{\bar\alpha}(x,x)=2\sum_{k=1}^{\nu}P_k^2-2\sum_{k=1}^{\nu}Q_k^2. \tag{62}$$

Since the rank of this quadratic form is 2ν, the P_k, Q_k $(k=1,2,\ldots,\nu)$ are linearly independent, so that by (62) for a complex α

$$\sigma\left[T_\alpha(x,x)+T_{\bar\alpha}(x,x)\right]=0. \tag{63}$$

From (59), (61), and (63) it follows that

$$\sigma\left[S_n(x,x)\right]=\sum_{\left(\substack{\alpha_j \text{ real} \\ \nu \text{ odd}}\right)}\operatorname{sign} A_\nu^{(j)}.$$

But on p. 175 we saw that the sum on the right-hand side of this equation is $I_{-\infty}^{+\infty}R(z)$. This completes the proof.

From this theorem we deduce:

COROLLARY 1: *If* $R(z)\sim S=\left\| s_{i+k}\right\|_0^\infty$ *and* m *is the rank of* S, *then all the quadratic forms* $S_n(x,x)=\sum_{i,k=0}^{n-1}s_{i+k}x_ix_k$ $(n=m,m+1,\ldots)$ *have one and the same signature.*

In Chapter X, § 10 (Vol. I, pp. 343-44) we established a rule for computing the signature of a Hankel form; moreover, Frobenius' investigations enabled us to formulate a rule that embraces all singular cases. By the

[45] Each of the products $x_0x_{\nu-1}$, $x_1x_{\nu-2}$, ... can be replaced by a difference of squares

$$\left(\frac{x_0+x_{\nu-1}}{2}\right)^2-\left(\frac{x_0-x_{\nu-1}}{2}\right)^2,\ \left(\frac{x_1+x_{\nu-2}}{2}\right)^2-\left(\frac{x_1-x_{\nu-2}}{2}\right)^2,\ldots.$$

All the squares so obtained are linearly independent.

theorem above we can apply this rule to compute the Cauchy index. Thus we obtain:

COROLLARY 2: *The index of an arbitrary rational function* $R(z)$ *whose corresponding matrix* $S = \| s_{i+k} \|_0^\infty$ *is of rank* m, *is determined by the formula*

$$I_{-\infty}^{+\infty} R(z) = m - 2V(1, D_1, D_2, \ldots, D_m), \tag{64}$$

where

$$D_f = |s_{i+k}|_0^{f-1} = \begin{vmatrix} s_0 & s_1 & \ldots & s_{f-1} \\ s_1 & s_2 & \ldots & s_f \\ \cdot & \cdot & \cdot & \cdot \\ s_{f-1} & s_f & \ldots & s_{2f-2} \end{vmatrix} \quad (f = 1, 2, \ldots, m); \tag{65}$$

if among $D_1, D_2, \ldots, D_m$ *there is a group of vanishing determinants*[46]

$$(D_h \neq 0) \quad D_{h+1} = \cdots = D_{h+p} = 0 \quad (D_{h+p+1} \neq 0),$$

then in the computation of $V(D_h, D_{h+1}, \ldots, D_{h+p+1})$ *we can take*

$$\operatorname{sign} D_{h+j} = (-1)^{\frac{j(j-1)}{2}} \operatorname{sign} D_h \quad (j = 1, 2, \ldots, p)$$

and this gives

$$V(D_h, D_{h+1}, \ldots, D_{h+p+1}) = \begin{cases} \dfrac{p+1}{2} & \text{for odd } p, \\ \dfrac{p+1-\varepsilon}{2} & \text{for even } p \text{ and } \varepsilon = (-1)^{\frac{p}{2}} \operatorname{sign} \dfrac{D_{h+p+1}}{D_h}. \end{cases} \tag{66}$$

In order to express the index of a rational function in terms of the coefficients of the numerator and denominator we shall require some additional relations.

First of all, we can always represent $R(z)$ in the form[47]

$$R(z) = Q(z) + \frac{g(z)}{h(z)},$$

where $Q(z)$, $g(z)$, $h(z)$ are polynomials and

$$h(z) = a_0 z^m + a_1 z^{m-1} + \cdots + a_m \ (a_0 \neq 0), \ g(z) = b_0 z^m + b_1 z^{m-1} + \cdots + b_m.$$

Obviously,

$$I_{-\infty}^{+\infty} R(z) = I_{-\infty}^{+\infty} \frac{g(z)}{h(z)}.$$

---

46 Here we always have $D_m \neq 0$ (p. 206).

47 It is not necessary to replace $R(z)$ by a proper fraction. For what follows it is sufficient that the degree of $g(z)$ does not exceed that of $h(z)$.

Let

$$\frac{g(z)}{h(z)} = s_{-1} + \frac{s_0}{z} + \frac{s_1}{z^2} + \cdots .$$

If we now get rid of the denominator and then equate equal powers of z on the two sides of the equation, we obtain:

$$\begin{aligned} &a_0 s_{-1} = b_0 , \\ &a_0 s_0 + a_1 s_{-1} = b_1 , \\ &\cdots\cdots\cdots \\ &a_0 s_{m-1} + a_1 s_{m-2} + \cdots + a_m s_{-1} = b_m , \\ &a_0 s_t + a_1 s_{t-1} + \cdots + a_m s_{t-m} = 0 \quad (t = m, m+1, \ldots) . \end{aligned} \tag{67}$$

Using (67), we find an expression for the following determinant of order $2p$ in which we put $a_j = 0$, $b_j = 0$ for $j > m$:

$$\begin{vmatrix} a_0 & a_1 & a_2 \ldots a_{2p-1} \\ b_0 & b_1 & b_2 \ldots b_{2p-1} \\ 0 & a_0 & a_1 \ldots a_{2p-2} \\ 0 & b_0 & b_1 \ldots b_{2p-2} \\ \cdots & \cdots & \cdots \\ \cdots & \cdots & \cdots \end{vmatrix} = \begin{vmatrix} 1 & 0 & 0 \ldots 0 \\ s_{-1} & s_0 & s_1 \ldots s_{2p-2} \\ 0 & 1 & 0 \ldots 0 \\ 0 & s_{-1} & s_0 \ldots s_{2p-3} \\ \cdots & \cdots & \cdots \\ \cdots & \cdots & \cdots \end{vmatrix} \cdot \begin{vmatrix} a_0 & a_1 & a_2 \ldots a_{2p-1} \\ 0 & a_0 & a_1 \ldots a_{2p-2} \\ 0 & 0 & a_0 \ldots a_{2p-3} \\ \cdots & \cdots & \cdots \\ \cdots & \cdots & \cdots \\ 0 & 0 & 0 \ldots a_0 \end{vmatrix}$$

$$= (-1)^{\frac{p(p-1)}{2}} a_0^{2p} \begin{vmatrix} s_{p-1} & s_p & \ldots s_{2p-2} \\ s_{p-2} & s_{p-1} & \ldots s_{2p-3} \\ \cdots & \cdots & \cdots \\ s_0 & s_1 & \ldots s_{p-1} \end{vmatrix} = a_0^{2p} \begin{vmatrix} s_0 & s_1 \ldots s_{p-1} \\ s_1 & s_2 \ldots s_p \\ \cdots & \cdots \\ s_{p-1} & s_p \ldots s_{2p-2} \end{vmatrix} = a_0^{2p} D_p . \tag{68}$$

We introduce the abbreviation

$$\nabla_{2p} = \begin{vmatrix} a_0 & a_1 \ldots a_{2p-1} \\ b_0 & b_1 \ldots b_{2p-1} \\ 0 & a_0 \ldots a_{2p-2} \\ 0 & b_0 \ldots b_{2p-2} \\ \cdots & \cdots \end{vmatrix} \qquad (p = 1, 2, \ldots;\ a_j = b_j = 0 \text{ for } j > m). \tag{69}$$

Then (68) can be written as follows:

$$\nabla_{2p} = a_0^{2p} D_p \quad (p = 1, 2, \ldots) . \tag{68'}$$

By this formula, Corollary 2 above leads to the following theorem:

THEOREM 10: *If* $\nabla_{2m} \neq 0$,[48] *then*

$$I_{-\infty}^{+\infty} \frac{b_0 z^m + b_1 z^{m-1} + \cdots + b_m}{a_0 z^m + a_1 z^{m-1} + \cdots + a_m} = m - 2V(1, \nabla_2, \nabla_4, \ldots, \nabla_{2m}) \quad (a_0 \neq 0), \tag{70}$$

where ∇_{2p} $(p = 1, 2, \ldots, m)$ *is determined by* (69); *if there is a group of zero determinants*

$$(\nabla_{2h} \neq 0) \quad \nabla_{2h+2} = \cdots = \nabla_{2h+2p} = 0 \quad (\nabla_{2h+2p+2} \neq 0),$$

then in computing $V(\nabla_{2h}, \nabla_{2h+2}, \ldots, \nabla_{2h+2p+2})$ *we have to set:*

$$\operatorname{sign} \nabla_{2h+2j} = (-1)^{\frac{j(j-1)}{2}} \operatorname{sign} \nabla_{2h} \quad (j = 1, 2, \ldots, p)$$

or, what is the same,

$$V(\nabla_{2h}, \ldots, \nabla_{2h+2p+2}) = \begin{cases} \dfrac{p+1}{2} & \text{for odd } p \\[2ex] \dfrac{p+1-\varepsilon}{2} & \text{for even } p \text{ and } \varepsilon = (-1)^{\frac{p}{2}} \operatorname{sign} \dfrac{\nabla_{2h+2p+2}}{\nabla_{2h}}. \end{cases}$$

Note. If $\nabla_{2m} \neq 0$, i.e., if the fraction under the index sign in (70) is reducible, then (70) must be replaced by another formula

$$I_{-\infty}^{+\infty} \frac{b_0 z^m + b_1 z^{m-1} + \cdots + b_m}{a_0 z^m + a_1 z^{m-1} + \cdots + a_m} = r - 2V(1, \nabla_2, \nabla_4, \ldots, \nabla_{2r}), \tag{70'}$$

where r is the number of poles (including multiplicities) of the rational fraction under the index sign (i.e., r is the degree of the denominator in the reduced fraction).

For in this case the index we are interested in is

$$r - 2V(1, D_1, D_2, \ldots, D_r),$$

since r is the rank of the corresponding matrix $S = \| s_{i+k} \|_0^\infty$. But the equation (68′) is of a formal character and also holds for reduced fractions. Therefore

$$V(1, D_1, D_2, \ldots, D_r) = V(1, \nabla_2, \nabla_4, \ldots, \nabla_{2r}),$$

and we have reached (70′).

Formula (70′) enables us to express the index of every rational fraction in which the degree of the numerator does not exceed that of the denominator in terms of the coefficients of numerator and denominator.

---

[48] The condition $\nabla_{2m} \neq 0$ means that $D_m \neq 0$, so that the fraction under the index sign in (70) is reduced.

§ 12. Another Proof of the Routh-Hurwitz Theorem

1. In § 6 we proved the Routh-Hurwitz theorem with the help of Sturm's theorem and the Routh algorithm. In this section we shall give an alternative proof based on Theorem 10 of § 11 and on properties of the Cauchy indices.

We mention a few properties of the Cauchy indices that will be required in what follows.

1. $I_a^b R(x) = -I_b^a R(x)$.[49]

2. $I_a^b R_1(x)R(x) = \text{sign } R_1(x)\, I_a^b R(x)$ *if* $R_1(x) \neq 0, \infty$ *within the interval* (a, b).

3. *If* $a < c < b$, *then* $I_a^b R(x) = I_a^c R(x) + I_c^b R(x) + \eta_c$, *where* $\eta_c = 0$ *if* $R(c)$ *is finite and* $\eta_c = \pm 1$ *if* $R(x)$ *becomes infinite at* c; *here* $\eta_c = +1$ *corresponds to a jump from* $-\infty$ *to* $+\infty$ *at* c *(for increasing* x*), and* $\eta_c = -1$ *to a jump from* $+\infty$ *to* $-\infty$.

4. *If* $R(-x) = -R(x)$, *then* $I_{-a}^0 R(x) = I_0^a R(x)$. *If* $R(-x) = R(x)$, *then* $I_{-a}^0 R(x) = -I_0^a R(x)$.

5. $I_a^b R(x) + I_a^b(1/R(x)) = \dfrac{\varepsilon_a - \varepsilon_b}{2}$, *where* ε_a *is the sign of* $R(x)$ *within* (a, b) *near* a *and* ε_b *is the sign of* $R(x)$ *within* (a, b) *near* b.

The first four properties follow immediately from the definition of the Cauchy index (see § 2). Property 5. follows from the fact that the sum of the indices $I_a^b R(x)$ and $I_a^b \dfrac{1}{R(x)}$ is equal to the difference $n_1 - n_2$, where n_1 is the number of times $R(x)$ changes from negative to positive when x changes from a to b, and n_2 the number of times $R(x)$ changes from positive to negative.

We consider a real polynomial[50]

$$f(z) = a_0 z^n + a_1 z^{n-1} + a_2 z^{n-2} + \cdots + a_{n-1} z + a_n \quad (a_0 > 0),$$

We can represent it in the form

$$f(z) = h(z^2) + z g(z^2),$$

where

$$h(u) = a_n + a_{n-2} u + \cdots, \quad g(u) = a_{n-1} + a_{n-3} u + \cdots.$$

---

[49] Here and in what follows the lower limit of the index may be $-\infty$ and the upper limit may be $+\infty$.

[50] We have here reverted to the usual notation for the coefficients of a polynomial.

We shall use the notation

$$\varrho = I_{-\infty}^{+\infty} \frac{a_1 z^{n-1} - a_3 z^{n-3} + \cdots}{a_0 z^n - a_2 z^{n-2} + \cdots}. \tag{71}$$

In § 3 we proved (see (20) on p. 180) that

$$\varrho = n - 2k - s, \tag{72}$$

where k is the number of roots of $f(z)$ with positive real parts and s the number of roots of $f(z)$ on the imaginary axis.

We shall transform the expression (71) for ϱ.

To begin with, we deal with the case where n is even. Let $n = 2m$. Then

$$h(u) = a_0 u^m + a_2 u^{m-1} + \cdots + a_n, \quad g(u) = a_1 u^{m-1} + a_3 u^{m-2} + \cdots + a_{n-1}.$$

Using the properties 1.-4. and setting $\eta = \pm 1$ if $\lim_{u \to 0-} \frac{g(u)}{h(u)} = \pm \infty$, respectively, and $\eta = 0$ otherwise, we have:

$$\varrho = -I_{-\infty}^{+\infty} \frac{zg(-z^2)}{h(-z^2)} = -(I_{-\infty}^{0} + I_{0}^{+\infty} + \eta) = -2I_{-\infty}^{0} \frac{zg(-z^2)}{h(-z^2)} - \eta$$

$$= 2I_{-\infty}^{0} \frac{g(-z^2)}{h(-z^2)} - \eta = 2I_{-\infty}^{0} \frac{g(u)}{h(u)} - \eta = I_{-\infty}^{0} \frac{g(u)}{h(u)} - I_{-\infty}^{0} \frac{ug(u)}{h(u)} - \eta$$

$$= I_{-\infty}^{+\infty} \frac{g(u)}{h(u)} - I_{-\infty}^{+\infty} \frac{ug(u)}{h(u)}.$$

Similarly we have for odd n, $n = 2m + 1$:

$$h(u) = a_1 u^m + a_3 u^{m-1} + \cdots + a_n, \quad g(u) = a_0 u^m + a_2 u^{m-1} + \cdots + a_{n-1}.$$

Setting[51] $\zeta = \operatorname{sign} \left[\frac{g(u)}{h(u)}\right]_{u=0-}$ if $\lim_{u \to 0-} \frac{g(u)}{h(u)} = 0$ and $\zeta = 0$ otherwise, we find:

$$\varrho = I_{-\infty}^{+\infty} \frac{h(-z^2)}{zg(-z^2)} = I_{-\infty}^{0} + I_{0}^{+\infty} + \zeta = 2I_{-\infty}^{0} \frac{h(-z^2)}{zg(-z^2)} + \zeta = 2I_{-\infty}^{0} \frac{h(u)}{ug(u)} + \zeta$$

$$= I_{-\infty}^{0} \frac{h(u)}{ug(u)} - I_{-\infty}^{0} \frac{h(u)}{g(u)} + \zeta = I_{-\infty}^{+\infty} \frac{h(u)}{ug(u)} - I_{-\infty}^{+\infty} \frac{h(u)}{g(u)}.$$

Thus[52]

---

[51] Here we mean by sign $[g(u)/h(u)]_{u=0-}$ the sign of $g(u)/h(u)$ for negative values of u of sufficiently small modulus.

[52] If $a_1 \neq 0$, then the two formulas (73′) and (73″) may be combined into the single formula

$$\varrho = I_{-\infty}^{+\infty} \frac{g(u)}{h(u)} + I_{-\infty}^{+\infty} \frac{h(u)}{ug(u)}. \tag{73‴}$$

$$\varrho = I_{-\infty}^{+\infty} \frac{g(u)}{h(u)} - I_{-\infty}^{+\infty} \frac{ug(u)}{h(u)} \quad (n = 2m), \tag{73'}$$

$$\varrho = I_{-\infty}^{+\infty} \frac{h(u)}{ug(u)} - I_{-\infty}^{+\infty} \frac{h(u)}{g(u)} \quad (n = 2m+1). \tag{73''}$$

As before, we denote by $\Delta_1, \Delta_2, \dots, \Delta_n$ the Hurwitz determinants of $f(z)$. We assume that $\Delta_n \neq 0$.[53]

1) $n = 2m$. By (70),[54]

$$I_{-\infty}^{+\infty} \frac{g(u)}{h(u)} = m - 2V(1, \Delta_1, \Delta_3, \dots, \Delta_{n-1}), \tag{74}$$

$$\begin{aligned} I_{-\infty}^{+\infty} \frac{ug(u)}{h(u)} &= m - 2V(1, -\Delta_2, +\Delta_4, -\Delta_6, \dots) \\ &= -m + 2V(1, \Delta_2, \Delta_4, \dots, \Delta_n). \end{aligned} \tag{75}$$

But then, by (73′),

$$\varrho = n - 2V(1, \Delta_1, \Delta_3, \dots, \Delta_{n-1}) - 2V(1, \Delta_2, \Delta_4, \dots, \Delta_n),$$

which in conjunction with $\varrho = n - 2k$ gives

$$k = V(1, \Delta_1, \Delta_3, \dots, \Delta_{n-1}) + V(1, \Delta_2, \Delta_4, \dots, \Delta_n). \tag{76}$$

2) $n = 2m+1$. By (70),[55]

$$I_{-\infty}^{+\infty} \frac{h(u)}{ug(u)} = m + 1 - 2V(1, \Delta_1, \Delta_3, \dots, \Delta_n), \tag{77}$$

$$\begin{aligned} I_{-\infty}^{+\infty} \frac{h(u)}{g(u)} &= m - 2V(1, -\Delta_2, +\Delta_4, - \cdots) \\ &= -m + 2V(1, \Delta_2, \Delta_4, \dots, \Delta_{n-1}). \end{aligned} \tag{78}$$

The equation $\varrho = 2m + 1 - 2k$ together with (73″), (77), and (78) again gives (76).

This proves the Routh-Hurwitz theorem (see p. 194).

[53] In this case $s = 0$, so that $\varrho = n - 2k$. Moreover, $\Delta_n \neq 0$ means that the fractions under the index signs in (73′) and (73″) are reduced.

[54] In computing $V_2, V_4, \dots, V_{2m}$ the values $a_0, a_1, \dots, a_m$ and $b_0, b_1, \dots, b_m$ must be replaced by $a_0, a_2, \dots, a_{2m}$ and $0, a_1, a_3, \dots, a_{2m-1}$ respectively in computing the first index and by $a_0, a_2, \dots, a_{2m}$ and $a_1, a_3, \dots, a_{2m-1}, 0$ respectively in computing the second index.

[55] In computing the first index in (70) we take $a_0, a_2, \dots, a_{2m}, 0$ and $0, a_1, a_3, \dots, a_{2m+1}$, respectively, instead of $a_0, a_1, \dots, a_{m+1}$ and $b_0, b_1, \dots, b_{m+1}$; and in computing the second index we take $a_1, a_3, \dots, a_{2m+1}$ and $a_0, a_2, \dots, a_{2m}$, respectively, instead of $a_0, a_1, \dots, a_m$ and $b_0, b_1, \dots, b_m$.

2. *Note* 1. If in the formula

$$k = V(1, \Delta_1, \Delta_3, \ldots) + V(1, \Delta_2, \Delta_4, \ldots)$$

some intermediate Hurwitz determinants are zero, then the formula remains valid, only in each group of successive zero determinants

$$(\Delta_l \neq 0) \quad \Delta_{l+2} = \Delta_{l+4} = \cdots = \Delta_{l+2p} = 0 \quad (\Delta_{l+2p+2} \neq 0)$$

the following signs must be attributed to these determinants (in accordance with Theorem 7)

$$\operatorname{sign} \Delta_{l+2j} = (-1)^{\frac{j(j-1)}{2}} \operatorname{sign} \Delta_l \quad (j = 1, 2, \ldots, p),$$

which yields:

$$V(\Delta_l, \Delta_{l+2}, \ldots, \Delta_{l+2p+2}) = \begin{cases} \dfrac{p+1}{2}, & \text{for odd } p, \\[2ex] \dfrac{p+1-\varepsilon}{2}, & \text{for even } p \text{ and } \varepsilon = (-1)^{\frac{p}{2}} \operatorname{sign} \dfrac{\Delta_{l+2p+2}}{\Delta_l}. \end{cases} \tag{79}$$

A careful comparison of this rule for computing k in the presence of vanishing Hurwitz determinants with the rule given in Theorem 5 (p. 201) shows that the two rules coincide.[56]

Note 2. If $\Delta_n = 0$, then the polynomials $ug(u)$ and $h(u)$ are not co-prime. We denote by $d(u)$ the greatest common divisor of $g(u)$ and $h(u)$ and by $u^\gamma d(u)$ that of $ug(u)$ and $h(u)$ ($\gamma = 0$ or 1). We denote the degree of $d(u)$ by δ and we set $h(u) = d(u)h_1(u)$ and $g(u) = d(u)g_1(u)$.

The irreducible rational fraction $g_1(u)/h_1(u)$ always corresponds to an infinite Hankel matrix $S = \| s_{i+k} \|_2^\infty$ of rank r, where r is the degree of $h_1(u)$. The corresponding determinant $D_r \neq 0$ and $D_{r+1} = D_{r+2} = \ldots = 0$. By (68′) $\nabla_{2r} \neq 0, \nabla_{2r+2} = \nabla_{2r+4} = \ldots = 0$. Moreover,

$$I_{-\infty}^{+\infty} \frac{g_1(u)}{h_1(u)} = r - 2V(1, \nabla_2, \ldots, \nabla_{2r}).$$

When we apply all this to the fractions under the index sign in (74), (75), (77), and (78) we easily find that for every n (even or odd) and $\varkappa = 2\delta + \gamma$

$$\Delta_{n-\varkappa-1} \neq 0, \; \Delta_{n-\varkappa} \neq 0, \; \overbrace{\Delta_{n-\varkappa+1} = \cdots = \Delta_n}^{\varkappa} = 0$$

and that the formulas (74), (75), (77), and (78) all remain valid in this case, provided we omit all the Δ_i with $i > n - \varkappa$ on the right-hand sides and replace the number m (in (77), $m+1$) by the degree of the corresponding

[56] We have to take account here of the remark made in footnote 36 (p. 201).

denominator of the fraction under the index, after reduction. We then obtain by taking (73′) and (73″) into account:

$$\varrho = n - \varkappa - 2V(1, \Delta_1, \Delta_3, \ldots) - 2V(1, \Delta_2, \Delta_4, \ldots).$$

Together with the formula $\varrho = n - 2k - s$ this gives:

$$k_1 = V(1, \Delta_1, \Delta_3, \ldots) + V(1, \Delta_2, \Delta_4, \ldots), \tag{80}$$

where $k_1 = k + s/2 - \varkappa/2$ is the number of all the roots of $f(z)$ in the right half-plane, excluding those that are also roots of $f(-z)$.[57]

§ 13. Some Supplements to the Routh-Hurwitz Theorem. Stability Criterion of Liénard and Chipart

1. Suppose given a polynomial with real coefficients

$$f(z) = a_0 z^n + a_1 z^{n-1} + \cdots + a_n \quad (a_0 > 0).$$

Then the Routh-Hurwitz conditions that are necessary and sufficient for all the roots of $f(z)$ to have negative real parts can be written in the form of the inequalities

$$\Delta_1 > 0, \Delta_2 > 0, \ldots, \Delta_n > 0, \tag{81}$$

where

$$\Delta_i = \begin{vmatrix} a_1 & a_3 & a_5 & \cdots & \\ a_0 & a_2 & a_4 & \cdots & \\ 0 & a_1 & a_3 & \cdots & \\ 0 & a_0 & a_2 & a_4 & \\ & & & \ddots & \\ & & & & a_i \end{vmatrix} \quad (a_k = 0 \text{ for } k > n)$$

is the Hurwitz determinant of order i $(i = 1, 2, \ldots, n)$.

If (81) is satisfied, then $f(z)$ can be represented in the form of a product of a_0 with factors of the form $z + u$, $z^2 + vz + w$ $(u > 0, v > 0, w > 0)$, so that all the coefficients of $f(z)$ are positive:[58]

---

[57] This follows from the fact that $\varkappa$ is the degree of the greatest common divisor of $h(u)$ and $ug(u)$; $\varkappa$ is the number of 'special' roots of $f(z)$, i.e., those roots z^* for which $-z^*$ is also a root of $f(z)$. The number of these special roots is equal to the number of determinants in the last uninterrupted sequence of vanishing Hurwitz determinants (including Δ_n): $\Delta_{n-\varkappa+1} = \cdots = \Delta_n = 0$.

[58] $a_0 > 0$, by assumption.

$$a_1 > 0,\ a_2 > 0,\ \ldots,\ a_n > 0. \tag{82}$$

Unlike (81), the conditions (82) are necessary but by no means sufficient for all the roots of $f(z)$ to lie in the left half-plane $\operatorname{Re} z < 0$.

However, when the conditions (82) hold, then the inequalities (81) are not independent. For example: For $n = 4$ the Routh-Hurwitz conditions reduce to the single inequality $\Delta_3 > 0$; for $n = 5$, to the two: $\Delta_2 > 0$, $\Delta_4 > 0$; for $n = 6$ to the two: $\Delta_3 > 0$, $\Delta_5 > 0$.[59]

This circumstance was investigated by the French mathematicians Liénard and Chipart[60] in 1914 and enabled them to set up a stability criterion different from the Routh-Hurwitz criterion.

THEOREM 11 (Stability Criterion of Liénard and Chipart): *Necessary and sufficient conditions for all the roots of the real polynomial* $f(z) = a_0z^n + a_1z^{n-1} + \cdots + a_n$ $(a_0 > 0)$ *to have negative real parts can be given in any one of the following four forms:*[61]

1) $a_n > 0,\ a_{n-2} > 0,\ \ldots;\ \Delta_1 > 0,\ \Delta_3 > 0,\ \ldots,$
2) $a_n > 0,\ a_{n-2} > 0,\ \ldots;\ \Delta_2 > 0,\ \Delta_4 > 0,\ \ldots,$
3) $a_n > 0;\ a_{n-1} > 0,\ a_{n-3} > 0,\ \ldots;\ \Delta_1 > 0,\ \Delta_3 > 0,\ \ldots,$
4) $a_n > 0;\ a_{n-1} > 0,\ a_{n-3} > 0,\ \ldots;\ \Delta_2 > 0,\ \Delta_4 > 0,\ \ldots.$

From Theorem 11 it follows that Hurwitz's determinant inequalities (81) are not independent for a real polynomial $f(z) = a_0z^n + a_1z^{n-1} + \cdots + a_n$ $(a_0 > 0)$ in which all the coefficients (or even only part of them: $a_n, a_{n-2}, \ldots$ or $a_n, a_{n-1}, a_{n-3}, \ldots$) are positive. In fact: *If the Hurwitz determinants of odd order are positive, then those of even order are also positive, and vice versa.*

Liénard and Chipart obtained the condition 1) in the paper [259] by means of special quadratic forms. We shall give a simpler derivation of the condition 1) (and also of 2), 3), 4)) based on Theorem 10 of § 11 and the theory of Cauchy indices and we shall obtain these conditions as a special case of a much more general theorem which we are now about to expound.

We again consider the polynomials $h(u)$ and $g(u)$ that are connected with $f(z)$ by the identity

---

[59] This fact has been established for the first few values of n in a number of papers on the theory of governors, independently of the general criterion of Liénard and Chipart, with which the authors of these papers were obviously not acquainted.

[60] See [259]. An account of some of the basic results of Liénard and Chipart can be found in the fundamental survey by M. G. Kreĭn and M. A. Naĭmark [25].

[61] Conditions 1), 2), 3), and 4) have a decided advantage over Hurwitz' conditions, because they involve only about half the number of determinantal inequalities.

$$f(z) = h(z^2) + z\,g(z^2).$$

If n is even, $n = 2m$, then

$$h(u) = a_0 u^m + a_2 u^{m-1} + \cdots + a_n, \quad g(u) = a_1 u^{m-1} + a_3 u^{m-2} + \cdots + a_{n-1};$$

if n is odd, $n = 2m + 1$, then

$$h(u) = a_1 u^m + a_3 u^{m-1} + \cdots + a_n, \; g(u) = a_0 u^m + a_2 u^{m-1} + \cdots + a_{n-1}.$$

The conditions $a_n > 0,\ a_{n-2} > 0, \ldots$ (or $a_{n-1} > 0,\ a_{n-3} > 0, \ldots$) can therefore be replaced by the more general condition: $h(u)$ (or $g(u)$) does not change sign for $u > 0$.[62]

Under these conditions we can deduce a formula for the number of roots of $f(z)$ in the right half-plane, using only Hurwitz determinants of odd order or of even order.

THEOREM 12: *If for the real polynomial*

$$f(z) = a_0 z^n + a_1 z^{n-1} + \cdots + a_n = h(z^2) + z g(z^2) \quad (a_0 > 0)$$

$h(u)$ (or $g(u)$) does not change sign for $u > 0$ and the last Hurwitz determinant $\Delta_n \neq 0$, then the number k of roots of $f(z)$ in the right half-plane is determined by the formulas

	$n = 2m$	$n = 2m + 1$
$h(u)$ does not change sign for $u > 0$	$k = 2V(1, \Delta_1, \Delta_3, \ldots, \Delta_{n-1})$ $= 2V(1, \Delta_2, \Delta_4, \ldots, \Delta_n)$	$k = 2V(1, \Delta_1, \Delta_3, \ldots, \Delta_n) - \frac{1-\varepsilon_\infty}{2}$ $= 2V(1, \Delta_2, \Delta_4, \ldots, \Delta_{n-1}) + \frac{1-\varepsilon_\infty}{2}$
$g(u)$ does not change sign for $u > 0$	$k = 2V(1, \Delta_1, \Delta_3, \ldots, \Delta_{n-1}) + \frac{\varepsilon_\infty - \varepsilon_0}{2}$ $= 2V(1, \Delta_2, \Delta_4, \ldots, \Delta_n) - \frac{\varepsilon_\infty - \varepsilon_0}{2}$	$k = 2V(1, \Delta_1, \Delta_3, \ldots, \Delta_n) - \frac{1-\varepsilon_0}{2}$ $= 2V(1, \Delta_2, \Delta_4, \ldots, \Delta_{n-1}) + \frac{1-\varepsilon_0}{2}$

(83)

where[63]

$$\varepsilon_\infty = \operatorname{sign}\left[\frac{g(u)}{h(u)}\right]_{u=+\infty}, \quad \varepsilon_0 = \operatorname{sign}\left[\frac{g(u)}{h(u)}\right]_{u=0+} \tag{84}$$

---

[62] I.e., $h(u) \geqq 0$ or $h(u) \leqq 0$ for $u > 0$ ($g(u) \geqq 0$ or $g(u) \leqq 0$ for $u > 0$).

[63] If $a_1 \neq 0$, then $\varepsilon_\infty = \operatorname{sign} a_1$; and, more generally, if $a_1 = a_3 = \ldots = a_{2\mu-1} = 0$, $a_{2\mu+1} \neq 0$, then $\varepsilon_\infty = \operatorname{sign} a_{2\mu+1}$. If $a_{n-1} \neq 0$, then $\varepsilon_0 = \operatorname{sign} a_{n-1}/a_n$; and, more generally, if $a_{n-1} = a_{n-3} = \ldots = a_{n-2\mu-1} = 0$ and $a_{n-2\mu-1} \neq 0$, then $\varepsilon_0 = \operatorname{sign} a_{n-2\mu-1}/a_n$.

Proof. Again we use the notation

$$\varrho = I_{-\infty}^{+\infty} \frac{a_1 z^{n-1} - a_3 z^{n-3} + \cdots}{a_0 z^n - a_2 z^{n-2} + \cdots}.$$

Corresponding to the table (83) we consider four cases:

1) $n = 2m$; $h(u)$ does not change sign for $u > 0$. Then[64]

$$I_0^{+\infty} \frac{g(u)}{h(u)} = I_0^{+\infty} \frac{ug(u)}{h(u)} = 0,$$

and so the obvious equation

$$I_{-\infty}^{0} \frac{g(u)}{h(u)} = - I_{-\infty}^{0} \frac{ug(u)}{h(u)}$$

implies that:[65]

$$I_{-\infty}^{+\infty} \frac{g(u)}{h(u)} = - I_{-\infty}^{+\infty} \frac{ug(u)}{h(u)}.$$

But then we have from (74) and (75):

$$V(1, \Delta_1, \Delta_3, \ldots) = V(1, \Delta_2, \Delta_4, \ldots),$$

and therefore the Routh-Hurwitz formula (76) gives:

$$k = 2V(1, \Delta_1, \Delta_3, \ldots, \Delta_{n-1}) = 2V(1, \Delta_2, \Delta_4, \ldots, \Delta_n).$$

2) $n = 2m$; $g(u)$ does not change sign for $u > 0$. In this case,

$$I_0^{+\infty} \frac{h(u)}{g(u)} = I_0^{+\infty} \frac{h(u)}{ug(u)} = 0,$$

$$I_{-\infty}^{0} \frac{h(u)}{g(u)} + I_{-\infty}^{0} \frac{h(u)}{ug(u)} = 0,$$

so that with the notation (84) we have:

$$I_{-\infty}^{+\infty} \frac{h(u)}{g(u)} + I_{-\infty}^{+\infty} \frac{h(u)}{ug(u)} - \varepsilon_0 = 0. \tag{85}$$

When we replace the functions under the index sign by their reciprocals, then we obtain by 5. (see p. 216):

$$I_{-\infty}^{+\infty} \frac{g(u)}{h(u)} + I_{-\infty}^{+\infty} \frac{ug(u)}{h(u)} = \varepsilon_\infty - \varepsilon_0.$$

---

[64] If $h(u_1) = 0$ $(u_1 > 0)$, then $g(u_1) \neq 0$, because $\Delta_n \neq 0$. Therefore $h(u) \geqq 0$ $(u > 0)$ implies that $g(u)/h(u)$ does not change sign in passing through $u = u_1$.

[65] From $\Delta_n = a_n \Delta_{n-1} \neq 0$ it follows that $h(0) = a_n \neq 0$.

But this by (74) and (75) gives:

$$V(1, \Delta_2, \Delta_4, \ldots) - V(1, \Delta_1, \Delta_3, \ldots) = \frac{\varepsilon_\infty - \varepsilon_0}{2}.$$

Hence, in conjunction with the Routh-Hurwitz formula (76), we obtain:

$$k = 2V(1, \Delta_1, \Delta_3, \ldots) + \frac{\varepsilon_\infty - \varepsilon_0}{2} = 2V(1, \Delta_2, \Delta_4, \ldots) - \frac{\varepsilon_\infty - \varepsilon_0}{2}.$$

3) $n = 2m + 1$, $g(u)$ does not change sign for $u > 0$.

In this case, as in the preceding one, (85) holds. When we substitute the expressions for the indices from (77) and (78) into (85), we obtain:

$$V(1, \Delta_1, \Delta_3, \ldots) - V(1, \Delta_2, \Delta_4, \ldots) = \frac{1 - \varepsilon_0}{2}.$$

In conjunction with the Routh-Hurwitz formula this gives:

$$k = 2V(1, \Delta_1, \Delta_3, \ldots) - \frac{1 - \varepsilon_0}{2} = 2V(1, \Delta_2, \Delta_4, \ldots) + \frac{1 - \varepsilon_0}{2}.$$

4) $n = 2m + 1$, $h(u)$ does not change sign for $u > 0$.

From the equations

$$I_0^\infty \frac{g(u)}{h(u)} = I_0^\infty \frac{u\,g(u)}{h(u)} = 0 \text{ and } I_{-\infty}^{0} \frac{g(u)}{h(u)} + I_{-\infty}^{0} \frac{u\,g(u)}{h(u)} = 0$$

we deduce:

$$I_{-\infty}^{+\infty} \frac{g(u)}{h(u)} + I_{-\infty}^{+\infty} \frac{ug(u)}{h(u)} = 0.$$

Taking the reciprocals of the functions under the index sign, we obtain:

$$I_{-\infty}^{+\infty} \frac{u(u)}{g(u)} + I_{-\infty}^{+\infty} \frac{h(u)}{ug(u)} = \varepsilon_\infty.$$

Again, when we substitute the expressions for the indices from (77) and (78), we have:

$$V(1, \Delta_1, \Delta_3, \ldots) - V(1, \Delta_2, \Delta_4, \ldots) = \frac{1 - \varepsilon_\infty}{2}.$$

From this and the Routh-Hurwitz formula it follows that:

$$k = 2V(1, \Delta_1, \Delta_3, \ldots) - \frac{1 - \varepsilon_\infty}{2} = 2V(1, \Delta_2, \Delta_4, \ldots) + \frac{1 - \varepsilon_\infty}{2}.$$

This completes the proof of Theorem 12.

From Theorem 12 we obtain Theorem 11 as a special case.

2. COROLLARY TO THEOREM 12: *If the real polynomial*

$$f(z) = a_0 z^n + a_1 z^{n-1} + \cdots + a_n \quad (a_0 > 0)$$

has positive coefficients

$$a_0 > 0, \quad a_1 > 0, \quad a_2 > 0, \ldots, a_n > 0,$$

and $\Delta_n \neq 0$, *then the number* k *of its roots in the right half-plane* $\mathrm{Re}\, z > 0$ *is determined by the formula*

$$k = 2V(1, \Delta_1, \Delta_3, \ldots) = 2V(1, \Delta_2, \Delta_4, \ldots).$$

Note. If in the last formula, or in (83), some of the intermediate Hurwitz determinants are zero, then in the computation of $V(1, \Delta_1, \Delta_3, \ldots)$ and $V(1, \Delta_2, \Delta_4, \ldots)$ the rule given in Note 1 on p. 219 must be followed.

But if $\Delta_n = \Delta_{n-1} = \ldots = \Delta_{n-\varkappa+1} = 0$, $\Delta_{n-\varkappa} \neq 0$, then we disregard the determinants $\Delta_{n-\varkappa+1}, \ldots, \Delta_n$ in (83)[66] and determine from these formulas the number k_1 of the 'non-singular' roots of $f(z)$ in the right half-plane, provided only that $h(u) \neq 0$ for $u > 0$ or $g(u) \neq 0$ for $u > 0$.[67]

§ 14. Some Properties of Hurwitz Polynomials. Stieltjes' Theorem. Representation of Hurwitz Polynomials by Continued Fractions

1. Let

$$f(z) = a_0 z^n + a_1 z^{n-1} + \cdots + a_n \qquad (a_0 \neq 0)$$

be a real polynomial. We represent it in the form

$$f(z) = h(z^2) + z g(z^2).$$

We shall investigate what conditions have to be imposed on $h(u)$ and $g(u)$ in order that $f(z)$ be a Hurwitz polynomial.

Setting $k = s = 0$ in (20) (p. 180), we obtain a necessary and sufficient condition for $f(z)$ to be a Hurwitz polynomial, in the form

$$\varrho = n,$$

where, as in the preceding sections,

$$\varrho = I_{-\infty}^{+\infty} \frac{a_1 z^{n-1} - a_3 z^{n-3} + \cdots}{a_0 z^n - a_2 z^{n-2} + \cdots}.$$

---

66 See p. 220.

67 In this case the polynomials $h_1(u)$ and $g_1(u)$ obtained from $h(u)$ and $g(u)$ by dividing them by their greatest common divisor $d(u)$ satisfy the conditions of Theorem 12.

Let $n=2m$. By (73′) (p. 218), this condition can be written as follows:

$$n=2m=I_{-\infty}^{+\infty}\frac{g(u)}{h(u)}-I_{-\infty}^{+\infty}\frac{ug(u)}{h(u)}. \tag{86}$$

Since the absolute value of the index of a rational fraction cannot exceed the degree of the denominator (in this case, m), the equation (86) can hold if and only if

$$I_{-\infty}^{+\infty}\frac{g(u)}{h(u)}=m \text{ and } I_{-\infty}^{+\infty}\frac{ug(u)}{h(u)}=-m \tag{87}$$

hold simultaneously.

For $n=2m+1$ the equation (73″) gives (on account of $\varrho=n$):

$$n=I_{-\infty}^{+\infty}\frac{h(u)}{ug(u)}-I_{-\infty}^{+\infty}\frac{h(u)}{g(u)}.$$

When we replace the fractions under the index signs by their reciprocals (see 5. on p. 216) and observe that $h(u)$ and $g(u)$ are of the same degree m, we obtain:[68]

$$n=2m+1=I_{-\infty}^{+\infty}\frac{g(u)}{h(u)}-I_{-\infty}^{+\infty}\frac{ug(u)}{h(u)}+\varepsilon_\infty. \tag{88}$$

Starting again from the fact that the absolute value of the index of a fraction cannot exceed the degree of the denominator we conclude that (88) holds if and only if

$$I_{-\infty}^{+\infty}\frac{g(u)}{h(u)}=m,\quad I_{-\infty}^{+\infty}\frac{ug(u)}{h(u)}=-m \text{ and } \varepsilon_\infty=1 \tag{89}$$

hold simultaneously.

If $n=2m$, the first of equations (87) indicates that $h(u)$ has m distinct real roots $u_1<u_2<\dots<u_m$ and that the proper fractions $g(u)/h(u)$ can be represented in the form

$$\frac{g(u)}{h(u)}=\sum_{i=1}^{m}\frac{R_i}{u-u_i}, \tag{90}$$

where

$$R_i=\frac{g(u_i)}{h'(u_i)}>0 \qquad (i=1,2,\dots,m). \tag{90′}$$

From this representation of $g(u)/h(u)$ it follows that between any two roots u_i, u_{i+1} of $h(u)$ there is a real root u_i' of $g(u)$ $(i=1,2,\dots,m-1)$ and that the highest coefficients of $h(u)$ and $g(u)$ are of like sign, i.e.,

68 As in the preceding section, $\varepsilon_\infty=\operatorname{sign}\left[\frac{g(u)}{h(u)}\right]_{u=+\infty}$.

$$h(u) = a_0 (u - u_1) \cdots (u - u_m), \qquad g(u) = a_1 (u - u_1') \cdots (u - u_{m-1}'),$$
$$u_1 < u_1' < u_2 < u_2' < \cdots < u_{m-1} < u_{m-1}' < u_m; \qquad a_0 a_1 > 0.$$

The second of equations (87) adds only one condition

$$u_m < 0.$$

By this condition all the roots of $h(u)$ and $g(u)$ must be negative.

If $n = 2m + 1$, then it follows from the first of equations (89) that $h(u)$ has m distinct real roots $u_1 < u_2 < \cdots < u_m$ and that

$$\frac{g(u)}{h(u)} = s_{-1} + \sum_{i=1}^{m} \frac{R_i}{u - u_i} \qquad (s_{-1} \neq 0), \tag{91}$$

where

$$R_i = \frac{g(u_i)}{h'(u_i)} > 0 \qquad (i = 1, 2, \ldots, m). \tag{91'}$$

The third of equations (89) implies that

$$s_{-1} > 0, \tag{92}$$

i.e., that the highest coefficients a_0 and a_1 are of like sign. Moreover, it follows from (91), (91′), and (92) that $g(u)$ has m real roots $u'_1 < u'_2 < \cdots < u'_m$ in the intervals $(-\infty, u_1), (u_1, u_2), \ldots, (u_{m-1}, u_m)$. In other words,

$$h(u) = a_1 (u - u_1) \cdots (u - u_m), \qquad g(u) = a_0 (u - u_1') \cdots (u - u_m'),$$
$$u_1' < u_1 < u_2' < u_2 < \cdots < u_m' < u_m; \qquad a_0 a_1 > 0.$$

The second of equations (89), as in the case $n = 2m$, only adds one further inequality

$$u_m < 0.$$

DEFINITION 3. *We shall say that two polynomials $h(u)$ and $g(u)$ of degree m (or the first of degree m and the second of degree $m - 1$) form a positive pair*[69] *if the roots $u_1, u_2, \ldots, u_m$ and $u_1', u_2', \ldots, u_m'$ (or $u_1', u_2', \ldots, u_{m-1}'$) are all distinct, real, and negative and they alternate as follows:*

$$u_1' < u_1 < u_2' < u_2 < \cdots < u_m' < u_m < 0$$
$$(\text{or } u_1 < u_1' < u_2 < \cdots < u_{m-1}' < u_m < 0)$$

and their highest coefficients are of like sign.[70]

---

[69] See [17], p. 333. The definition of a positive pair of polynomials given here differs slightly from that given in the book [17].

[70] If we omit the condition that the roots be negative, we obtain a real pair of polynomials. For the application of this concept to the Routh-Hurwitz problem, see [36].

When we introduce the positive numbers $v_i = -u_i$ and $v_i' = -u_i'$ and multiply $h(u)$ and $g(u)$ by $+1$ or -1 so that their highest coefficients are positive, then we can write the polynomials of this positive pair in the form

$$h(u) = a_1 \prod_{i=1}^{m} (u + v_i), \qquad g(u) = a_0 \prod_{i=1}^{m} (u + v_i'), \tag{93}$$

where

$$a_1 > 0,\ a_0 > 0, \quad 0 < v_m < v_m' < v_{m-1} < v_{m-1}' < \cdots < v_1 < v_1',$$

in case both $h(u)$ and $g(u)$ are of degree m, and in the form

$$h(u) = a_0 \prod_{i=1}^{m} (u + v_i), \qquad g(u) = a_1 \prod_{i=1}^{m-1} (u + v_i'), \tag{93'}$$

where

$$a_0 > 0,\ a_1 > 0, \quad 0 < v_m < v_{m-1}' < v_{m-1} < \cdots < v_1' < v_1,$$

in case $h(u)$ is of degree m and $g(u)$ of degree $m-1$.

By our earlier arguments we have proved the following two theorems:

THEOREM 13: *The polynomial $f(z) = h(z^2) + zg(z^2)$ is a Hurwitz polynomial if and only if $h(u)$ and $g(u)$ form a positive pair.*[71]

THEOREM 14: *Two polynomials $h(u)$ and $g(u)$ the first of which is of degree m and the second of degree m or $m-1$ form a positive pair if and only if the equations*

$$I_{-\infty}^{+\infty} \frac{g(u)}{h(u)} = m, \qquad I_{-\infty}^{+\infty} \frac{ug(u)}{h(u)} = -m \tag{94}$$

hold and, when $h(u)$ and $g(u)$ are of equal degree, the additional condition

$$\varepsilon_\infty = \operatorname{sign} \left[\frac{g(u)}{h(u)}\right]_{+\infty} = 1 \tag{95}$$

holds.

2. Using properties of the Cauchy indices we can easily deduce from the last theorem a theorem of Stieltjes on the representation of a fraction $g(u)/h(u)$ as a continued fraction of a special type, provided $h(u)$ and $g(u)$ form a positive pair of polynomials.

The proof of Stieltjes' theorem will be based on the following lemma:

---

[71] This theorem is a special case of the so-called Hermite-Biehler theorem (see [7], p. 21).

LEMMA. *If the polynomials $h(u)$ and $g(u)$ ($h(u)$ of degree m) form a positive pair and*

$$\frac{g(u)}{h(u)} = c + \cfrac{1}{du + \cfrac{h_1(u)}{g_1(u)}}, \tag{96}$$

where c, d are constants and $h_1(u)$, $g_1(u)$ are polynomials of degree not exceeding $m-1$, then

1. $c \geqq 0$, $d > 0$;
2. *$h_1(u)$, $g_1(u)$ are of degree $m-1$;*
3. *$h_1(u)$ and $g_1(u)$ form a positive pair.*

Given $h(u)$ and $g(u)$, the polynomials $h_1(u)$ and $g_1(u)$ are uniquely determined (to within a common constant factor) and so are c and d.

Conversely, from (96) and 1., 2., 3. it follows that $h(u)$ and $g(u)$ form a positive pair, that $h(u)$ is of degree m, and $g(u)$ is of degree m or $m-1$ according as $c > 0$ or $c = 0$.

Proof. Let $h(u)$, $g(u)$ be a positive pair. Then it follows from (94) and (96) that:

$$m = I_{-\infty}^{+\infty} \frac{g(u)}{h(u)} = I_{-\infty}^{+\infty} \cfrac{1}{du + \cfrac{h_1(u)}{g_1(u)}}. \tag{97}$$

This equation implies that $g_1(u)$ is of degree $m-1$ and that $d \neq 0$.

Further, from (97) we find:

$$m = -I_{-\infty}^{+\infty}\left[du + \frac{h_1(u)}{g_1(u)}\right] + \operatorname{sign} d = -I_{-\infty}^{+\infty}\frac{h_1(u)}{g_1(u)} + \operatorname{sign} d.$$

Hence it follows that $d > 0$ and that

$$I_{-\infty}^{+\infty}\frac{h_1(u)}{g_1(u)} = -(m-1). \tag{98}$$

The second of equations (94) now gives:

$$\begin{aligned} -m = I_{-\infty}^{+\infty}\frac{ug(u)}{h(u)} &= I_{-\infty}^{+\infty}\left[cu + \cfrac{1}{d + \cfrac{h_1(u)}{ug_1(u)}}\right] \\ = I_{-\infty}^{+\infty} \cfrac{1}{d + \cfrac{h_1(u)}{ug_1(u)}} &= -I_{-\infty}^{+\infty}\left[d + \frac{h_1(u)}{ug_1(u)}\right] = -I_{-\infty}^{+\infty}\frac{h_1(u)}{ug_1(u)}. \end{aligned} \tag{99}$$

Hence it follows that $h_1(u)$ is of degree $m-1$.

Condition (95) yields, by (96) : $c > 0$. But if $g(u)$ is of smaller degree than $h(u)$, then it follows from (96) that $c = 0$.

(98) and (99) imply:

$$I_{-\infty}^{+\infty}\frac{g_1(u)}{h_1(u)}=m-1, \quad I_{-\infty}^{+\infty}\frac{ug_1(u)}{h_1(u)}=-m+\varepsilon_\infty^{(1)}, \tag{100}$$

where

$$\varepsilon_\infty^{(1)}=\operatorname{sign}\left[\frac{g_1(u)}{h_1(u)}\right]_{u=+\infty}.$$

Since the second of the indices (100) is in absolute value less than $m-1$, we have

$$\varepsilon_\infty^{(1)}=1, \tag{101}$$

and then we conclude from (100) and (101), by Theorem 12, that the polynomials $h_1(u)$ and $g_1(u)$ form a positive pair.

From (96) it follows that

$$c=\lim_{u\to\infty}\frac{g(u)}{h(u)}, \quad \lim_{u\to\infty}\left[\frac{g(u)}{h(u)}-c\right]u=\frac{1}{d}.$$

After c and d have been found, the ratio $\frac{h_1(u)}{g_1(u)}$ is determined by (96).

The relations (97), (98), (99), (100), and (101) applied in the reverse order, establish the second part of the lemma. Thus the proof of the lemma is complete.

Suppose given a positive pair of polynomials $h(u)$, $g(u)$, with $h(u)$ of degree m. Then when we divide $g(u)$ by $h(u)$ and denote the quotient by c_0 and the remainder by $g_1(u)$, we obtain:

$$\frac{g(u)}{h(u)}=c_0+\frac{g_1(u)}{h(u)}=c_0+\cfrac{1}{\cfrac{h(u)}{g_1(u)}}.$$

$\frac{h(u)}{g_1(u)}$ can be represented in the form $d_0u+\frac{h_1(u)}{g_1(u)}$, where $h_1(u)$, like $g_1(u)$, is of degree less than m. Hence

$$\frac{g(u)}{h(u)}=c_0+\cfrac{1}{d_0u+\cfrac{h_1(u)}{g_1(u)}}. \tag{102}$$

Thus, the representation (96) always holds for a positive pair $h(u)$ and $g(u)$. By the lemma

$$c_0 \geqq 0,\ d_0 > 0,$$

and the polynomials $h_1(u)$ and $g_1(u)$ are of degree $m-1$ and form a positive pair.

When we apply the same arguments to the positive pair $h_1(u)$, $g_1(u)$, we obtain

$$\frac{g_1(u)}{h_1(u)} = c_1 + \cfrac{1}{d_1 u + \cfrac{h_2(u)}{g_2(u)}}, \tag{102'}$$

where

$$c_1 > 0,\ d_1 > 0,$$

and the polynomials $h_2(u)$ and $g_2(u)$ are of degree $m-2$ and form a positive pair. Continuing the process, we finally end up with a positive pair h_m and g_m, where h_m and g_m are constants of like sign. We set:

$$\frac{g_m}{h_m} = c_m. \tag{102$^{(m)}$}$$

Then it follows from (102), (102′), ..., (102$^{(m)}$) that:

$$\frac{g(u)}{h(u)} = c_0 + \cfrac{1}{d_0 u + \cfrac{1}{c_1 + \cfrac{1}{d_1 u + \cfrac{1}{c_2 + \cdots + \cfrac{1}{d_{m-1} u + \cfrac{1}{c_m}}}}}}.$$

Using the second part of the lemma, we show similarly that for arbitrary $c_0 \geqq 0$, $c_1 > 0, \ldots, c_m > 0$, $d_0 > 0$, $d_1 > 0, \ldots, d_{m-1} > 0$ the above continued fraction determines uniquely (to within a common constant factor) a positive pair of polynomials $h(u)$ and $g(u)$, where $h(u)$ is of degree m and $g(u)$ is of degree m when $c_0 > 0$ and of degree $m-1$ when $c_0 = 0$.

Thus we have proved the following theorem.[72]

[72] A proof of Stieltjes' theorem that is not based on the theory of Cauchy indices can be found in the book [17], pp. 333-37.

THEOREM 15 (Stieltjes): *If $h(u)$, $g(u)$ is a positive pair of polynomials and $h(u)$ is of degree m, then*

$$\frac{g(u)}{h(u)} = c_0 + \cfrac{1}{d_0 u + \cfrac{1}{c_1 + \cfrac{1}{d_1 u + \cfrac{1}{c_2 + \cfrac{}{\ddots + \cfrac{1}{d_{m-1} u + \cfrac{1}{c_m}}}}}}}, \tag{103}$$

where

$$c_0 \geqq 0,\ c_1 > 0, \ldots, c_m > 0, \quad d_0 > 0, \ldots, d_{m-1} > 0 .$$

Here $c_0 = 0$ if $g(u)$ is of degree $m - 1$ and $c_0 > 0$ if $g(u)$ is of degree m. The constants c_i, d_k are uniquely determined by $h(u)$, $g(u)$.

Conversely, for arbitrary $c_0 \geqq 0$ and arbitrary positive $c_1, \ldots, c_m, d_0, \ldots, d_{m-1}$, the continued fraction (103) *determines a positive pair of polynomials $h(u)$, $g(u)$, where $h(u)$ is of degree m.*

From Theorem 13 and Stieltjes' Theorem we deduce:

THEOREM 16: *A real polynomial of degree n $f(z) = h(z^2) + zg(z^2)$ is a Hurwitz polynomial if and only if the formula* (103) *holds with non-negative c_0 and positive $c_1, \ldots, c_m, d_0, \ldots, d_{m-1}$. Here $c_0 > 0$ when n is odd and $c_0 = 0$ when n is even.*

§ 15. Domain of Stability. Markov Parameters

1. With every real polynomial of degree n we can associate a point of an n-dimensional space whose coordinates are the quotients of the coefficients divided by the highest coefficient. In this 'coefficient space' all the Hurwitz polynomials form a certain n-dimensional domain which is determined[73] by the Hurwitz inequalities $\Delta_1 > 0, \Delta_2 > 0, \ldots, \Delta_n > 0$, or, for example, by the Liénard-Chipart inequalities $a_n > 0, a_{n-2} > 0, \ldots, \Delta_1 > 0, \Delta_3 > 0, \ldots$. We shall call it the *domain of stability*. If the coefficients are given as functions of p parameters, then the domain of stability is constructed in the space of these parameters.

[73] For $a_0 = 1$.

The study of the domain of stability is of great practical interest; for example, it is essential in the design of new systems of governors.[74]

In § 17 we shall show that two remarkable theorems which were found by Markov and Chebyshev in connection with the expansion of continued fractions in power series with negative powers of the argument are closely connected with the investigation of the domain of stability. In formulating and proving these theorems it is convenient to give the polynomial not by its coefficients, but by special parameters, which we shall call *Markov parameters.*

Suppose that

$$f(z) = a_0 z^n + a_1 z^{n-1} + \cdots + a_n \qquad (a_0 \neq 0)$$

is a real polynomial. We represent it in the form

$$f(z) = h(z^2) + z g(z^2).$$

We may assume that $h(u)$ and $g(u)$ are co-prime ($\Delta_n \neq 0$). We expand the irreducible rational fraction $\frac{g(u)}{h(u)}$ in a series of decreasing powers of u:[75]

$$\frac{g(u)}{h(u)} = s_{-1} + \frac{s_0}{u} - \frac{s_1}{u^2} + \frac{s_2}{u^3} - \frac{s_3}{u^4} + \cdots. \tag{104}$$

The sequence $s_0, s_1, s_2, \ldots$ determines an infinite Hankel matrix $S = \| s_{i+k} \|_0^\infty$. We define a rational function $R(v)$ by

$$R(v) = -\frac{g(-v)}{h(-v)}. \tag{105}$$

Then

$$R(v) = -s_{-1} + \frac{s_0}{v} + \frac{s_1}{v^2} + \frac{s_2}{v^3} + \cdots, \tag{106}$$

so that we have the relation (see p. 208)

$$R(v) \sim S. \tag{107}$$

Hence it follows that the matrix S is of rank $m = [n/2]$, since m, being the degree of $h(u)$, is equal to the number of poles of $R(v)$.[76]

For $n = 2m$ (in this case, $s_{-1} = 0$), the matrix S determines the irreducible fraction $\frac{g(u)}{h(u)}$ uniquely and therefore determines $f(z)$ to within a

[74] A number of papers by Y. I. Naĭmark deal with the investigation of the domain of stability and also of the domains corresponding to various values of k (k is the number of roots in the right half-plane). (See the monograph [41].)

[75] In what follows it is convenient to denote the coefficients of the even negative powers of u by $-s_1$, $-s_3$, etc.

[76] See Theorem 8 (p. 207).

constant factor. For $n = 2m + 1$, in order to give $f(z)$ by means of S it is necessary also to know the coefficient s_{-1}.

On the other hand, in order to give the infinite Hankel matrix S of rank m it is sufficient to know the first $2m$ numbers $s_0, s_1, \ldots, s_{2m-1}$. These numbers may be chosen arbitrarily subject to only one restriction

$$D_m = | s_{i+k} |_0^m \neq 0 ; \tag{108}$$

all the subsequent coefficients $s_{2m}, s_{2m+1}, \ldots$ of (104) are uniquely (and rationally) expressible in terms of the first $2m$: $s_0, s_1, \ldots, s_{2m-1}$. For in the infinite Hankel matrix S of rank m the elements are connected by a recurrence relation (see Theorem 7 on p. 205)

$$s_q = \sum_{g=1}^{m} \alpha_g s_{q-g} \qquad (q = m, m+1, \ldots). \tag{109}$$

If the numbers $s_0, s_1, \ldots, s_{m-1}$ satisfy (108), then the coefficients $\alpha_1, \alpha_2, \ldots, \alpha_m$ in (109) are uniquely determined by the first m relations; the subsequent relations then determine $s_{2m}, s_{2m+1}, \ldots$.

Thus, a real polynomial $f(z)$ of degree $n = 2m$ with $\Delta_n \neq 0$ can be given uniquely[77] by $2m$ numbers $s_0, s_1, \ldots, s_{2m-1}$ satisfying (108). When $n = 2m + 1$, we have to add s_{-1} to these numbers.

We shall call the n values $s_0, s_1, \ldots, s_{2m-1}$ (for $n = 2m$) or $s_{-1}, s_0, \ldots, s_{2m-1}$ (for $n = 2m + 1$) the *Markov parameters* of the polynomial $f(z)$. These parameters may be regarded as the coordinates in an n-dimensional space of a point that represents the given polynomial $f(z)$.

We shall find out what conditions must be imposed on the Markov parameters in order that the corresponding polynomial be a Hurwitz polynomial. In this way we shall determine the domain of stability in the space of Markov parameters.

A Hurwitz polynomial is characterized by the conditions (94) and the additional condition (95) for $n = 2m + 1$. Introducing the function $R(v)$ (see (105)), we write (94) as follows:

$$I_{-\infty}^{+\infty} R(v) = m, \qquad I_{-\infty}^{+\infty} vR(v) = m. \tag{110}$$

The additional condition (95) for $n = 2m + 1$ gives:

$$s_{-1} > 0.$$

Apart from the matrix $S = \| s_{i+k} \|_0^\infty$ we introduce the infinite Hankel matrix $S^{(1)} = \| s_{i+k+1} \|_0^\infty$. Then, since by (106)

[77] To within a constant factor.

$$vR(v) = -s_{-1}v + s_0 + \frac{s_1}{v} + \frac{s_2}{v^2} + \cdots ,$$

the following relation holds:

$$vR(v) \sim S^{(1)}. \tag{111}$$

The matrix $S^{(1)}$, like S, is of finite rank m, since the function $vR(v)$, like $R(v)$, has m poles. Therefore the forms

$$S_m(x, x) = \sum_{i,\,k=0}^{m-1} s_{i+k}x_ix_k, \quad S_m^{(1)}(x, x) = \sum_{i,\,k=0}^{m-1} s_{i+k+1}x_ix_k$$

are of rank m. But by Theorem 9 (p. 190) the signatures of these forms, in virtue of (107) and (111), are equal to the indices (110) and hence also to m. Thus, the conditions (110) mean that the quadratic forms $S_m(x, x)$ and $S_m^{(1)}(x, x)$ are positive definite. Hence:

THEOREM 17: *A real polynomial $f(z) = h(z^2) + zg(z^2)$ of degree $n = 2m$ or $n = 2m + 1$ is a Hurwitz polynomial if and only if:*[78]

1. *The quadratic forms*

$$S_m(x, x) = \sum_{i,\,k=0}^{m-1} s_{i+k}x_ix_k, \quad S_m^{(1)}(x, x) = \sum_{i,\,k=0}^{m-1} s_{i+k+1}x_ix_k \tag{112}$$

are positive definite; and

2. *(For $n = 2m + 1$)*

$$s_{-1} > 0. \tag{113}$$

Here $s_{-1}, s_0, s_1, \ldots, s_{2m-1}$ are the coefficients of the expansion

$$\frac{g(u)}{h(u)} = s_{-1} + \frac{s_0}{u} - \frac{s_2}{u^2} + \frac{s_2}{u^3} - \frac{s_3}{u^4} + \cdots.$$

---

[78] We do not mention the inequality $\Delta_n \neq 0$ expressly, because it follows automatically from the conditions of the theorem. For if $f(z)$ is a Hurwitz polynomial, then it is known that $\Delta_n \neq 0$. But if the conditions 1., 2. are given, then the fact that the form $S_m^{(1)}(x, x)$ is positive definite implies that

$$-I_{-\infty}^{+\infty}\frac{ug(u)}{h(u)} - I_{-\infty}^{+\infty} vR(v) = m,$$

and from this it follows that the fraction $ug(u)/h(u)$ is reduced, which can be expressed by the inequality $\Delta_n \neq 0$.

In exactly the same way, it follows automatically from the conditions of the theorem that $D_m = |\, s_{i+k} \,|_0^{m-1} \neq 0$, i.e., that the numbers $s_0, s_1, \ldots, s_{2m-1}$, and (for $n = 2m + 1$) s_{-1} are the Markov parameters of $f(z)$.

We introduce a notation for the determinants

$$D_p = \left| s_{i+k} \right|_0^{p-1}, \quad D_p^{(1)} = \left| s_{i+k+1} \right|_0^{p-1} \quad (p=1, 2, \ldots, m). \tag{114}$$

Then condition 1. is equivalent to the system of determinantal inequalities

$$\left.\begin{aligned} D_1 = s_0 > 0,\ D_2 = \begin{vmatrix} s_0 & s_1 \\ s_1 & s_2 \end{vmatrix} > 0, \ldots, D_m = \begin{vmatrix} s_0 & s_1 & \cdots & s_{m-1} \\ s_1 & s_2 & \cdots & s_m \\ \cdot & \cdot & \cdots & \cdot \\ s_{m-1} & s_m & \cdots & s_{2m-2} \end{vmatrix} > 0, \\ D_1^{(1)} = s_1 > 0, D_2^{(1)} = \begin{vmatrix} s_1 & s_2 \\ s_2 & s_3 \end{vmatrix} > 0, \ldots, D_m^{(1)} = \begin{vmatrix} s_1 & s_2 & \cdots & s_m \\ s_2 & s_3 & \cdots & s_{m+1} \\ \cdot & \cdot & \cdots & \cdot \\ s_m & s_{m+1} & & s_{2m-1} \end{vmatrix} > 0. \end{aligned}\right\} \tag{115}$$

If $n=2m$, the inequalities (115) determine the domain of stability in the space of Markov parameters. If $n=2m+1$, we have to add the further inequality:

$$s_{-1} > 0. \tag{116}$$

In the next section we shall find out what properties of S follow from the inequalities (115) and, in so doing, shall single out the special class of infinite Hankel matrices S that correspond to Hurwitz polynomials.

§ 16. Connection with the Problem of Moments

1. We begin by stating the following problem:

PROBLEM OF MOMENTS FOR THE POSITIVE AXIS $0 < v < \infty$:[79] *Given a sequence $s_0, s_1, \ldots$ of real numbers, it is required to determine positive numbers*

$$\mu_1 > 0, \quad \mu_2 > 0, \ldots, \quad \mu_m > 0, \quad 0 < v_1 < v_2 < \cdots < v_m \tag{117}$$

such that the following equations hold:

$$s_p = \sum_{j=1}^{m} \mu_j v_j^p \qquad (p=0, 1, 2, \ldots). \tag{118}$$

It is not difficult to see that the system (118) of equations is equivalent to the following expansion in a series of negative powers of u:

---

[79] This problem of moments ought to be called discrete in contrast to the general exponential problem of moments, in which the sums $\sum_{j=1}^{m} \mu_j v_j^p$ are replaced by Stieltjes integrals $\int_0^\infty v^p \, d\mu(v)$ (see [55]).

$$\sum_{j=1}^{m} \frac{\mu_j}{u+v_j} = \frac{s_0}{u} - \frac{s_1}{u^2} + \frac{s_2}{u^3} - \cdots. \tag{119}$$

In this case the infinite Hankel matrix $S = \| s_{i+k} \|_0^\infty$ is of finite rank m and by (117) in the irreducible proper fraction

$$\frac{g(u)}{h(u)} = \sum_{j=1}^{m} \frac{\mu_j}{u+v_j} \tag{120}$$

(we choose the highest coefficients of $h(u)$ and $g(u)$ to be positive) the polynomials $h(u)$ and $g(u)$ form a positive pair (see (91) and (91′)).

Therefore (see Theorem 14), our problem of moments has a solution if and only if the sequence $s_0, s_1, s_2, \ldots$ determines by means of (119) and (120) a Hurwitz polynomial $f(z) = h(z^2) + zg(z^2)$ of degree $2m$.

The solution of the problem of moments is unique, because the positive numbers v_j and μ_j $(j = 1, 2, \ldots, m)$ are uniquely determined from the expansion (119).

Apart from the 'infinite' problem of moments (118) we also consider the 'finite' problem of moments given by the first $2m$ equations (118)

$$s_p = \sum_{j=1}^{m} \mu_j v_j^p \qquad (p = 0, 1, \ldots, 2m-1). \tag{121}$$

These relations already determine the following expressions for the Hankel quadratic forms:

$$\begin{aligned} \sum_{i,k=0}^{m-1} s_{i+k} x_i x_k &= \sum_{j=1}^{m} \mu_j (x_0 + x_1 v_j + \cdots + x_{m-1} v_j^{m-1})^2, \\ \sum_{i,k=0}^{m-1} s_{i+k+1} x_i x_k &= \sum_{j=1}^{m} \mu_j v_j (x_0 + x_1 v_j + \cdots + x_{m-1} v_j^{m-1})^2. \end{aligned} \tag{122}$$

Since the linear forms in the variables $x_0, x_1, \ldots, x_{m-1}$

$$x_0 + x_1 v_j + \cdots + x_{m-1} v_j^{m-1} \qquad (j = 1, 2, \ldots, m)$$

are independent (their coefficients form a non-vanishing Vandermonde determinant), the quadratic forms (122) are positive definite. But then by Theorem 17 the numbers $s_0, s_1, \ldots, s_{2m-1}$ are the Markov parameters of a certain Hurwitz polynomial $f(z)$. They are the first $2m$ coefficients of the expansion (119). Together with the remaining coefficients $s_{2m}, s_{2m+1}, \ldots$ they determine the infinite solvable problem of moments (118), which has the same solution as the finite problem (121).

Thus we have proved the following theorem:

THEOREM 18: 1) *The finite problem of moments*

$$s_p = \sum_{j=1}^{m} \mu_j v_j^p \tag{123}$$

$(p=0, 1, \ldots, 2m-1; \mu_1 > 0, \ldots, \mu_m > 0; 0 < v_1 < v_2 < \ldots < v_m)$, *where* s_p *are given real numbers and* v_j *and* μ_j *are unknown real numbers* $(p=0, 1, \ldots, 2m-1;\ j=1, 2, \ldots, m)$ *has a solution if and only if the quadratic forms*

$$\sum_{i,k=0}^{m-1} s_{i+k} x_i x_k, \quad \sum_{i,k=0}^{m-1} s_{i+k+1} x_i x_k \tag{124}$$

are positive definite, i.e., if the numbers $s_0, s_1, \ldots, s_{2m-1}$ *are the Markov parameters of some Hurwitz polynomial of degree* $2m$.

2) *The infinite problem of moments*

$$s_p = \sum_{j=1}^{m} \mu_j v_j^p \tag{125}$$

$(p=0, 1, 2, \ldots; \mu_1 > 0, \ldots, \mu_m > 0; 0 < v_1 < v_2 < \ldots < v_m)$, *where* s_p *are given real numbers and* v_j *and* μ_j *are unknown real numbers* $(p=0, 1, \ldots;\ j=1, 2, \ldots, m)$ *has a solution if and only if* 1. *the quadratic forms* (124) *are positive definite and* 2. *the infinite Hankel matrix* $S = \| s_{i+k} \|_0^\infty$ *is of rank* m, *i.e., if the series*

$$\frac{s_0}{u} - \frac{s_1}{u^2} + \frac{s_2}{u^3} - \cdots = \frac{g(u)}{h(u)} \tag{126}$$

determines a Hurwitz polynomial $f(z) = h(z^2) = zg(z^2)$ *of degree* $2m$.

3) *The solution of the problem of moments, both the finite* (123) *and the infinite* (124) *problem, is always unique.*

2. We shall use this theorem in investigating the minors of an infinite Hankel matrix $S = \| s_{i+k} \|_0^\infty$ of rank m corresponding to some Hurwitz polynomial, i.e., one for which the quadratic form (124) is positive definite. In this case the generating numbers $s_0, s_1, s_2, \ldots$ of S can be represented in the form (123), so that for an arbitrary minor of S of order $h \leq m$ we have:

$$\left\| \begin{matrix} s_{i_1+k_1} & \cdots & s_{i_1+k_h} \\ \cdots & \cdots & \cdots \\ s_{i_h+k_1} & \cdots & s_{i_h+k_h} \end{matrix} \right\| = \left\| \begin{matrix} \mu_1 v_1^{i_1} & \mu_2 v_2^{i_1} & \ldots & \mu_m v_m^{i_1} \\ \cdots & \cdots & \cdots & \cdots \\ \mu_1 v_1^{i_h} & \mu_2 v_2^{i_h} & \ldots & \mu_m v_m^{i_h} \end{matrix} \right\| \cdot \left\| \begin{matrix} v_1^{k_1} & \ldots & v_1^{k_h} \\ v_2^{k_1} & \ldots & v_2^{k_h} \\ \cdots & \cdots & \cdots \\ \cdots & \cdots & \cdots \\ \cdots & \cdots & \cdots \\ v_m^{k_1} & \ldots & v_m^{k_h} \end{matrix} \right\|$$

and therefore

$$S\begin{pmatrix} i_1 & i_2 & \dots & i_h \\ k_1 & k_2 & \dots & k_h \end{pmatrix}$$

$$= \sum_{1 \leqq \alpha_1 < \alpha_2 < \cdots < \alpha_h \leqq m} \mu_{\alpha_1} \mu_{\alpha_2} \dots \mu_{\alpha_h} \begin{vmatrix} v_{\alpha_1}^{i_1} & v_{\alpha_2}^{i_1} & \dots & v_{\alpha_h}^{i_1} \\ v_{\alpha_1}^{i_2} & v_{\alpha_2}^{i_2} & \dots & v_{\alpha_h}^{i_2} \\ \dots & \dots & \dots & \dots \\ v_{\alpha_1}^{i_h} & v_{\alpha_1}^{i_h} & \dots & v_{\alpha_h}^{i_h} \end{vmatrix} \begin{vmatrix} v_{\alpha_1}^{k_1} & v_{\alpha_1}^{k_2} & \dots & v_{\alpha_1}^{k_h} \\ v_{\alpha_2}^{k_1} & v_{\alpha_2}^{k_2} & \dots & v_{\alpha_2}^{k_h} \\ \dots & \dots & \dots & \dots \\ v_{\alpha_h}^{k_1} & v_{\alpha_h}^{k_2} & \dots & v_{\alpha_h}^{k_h} \end{vmatrix}. \tag{127}$$

But from the inequalities

$$0 < v_1 < v_2 < \cdots < v_m, \quad i_1 < i_2 < \cdots < i_h, \quad k_1 < k_2 < \cdots < k_h$$

it follows that the generalized Vandermonde determinants[80]

$$\begin{vmatrix} v_{\alpha_1}^{i_1} & v_{\alpha_2}^{i_1} & \dots & v_{\alpha_h}^{i_1} \\ v_{\alpha_1}^{i_2} & v_{\alpha_2}^{i_2} & \dots & v_{\alpha_h}^{i_2} \\ \dots & \dots & \dots & \dots \\ v_{\alpha_1}^{i_h} & v_{\alpha_2}^{i_h} & \dots & v_{\alpha_h}^{i_h} \end{vmatrix} > 0, \qquad \begin{vmatrix} v_{\alpha_1}^{k_1} & v_{\alpha_1}^{k_2} & \dots & v_{\alpha_1}^{k_h} \\ v_{\alpha_2}^{k_1} & v_{\alpha_2}^{k_2} & \dots & v_{\alpha_2}^{k_h} \\ \dots & \dots & \dots & \dots \\ v_{\alpha_h}^{k_1} & v_{\alpha_h}^{k_2} & \dots & v_{\alpha_h}^{k_h} \end{vmatrix} > 0$$

are positive.

Since the numbers μ_j are positive ($j = 1, 2, \dots, m$), it therefore follows from (127) that

$$S\begin{pmatrix} i_1 & i_2 & \dots & i_h \\ k_1 & k_2 & \dots & k_h \end{pmatrix} > 0 \quad \left(0 \leqq \begin{matrix} i_1 < i_2 < \cdots < i_h \\ k_1 < k_2 < \cdots < k_h, \end{matrix} \quad h = 1, 2, \dots, m\right). \tag{128}$$

Conversely, if in an infinite Hankel matrix $S = \| s_{i+k} \|_0^\infty$ of rank m all the minors of every order $h \leqq m$ are positive, then the quadratic forms (124) are positive definite.

DEFINITION 4: *An infinite matrix $A = \| a_{ik} \|_1^n$ will be called totally positive of rank m if and only if all the minors of A of order $h \leqq m$ are positive and all the minors of order $h > m$ are zero.*

The property of S that we have found can now be expressed in the following theorem:[81]

THEOREM 19: *An infinite Hankel matrix $S = \| s_{i+k} \|_0^\infty$ is totally positive of rank m if and only if* 1) *S is of rank m and* 2) *the quadratic forms*

$$\sum_{i,k=0}^{m-1} s_{i+k} x_i x_k, \quad \sum_{i,k=0}^{m-1} s_{i+k+1} x_i x_k$$

are positive definite.

---

[80] See p. 99, Example 1.

[81] See [173].

From this theorem and Theorem 17 we obtain:

THEOREM 20: *A real polynomial $f(z)$ of degree n is a Hurwitz polynomial if and only if the corresponding infinite Hankel matrix $S = \| s_{i+k} \|_0^\infty$ is totally positive of degree $m = [n/2]$ and if, in addition, $s_{-1} > 0$ when n is odd.*

Here the elements $s_0, s_1, s_2, \ldots$ of S and s_{-1} are determined by the expansion

$$\frac{g(u)}{h(u)} = s_{-1} + \frac{s_0}{u} - \frac{s_1}{u^2} + \frac{s_2}{u^3} - \cdots, \tag{129}$$

where

$$f(z) = h(z^2) + z g(z^2).$$

§ 17. Theorems of Markov and Chebyshev

1. In a notable memoir 'On functions obtained by converting series into continued fractions'[82] Markov proved two theorems, the second of which had been established in 1892 by Chebyshev by other methods, and not in the same generality.[83]

In this section we shall show that these theorems have an immediate bearing on the study of the domain of stability in the Markov parameters and shall give a comparatively simple proof (without reference to continued fractions) which is based on Theorem 19 of the preceding section.

In proceeding to state the first theorem, we quote the corresponding passage from the above-mentioned memoir of Markov:[84]

> On the basis of what has preceded it is not difficult to prove two remarkable theorems with which we conclude our paper.
>
> One is concerned with the determinants[85]
>
> $$\Delta_1, \Delta_2, \ldots, \Delta_m, \quad \Delta^{(1)}, \Delta^{(2)}, \ldots, \Delta^{(m)}$$
>
> and the other with the roots of the equation[86]
>
> $$\psi_m(x) = 0.$$

[82] Zap. Petersburg Akad. Nauk, Petersburg, 1894 [in Russian]; see also [38], pp. 78-105.

[83] This theorem was first published in Chebyshev's paper 'On the expansion in continued fractions of series in descending powers of the variable' [in Russian]. See [8], pp. 307-62.

[84] [38], p. 95, beginning with line 3 from below.

[85] In our notation, $D_1, D_2, \ldots, D_m, D_1^{(1)}, D_2^{(1)}, \ldots, D_m^{(1)}$. (See p. 236.)

[86] In our notation, $h(-x) = 0$.

THEOREM ON DETERMINANTS: *If we have for the numbers*

$$s_0, s_1, s_2, \ldots, s_{2m-2}, s_{2m-1}$$

two sets of values

$$1.\ s_0 = a_0,\ s_1 = a_1,\ s_2 = a_2,\ \ldots,\ s_{2m-2} = a_{2m-2},\ s_{2m-1} = a_{2m-1},$$
$$2.\ s_0 = b_0,\ s_1 = b_1,\ s_2 = b_2,\ \ldots,\ s_{2m-2} = b_{2m-2},\ s_{2m-1} = b_{2m-1}$$

for which all the determinants

$$\Delta_1 = s_0,\ \Delta_2 = \begin{vmatrix} s_0 & s_1 \\ s_1 & s_2 \end{vmatrix},\ \ldots,\ \Delta_m = \begin{vmatrix} s_0 & s_1 \ldots s_{m-1} \\ s_1 & s_2 \ldots s_m \\ \cdots & \cdots \\ s_{m-1} & s_m \quad s_{2m-2} \end{vmatrix},$$

$$\Delta^{(1)} = s_1,\ \Delta^{(2)} = \begin{vmatrix} s_1 & s_2 \\ s_2 & s_3 \end{vmatrix},\ \ldots,\ \Delta^{(m)} = \begin{vmatrix} s_1 & s_2 \ldots s_m \\ s_2 & s_3 \ldots s_{m+1} \\ \cdots & \cdots \\ s_m & s_{m+1} \quad s_{2m-1} \end{vmatrix}$$

turn out to be positive numbers satisfying the inequalities

$$a_0 \geqq b_0,\ b_1 \geqq a_1,\ a_2 \geqq b_2,\ b_3 \geqq a_3,\ \ldots,\ a_{2m-2} \geqq b_{2m-2},\ b_{2m-1} \geqq a_{2m-1},$$

then our determinant

$$\Delta_1, \Delta_2, \ldots, \Delta_m;\ \Delta^{(1)}, \Delta^{(2)}, \ldots, \Delta^{(m)}$$

must be positive for all values

$$s_0, s_1, s_2, \ldots, s_{2m-1}$$

satisfying the inequalities

$$a_0 \geqq s_0 \geqq b_0,\ b_1 \geqq s_1 \geqq a_1,\ a_2 \geqq s_2 \geqq b_2,\ \ldots,$$
$$a_{2m-2} \geqq s_{2m-2} \geqq b_{2m-2},\ b_{2m-1} \geqq s_{2m-1} \geqq a_{2m-1}.$$

Under these conditions we have

$$\begin{vmatrix} a_0 & a_1 \ldots a_{k-1} \\ a_1 & a_2 \ldots a_k \\ \cdots & \cdots \\ a_{k-1} & a_k \ldots a_{2k-2} \end{vmatrix} \geqq \begin{vmatrix} s_0 & s_1 \ldots s_{k-1} \\ s_1 & s_2 \ldots s_k \\ \cdots & \cdots \\ s_{k-1} & s_k \ldots s_{2k-2} \end{vmatrix} \geqq \begin{vmatrix} b_0 & b_1 \ldots b_{k-1} \\ b_1 & b_2 \ldots b_k \\ \cdots & \cdots \\ b_{k-1} & b_k \ldots b_{2k-2} \end{vmatrix}$$

and

$$\begin{vmatrix} b_1 & b_2 & \dots & b_k \\ b_2 & b_3 & \dots & b_{k+1} \\ \dots & \dots & \dots & \dots \\ b_k & b_{k+1} & \dots & b_{2k-1} \end{vmatrix} \geqq \begin{vmatrix} s_1 & s_2 & \dots & s_k \\ s_2 & s_3 & \dots & s_{k+1} \\ \dots & \dots & \dots & \dots \\ s_k & s_{k+1} & \dots & s_{2k-1} \end{vmatrix} \geqq \begin{vmatrix} a_1 & a_2 & \dots & a_k \\ a_2 & a_3 & \dots & a_{k+1} \\ \dots & \dots & \dots & \dots \\ a_k & a_{k+1} & \dots & a_{2k-1} \end{vmatrix}$$

for $k = 1, 2, \dots, m$.

In order to give another statement of this theorem in connection with the problem of stability, we introduce some concepts and notations.

The Markov parameters $s_0, s_1, \dots, s_{2m-1}$ (for $n = 2m$) or $s_{-1}, s_0, s_1, \dots, s_{2m-1}$ (for $n = 2m + 1$) will be regarded as the coordinates of some point $\boldsymbol{P}$ in an n-dimensional space. The domain of stability in this space will be denoted by $\boldsymbol{G}$. The domain $\boldsymbol{G}$ is characterized by the inequalities (115) and (116) (p. 236).

We shall say that a point $\boldsymbol{P} = \{s_i\}$ 'precedes' a point $\boldsymbol{P}^* = \{s_i^*\}$ and shall write $\boldsymbol{P} \prec \boldsymbol{P}^*$ if

$$\left.\begin{array}{l} s_0 \leqq s_0^*,\ s_1^* \leqq s_1,\ s_2 \leqq s_2^*,\ s_3^* \leqq s_3,\ \dots,\ s_{2m-1}^* \leqq s_{2m-1} \\ \text{and (for } n = 2m+1) \\ s_{-1} \leqq s_{-1}^* \end{array}\right\} \tag{130}$$

and the sign $<$ holds in at least one of these relations.

If only the relations (130) hold, without the last clause, then we shall write:.

$$\mathfrak{P} \preceq \mathfrak{P}^*.$$

We shall say that a point $\boldsymbol{Q}$ lies 'between' $\boldsymbol{P}$ and $\boldsymbol{R}$ if $\boldsymbol{P} \prec \boldsymbol{Q} \prec \boldsymbol{R}$.

To every point $\boldsymbol{P}$ there corresponds an infinite Hankel matrix of rank m: $S = \| s_{i+k} \|_0^\infty$. We shall denote this matrix by $S_{\mathfrak{p}}$.

Now we can state Markov's theorem in the following way:

THEOREM 21 (Markov): *If two points* **P** *and* **R** *belong to the domain of stability* **G** *and if* **P** *precedes* **R**, *then every point* **Q** *between* **P** *and* **R** *also belongs to* **G**, *i.e.,*

$$\textit{from } \boldsymbol{P}, \boldsymbol{R} \in \boldsymbol{G},\ \boldsymbol{P} \preceq \boldsymbol{Q} \preceq \boldsymbol{R} \textit{ it follows that } \boldsymbol{Q} \in \boldsymbol{G}.$$

Proof. From $\boldsymbol{P} \preceq \boldsymbol{Q} \preceq \boldsymbol{R}$ it follows that $\boldsymbol{P}$ and $\boldsymbol{Q}$ can be connected by an arc of a curve

$$s_i = (-1)^i \varphi_i(t)\ [\alpha \leqq t \leqq \gamma;\ i = 0, 1, \dots, 2m-1 \text{ and (for } n = 2m+1)\ i = -1] \tag{131}$$

passing through $\boldsymbol{Q}$ such that: 1) the functions $\varphi_i(t)$ are continuous, monotonic increasing, and differentiable when t varies from $t=\alpha$ to $t=\gamma$; and 2) the values α, β, γ $(\alpha<\beta<\gamma)$ of t correspond to the points $\boldsymbol{P}$, $\boldsymbol{Q}$, $\boldsymbol{R}$ on the curve.

From the values (131) we form the infinite Hankel matrix $S=S(t)=\| s_{i+k}(t) \|_0^\infty$ of rank m. We consider part of this matrix, namely the rectangular matrix

$$\left\| \begin{matrix} s_0 & s_1 & \dots & s_{m-1} & s_m \\ s_1 & s_2 & \dots & s_m & s_{m+1} \\ \dots & \dots & \dots & \dots & \dots \\ s_{m-1} & s_m & \dots & s_{2m-2} & s_{2m-1} \end{matrix} \right\| . \tag{132}$$

By the conditions of the theorem, the matrix $S(t)$ is totally positive of rank m for $t=\alpha$ and $t=\gamma$, so that all the minors of (132) of order $p=1, 2, 3, \dots, m$ are positive.

We shall now show that this property also holds for every intermediate value of t $(\alpha<t<\gamma)$.

For $p=1$, this is obvious. Let us prove the statement for the minors of order p, on the assumption that it is true for those of order $p-1$. We consider an arbitrary minor of order p formed from successive rows and columns of (132):

$$D_p^{(q)}=\left| \begin{matrix} s_q & s_{q+1} & \dots & s_{q+p-1} \\ s_{q+1} & s_{q+2} & \dots & s_{q+p} \\ \dots & \dots & \dots & \dots \\ s_{q+p-1} & s_{q+p} & \dots & s_{q+2p-2} \end{matrix} \right| \qquad [q=0, 1, \dots, 2(m-p)+1]. \tag{133}$$

We compute the derivative of this minor

$$\frac{d}{dt} D_p^{(q)} = \sum_{i,k=0}^{p-1} \frac{\partial D_p^{(q)}}{\partial s_{q+i+k}} \frac{d s_{q+i+k}}{dt}. \tag{134}$$

$\dfrac{\partial D_p^{(q)}}{\partial s_{q+i+k}}$ $(i, k=0, 1, \dots, p-1)$ are the algebraic complements (adjoints) of the elements of (133). Since by assumption all the minors of this determinant are positive, we have

$$(-1)^{i+k} \frac{\partial D_p^{(q)}}{\partial s_{q+i+k}} > 0 \qquad (i, k=0, 1, \dots, p-1). \tag{135}$$

On the other hand, we find from (131):

$$(-1)^{q+i+k}\frac{ds_{q+i+k}}{dt}=\frac{d\varphi_{q+i+k}}{dt}\geqq 0 \qquad (i,\ k=0,\ 1,\ \ldots\ ,\ p-1). \quad \textbf{(136)}$$

From (134), (135), and (136) it follows that

$$(-1)^q\frac{d}{dt}D_p^{(q)}\geqq 0 \qquad \begin{pmatrix} q=0,\ 1,\ \ldots\ ,\ 2(m-p)+1, \\ p=1,\ 2,\ \ldots\ ,\ m, \\ \alpha\leqq t\leqq\gamma \end{pmatrix}. \quad \textbf{(137)}$$

Thus, when the argument increases from $t=\alpha$, to $t=\gamma$, then every minor (133) with even q is a monotone non-decreasing function and with odd q is a monotone non-increasing function; but since the minor is positive for $t=\alpha$ and $t=\gamma$, it is also positive for every intermediate value of t $(\alpha<t<\gamma)$.

From the fact that the minors of (132) of order $p-1$ and those of order p that are formed from successive rows and columns are positive, it now follows that *all* the minors of (131) of order p are positive.[87]

What we have proved implies that for every t $(\alpha\leqq t\leqq\gamma)$ the values $s_0,\ s_1,\ \ldots,\ s_{2m-1}$ and (for $n=2m+1$) s_{-1} satisfy the inequalities (115) and (116), i.e., that for every t these values are the Markov parameters of a certain Hurwitz polynomial. In other words, the whole arc (131) and, in particular, the point $\boldsymbol{Q}$ lies in the domain of stability $\boldsymbol{G}$.

This completes the proof of Markov's Theorem.

Note. Since we have proved that every point of the arc (131) belongs to $\boldsymbol{G}$, the values of (131) for every t $(\alpha\leqq t\leqq\gamma)$ determine a totally positive matrix $S(t)=\|\ s_{i+k}(t)\ \|_0^\infty$ of rank m. Therefore the inequalities (135) and consequently (137) as well hold for every t $(\alpha\leqq t\leqq\gamma)$, i.e., with increasing t every $D_p^{(q)}$ increases for even q and decreases for odd q $(q=0,\ 1,\ 2,\ \ldots,\ 2(m-p)+1;\ p=1,\ \ldots,\ m)$. In other words, from $\boldsymbol{P}\preceq\boldsymbol{Q}\preceq\boldsymbol{R}$ it follows that

$$(-1)^qD_p^{(q)}(\mathfrak{P})\leqq(-1)^qD_p^{(q)}(\mathfrak{Q})\leqq(-1)^qD_p^{(q)}(\mathfrak{R})$$
$$(q=0,\ 1,\ \ldots,\ 2(m-p)+1;\ p=1,\ \ldots,\ m).$$

These inequalities for $q=0,\ 1$ give Markov's inequalities (pp. 241).

We now come to the Chebyshev-Markov theorem mentioned at the beginning of this section. Again we quote from Markov's memoir:[88]

---

[87] This follows from Fekete's determinant indentity (see [17], pp. 306-7).

[88] See [38], p. 103, beginning with line 5.

THEOREM ON ROOTS: *If the numbers*

$$a_0, a_1, a_2, \ldots, a_{2m-2}, a_{2m-1},$$
$$s_0, s_1, s_2, \ldots, s_{2m-2}, s_{2m-1},$$
$$b_0, b_1, b_2, \ldots, b_{2m-2}, b_{2m-1}$$

satisfy all the conditions of the preceding theorem,[89] *then the equations*

$$\begin{vmatrix} a_0 & a_1 & \ldots & a_{m-1} & 1 \\ a_1 & a_2 & \ldots & a_m & x \\ a_2 & a_3 & \ldots & a_{m+1} & x^2 \\ \cdot & \cdot & \cdot & \cdot & \cdot \\ a_m & a_{m+1} & \ldots & a_{2m-1} & x^m \end{vmatrix} = 0,$$

$$\begin{vmatrix} s_0 & s_1 & \ldots & s_{m-1} & 1 \\ s_1 & s_2 & \ldots & s_m & x \\ s_2 & s_3 & \ldots & s_{m+1} & x^2 \\ \cdot & \cdot & \cdot & \cdot & \cdot \\ s_m & s_{m+1} & \ldots & s_{2m-1} & x^m \end{vmatrix} = 0,$$

$$\begin{vmatrix} b_0 & b_1 & \ldots & b_{m-1} & 1 \\ b_1 & b_2 & \ldots & b_m & x \\ b_2 & b_3 & \ldots & b_{m+1} & x^2 \\ \cdot & \cdot & \cdot & \cdot & \cdot \\ b_m & b_{m+1} & \ldots & b_{2m-1} & x^m \end{vmatrix} = 0$$

of degree m in the unknown x do not have multiple or imaginary or negative roots.

And the roots of the second equation are larger than the corresponding roots of the first equation and smaller than the corresponding roots of the last equation.

Let us find out the connection of this theorem with the domain of stability in the space of the Markov parameters. Setting $f(z) = h(z^2) + zg(z^2)$ and

$$h(-v) = c_0 v^m + c_1 v^{m-1} + \cdots + c_m \quad (c_0 \neq 0),$$

we obtain from the expansion (105)

$$R(v) = -\frac{g(-v)}{h(-v)} = -s_{-1} + \frac{s_0}{v} + \frac{s_1}{v^2} + \cdots$$

the identity

---

[89] He refers to the preceding theorem, Markov's theorem on determinants (pp. 241).

$$-g(-v)=\left(-s_{-1}+\frac{s_0}{v}+\frac{s_1}{v^2}+\cdots\right)(c_0v^m+c_1v^{m-1}+\cdots+c_m).$$

Equating to zero the coefficients of the powers $v^{-1}, v^{-2}, \ldots, v^{-m}$, we find:

$$\left.\begin{aligned} s_0c_m &+ s_1c_{m-1} + \cdots + s_mc_0 = 0,\\ s_1c_m &+ s_2c_{m-1} + \cdots + s_{m+1}\,c_0 = 0,\\ &\cdots\cdots\cdots\cdots\cdots\\ s_{m-1}c_m &+ s_mc_{m-1} + \cdots + s_{2m-1}c_0 = 0; \end{aligned}\right\};\tag{138}$$

to these relations we add the equation

$$h(-v)=0,\tag{139}$$

written as

$$c_m+vc_{m-1}+\cdots+v^mc_0=0.\tag{139'}$$

Eliminating from (138) and (139′) the coefficients $c_0, c_1, \ldots, c_m$, we represent the equation (139) in the form

$$\begin{vmatrix} s_0 & s_1 & \ldots & s_{m-1} & 1\\ s_1 & s_2 & \ldots & s_m & v\\ s_2 & s_3 & \ldots & s_{m+1} & v^2\\ \cdot & \cdot & \cdot & \cdot & \cdot\\ s_m & s_{m+1} & \ldots & s_{2m-1} & v^m \end{vmatrix}=0.\tag{139''}$$

Thus, the algebraic equation in the Chebyshev-Markov theorem coincides with (139) and the inequalities imposed on $s_0, s_1, \ldots, s_{2m-1}$ coincide with the inequalities (115) that determine the domain of stability in the space of the Markov parameters.

The Chebyshev-Markov theorem shows how the roots $u_1=-v_1$, $u_2=-v_2, \ldots, u_m=-v_m$ of $h(u)$ change when the corresponding Markov parameters $s_0, s_1, \ldots, s_{2m-1}$ vary in the domain of stability.

The first part of the theorem states something we already know: When the inequalities (115) are satisfied, then all the roots $u_1, u_2, \ldots, u_m$ of $h(u)$ are simple, real, and negative.[90] We denote them as follows:

$$u_1(\boldsymbol{P}),\ u_2(\boldsymbol{P}),\ \ldots,\ u_m(\boldsymbol{P}),$$

where $\boldsymbol{P}$ is the corresponding point of $\boldsymbol{G}$.

The second (fundamental) part of the Chebyshev-Markov theorem can be stated as follows:

---

[90] See Theorem 13, on p. 228.

THEOREM 22 (Chebyshev-Markov): *If **P** and **Q** are two points of **G** and **P** 'precedes' **Q**,*

$$P \prec Q, \tag{140}$$

then[91]

$$u_1(P) < u_1(Q),\ u_2(P) < u_2(Q), \ldots, u_m(P) < u_m(Q). \tag{141}$$

Proof. The coefficients of $h(u)$ can be expressed rationally in terms of the parameters $s_0, s_1, \ldots, s_{2m-1}$.[92] Then

$$h(u_i) = 0 \qquad (i = 1, 2, \ldots, m)$$

implies that:[93]

$$\frac{\partial h(u_i)}{\partial s_l} + h'(u_i)\frac{du_i}{ds_l} = 0 \qquad (i = 1, 2, \ldots, m; \quad l = 0, 1, \ldots, 2m-1). \tag{142}$$

On the other hand, when we differentiate the expansion

$$\frac{g(u)}{h(u)} = s_{-1} + \frac{s_0}{u} - \frac{s_1}{u^2} + \frac{s_2}{u^3} - \cdots$$

term by term with respect to s, we find:

$$\frac{h(u)\dfrac{\partial g(u)}{\partial s_l} - g(u)\dfrac{\partial h(u)}{\partial s_l}}{h^2(u)} = \frac{(-1)^l}{u^{l+1}} + \frac{1}{u^{2m+1}}(*). \tag{143}$$

Multiplying both sides of this equation by $\dfrac{h^2(u)}{u-u_i}$ and denoting the coefficient of u^l in this polynomial by C_{il}, we obtain:

$$\frac{h(u)}{u-u_i}\frac{\partial g(u)}{\partial s_l} - \frac{g(u)\dfrac{\partial h(u)}{\partial s_l}}{u-u_i} = \frac{(-1)^l C_{il}}{u} + \cdots. \tag{144}$$

Comparing the coefficients of $1/u$ (the residues) on the two sides of (144), we find:

$$(-1)^{l-1} g(u_i)\frac{\partial h(u_i)}{\partial s_l} = C_{il}, \tag{145}$$

which gives in conjunction with (142):

$$\frac{du_i}{ds_l} = \frac{(-1)^l C_{il}}{g(u_i)\, h'(u_i)}.$$

---

[91] In other words, the roots $u_1, u_2, \ldots, u_m$ increase with increasing $s_0, s_2, \ldots, s_{2m-2}$ and with decreasing $s_1, s_3, \ldots, s_{2m-1}$,

[92] For example, by the equations (138) if, for simplicity, we set $c_0 = 1$ in these equations

[93] Here $\dfrac{\partial h(u_i)}{\partial s_l} = \left[\dfrac{\partial h(u)}{\partial s_l}\right]_{u=u_i}$.

Introducing the values

$$R_i = \frac{g(u_i)}{h'(u_i)} \quad (l = 1, 2, \ldots, m), \tag{146}$$

we obtain the *formula of Chebyshev-Markov*:

$$\frac{du_i}{ds_l} = \frac{(-1)_l C_{il}}{R_i[h'(u_i)]^2} \quad (i = 1, 2, \ldots, m; \quad l = 0, 1, \ldots, 2m-1). \tag{147}$$

But in the domain of stability the values R_i $(i = 1, 2, \ldots, m)$ are positive (see (90′) on p. 226). The same can be said of the coefficients C_{il}. For

$$\frac{h^2(u)}{u - u_i} = c_0^2 (u + v_1)^2 \cdots (u + v_{i-1})^2 (u + v_i)(u + v_{i+1})^2 \cdots (u + v_m)^2, \tag{148}$$

where

$$v_i = -u_i > 0 \quad (i = 1, 2, \ldots, m),$$

From (148) it is clear that all the coefficients C_{il} in the expansion of $\frac{h^2(u)}{u - u_i}$ in powers of u are positive. Thus, we obtain from the Chebyshev-Markov formula:

$$(-1)^l \frac{du_i}{ds_l} > 0. \tag{149}$$

In the proof of Markov's theorem we have shown that any two points $\boldsymbol{P} \prec \boldsymbol{Q}$ of $\boldsymbol{G}$ can be joined by an arc $s_l = (-1)^l \varphi_l(t)$ $(l = 0, 1, \ldots, 2m-1)$, where $\varphi_l(t)$ is a monotonic increasing differentiable function of t (t varies within the limits α and β $(\alpha < \beta)$ and $t = \alpha$ corresponds to $\boldsymbol{P}$, $t = \beta$ to $\boldsymbol{Q}$). Then along this arc we have, by (149):[94]

$$\frac{du_i}{dt} = \sum_{l=0}^{2m-1} \frac{du_i}{ds_l}\frac{ds_l}{dt} \geqq 0, \quad \frac{du_i}{dt} \not\equiv 0 \quad (\alpha \leqq t \leqq \beta). \tag{150}$$

Hence by integrating we obtain:

$$u_{i(t=\alpha)} = u_i(\boldsymbol{P}) < u_{i(t=\beta)} = u_i(\boldsymbol{Q}) \qquad (i = 1, 2, \ldots, m).$$

This completes the proof of the Chebyshev-Markov theorem.

§ 18. The Generalized Routh-Hurwitz Problem

1. In this section we shall give a rule to determine the number of roots in the right half-plane of a polynomial $f(z)$ with complex coefficients.

[94] Since $(-1)^l \frac{ds_l}{dt} = \frac{d\varphi_l}{dt} \geqq 0$ $(\alpha \leqq t \leqq \beta)$ and for at least one l there exist values of t for which $(-1)^l \frac{ds_l}{dt} > 0$.

Suppose that

$$f(iz) = b_0 z^n + b_1 z^{n-1} + \cdots + b_n + i(a_0 z^n + a_1 z^{n-1} + \cdots + a_n), \tag{151}$$

where $a_0, a_1, \ldots, a_n, b_0, b_1, \ldots, b_n$ are real numbers. If the degree of $f(z)$ is n, then $b_0 + ia_0 \neq 0$. Without loss of generality we may assume that $a_0 \neq 0$ (otherwise we could replace $f(z)$ by $if(z)$).

We shall assume that the real polynomials

$$a_0 z^n + a_1 z^{n-1} + \cdots + a_n \quad \text{and} \quad b_0 z^n + b_1 z^{n-1} + \cdots + b_n \tag{152}$$

are co-prime, i.e., that their resultant does not vanish:[95]

$$\nabla_{2n} + \begin{vmatrix} a_0 & a_1 & \ldots & a_n & 0 & \ldots & 0 \\ b_0 & b_1 & \ldots & b_n & 0 & \ldots & 0 \\ 0 & a_0 & \ldots & a_{n-1} & a_n & \ldots & 0 \\ 0 & b_0 & \ldots & b_{n-1} & b_n & \ldots & 0 \\ \cdot & \cdot & \cdot & \cdot & \cdot & \cdot & \cdot \\ \cdot & \cdot & \cdot & \cdot & \cdot & \cdot & \cdot \end{vmatrix} \neq 0. \tag{153}$$

Hence it follows, in particular, that the polynomials (152) have no roots in common and that, therefore, $f(z)$ has no roots on the imaginary axis.

We denote by k the number of roots of $f(z)$ with positive real parts. By considering the domain in the right half-plane bounded by the imaginary axis and the semi-circle of radius R $(R \to \infty)$ and by repeating verbatim the arguments used on p. 177 for the real polynomial $f(z)$, we obtain the formula for the increment of $\arg f(z)$ along the imaginary axis

$$\Delta_{-\infty}^{+\infty} \arg f(z) = (n - 2k)\pi. \tag{154}$$

Hence we obtain, by (151), in view of $a_0 \neq 0$:

$$I_{-\infty}^{+\infty} \frac{b_0 z^n + b_1 z^{n-1} + \cdots + b_n}{a_0 z^n + a_1 z^{n-1} + \cdots + z_n} = n - 2k. \tag{155}$$

Using Theorem 10 of § 11 (p. 215), we now obtain:

$$k = V(1, \nabla_2, \nabla_4, \ldots, \nabla_{2n}), \tag{156}$$

where

[95] ∇_{2n} is a determinant of order $2n$.

$$\nabla_{2p}=\begin{vmatrix} a_0 & a_1 & \dots & a_{2p-1} \\ b_0 & b_1 & \dots & b_{2p-1} \\ 0 & a_0 & \dots & a_{2p-2} \\ 0 & b_0 & \dots & a_{2p-2} \\ \cdot & \cdot & \cdot & \cdot \\ \cdot & \cdot & \cdot & \cdot \end{vmatrix} \quad (p=1,\ 2,\ \dots,\ n;\ a_k=b_k=0 \text{ for } k>n). \tag{157}$$

We have thus arrived at the following theorem.

Theorem 23: *If a complex polynomial $f(z)$ is given for which*

$$f(iz)=b_0z^n+b_1z^{n-1}+\cdots+b_n+i\,(a_0z^n+a_1z^{n-1}+\cdots+a_n) \qquad (a_0\neq 0)$$

and if the polynomials $a_0z^n+\ldots+a_n$ and $b_0z^n+\ldots+b_n$ are co-prime ($\nabla_{2n}\neq 0$), then the number of roots of $f(z)$ in the right half-plane is determined by the formulas (156) *and* (157).

Moreover, if some of the determinants (157) *vanish, then for each group of successive zeros*

$$(\nabla_{2h}\neq 0)\ \nabla_{2h+2}=\cdots=\nabla_{2h+2p}=0\ \ (\nabla_{2h+2p+2}\neq 0) \tag{158}$$

in the calculation of $V(1,\nabla_2,\nabla_4,\dots,\nabla_{2n})$ we must set:

$$\operatorname{sign}\nabla_{2h+2j}=(-1)^{\frac{j(j-1)}{2}}\operatorname{sign}\nabla_{2h} \qquad (j=1,\ 2,\ \dots,\ p) \tag{159}$$

or, what is the same,

$$V(\nabla_{2h},\ \nabla_{2h+2},\ \dots,\ \nabla_{2h+2p},\ \nabla_{2h+2p+2}) = \begin{cases} \dfrac{p+1}{2} & \textit{for odd } p, \\[2ex] \dfrac{p+1-\varepsilon}{2} & \textit{for even } p \textit{ and } \varepsilon=(-1)^{\frac{p}{2}}\operatorname{sign}\dfrac{\nabla_{2h+2p+2}}{\nabla_{2h}}. \end{cases} \tag{160}$$

We leave it to the reader to verify that in the special case where $f(z)$ is a real polynomial we can obtain the Routh-Hurwitz theorem (see § 6) from Theorem 23.[96]

In conclusion, we mention that in this chapter we have dealt with the application of quadratic forms (in particular, Hankel forms) to one problem of the disposition of the roots of a polynomial in the complex plane. Quadratic and hermitian forms also have interesting applications to other problems of the disposition of roots. We refer the reader who is interested in these questions to the survey, already quoted, of M. G. Kreĭn and M. A. Naĭmark 'The method of symmetric and hermitian forms in the theory of separation of roots of algebraic equations,' (Kharkov, 1936).

---

[96] Suitable algorithms for the solution of the generalized Routh-Hurwitz problem can be found in the monograph [41] and in the paper [39]. See also [7] and [37].

BIBLIOGRAPHY

BIBLIOGRAPHY

Items in the Russian language are indicated by *

PART A. Textbooks, Monographs, and Surveys

[1] ACHIESER (Akhieser), N. J., *Theory of Approximation*. New York: Ungar, 1956. [Translated from the Russian.]

[2] AITKEN, A. C., *Determinants and matrices*. 9th ed., Edinburgh: Oliver and Boyd, 1956.

[3] BELLMAN, R., *Stability Theory of Differential Equations*. New York: McGraw-Hill, 1953.

*[4] BERNSTEIN, S. N., *Theory of Probability*. 4th ed., Moscow: Gostekhizdat, 1946.

[5] BODEWIG, E., *Matrix Calculus*. 2nd ed., Amsterdam: North Holland, 1959.

[6] CAHEN, G., *Éléments du calcul matriciel*. Paris: Dunod, 1955.

*[7] CHEBOTARËV, N. G., and MEĬMAN, N. N., *The problem of Routh-Hurwitz for polynomials and integral functions*. Trudy Mat. Inst. Steklov., vol. 26 (1949).

*[8] CHEBYSHEV, P. L., *Complete collected works*. vol. III. Moscow: Izd. AN SSSR, 1948.

*[9] CHETAEV, N. G., *Stability of motion*. Moscow: Gostekhizdat, 1946.

[10] COLLATZ, L., *Eigenwertaufgaben mit technischen Anwendungen*. Leipzig: Akad. Velags., 1949.

[11] ——— *Eigenwertprobleme und ihre numerische Behandlung*. New York: Chelsea, 1948.

[12] COURANT, R. and HILBERT, D., *Methods of Mathematical Physics*, vol. I. Trans. and revised from the German original. New York: Interscience, 1953.

*[13] ERUGIN, N. R., *The method of Lappo-Danilevskiĭ in the theory of linear differential equations*. Leningrad: Leningrad University, 1956.

*[14] FADDEEV, D. K. and SOMINSKIĬ, I. S., *Problems in higher algebra*. 2nd ed., Moscow, 1949; 5th ed. Moscow: Gostekhizdat, 1954.

[15] FADDEEVA, V. N., *Computational methods of linear algebra*. New York: Dover Publications, 1959. [Translated from the Russian.]

[16] FRAZER, R. A., DUNCAN, W. J., and COLLAR, A., *Elementary Matrices and Some Applications to Dynamics and Differential Equations*. Cambridge: Cambridge University Press, 1938.

*[17] GANTMACHER (Gantmakher), F. R. and KREĬN, M. G., *Oscillation matrices and kernels and small vibrations of dynamical systems*. 2nd ed., Moscow: Gostekhizdat, 1950. [A German translation is in preparation.]

[18] GRÖBNER, W., *Matrizenrechnung*. Munich: Oldenburg, 1956.

[19] HAHN, W., *Theorie und Anwendung der direkten Methode von Lyapunov* (Ergebnisse der Mathematik, Neue Folge, Heft 22). Berlin: Springer, 1959. [Contains an extensive bibliography.]

[20] INCE, E. L., *Ordinary Differential Equations.* New York: Dover, 1948.

[21] JUNG, H., *Matrizen und Determinanten. Eine Einführung.* Leipzig, 1953.

[22] KLEIN, F., *Vorlesungen über höhere Geometrie.* 3rd ed., New York: Chelsea, 1949.

[23] KOWALEWSKI, G., *Einführung in die Determinantentheorie.* 3rd ed., New York: Chelsea, 1949.

*[24] KREĬN, M. G., *Fundamental propositions in the theory of λ-zone stability of a canonical system of linear differential equations with periodic coefficients.* Moscow: Moscow Academy, 1955.

*[25] KREĬN, M. G. and NAĬMARK, M. A., *The method of symmetric and hermitian forms in the theory of separation of roots of algebraic equations.* Kharkov: GNTI, 1936.

*[26] KREĬN, M. G. and RUTMAN, M. A., *Linear operators leaving a cone in a Banach space invariant.* Uspehi Mat. Nauk, vol. 3 no. 1, (1948).

*[27] KUDRYAVCHEV, L. D., *On some mathematical problems in the theory of electrical networks.* Uspehi Mat. Nauk, vol. 3 no. 4 (1948).

*[28] LAPPO-DANILEVSKIĬ, I. A., *Theory of functions of matrices and systems of linear differential equations.* Moscow, 1934.

[29] ——— *Mémoires sur la théorie des systèmes des équations différentielles linéaires.* 3 vols., Trudy Mat. Inst. Steklov. vols. 6-8 (1934-1936). New York: Chelsea, 1953.

[30] LEFSCHETZ, S., *Differential Equations: Geometric Theory.* New York: Interscience, 1957.

[31] LICHNEROWICZ, A., *Algèbre et analyse linéaires.* 2nd ed., Paris: Masson, 1956.

[32] LYAPUNOV (Liapounoff), A. M., *Le Problème général de la stabilité du mouvement* (Annals of Mathematics Studies, No. 17). Princeton: Princeton Univ. Press, 1949.

[33] MACDUFFEE, C. C., *The Theory of Matrices.* New York: Chelsea, 1946.

[34] ——— *Vectors and matrices.* La Salle: Open Court, 1943.

*[35] MALKIN, I. G., *The method of Lyapunov and Poincaré in the theory of non-linear oscillations.* Moscow: Gostekhizdat, 1949.

[36] ——— *Theory of stability of motion.* Moscow: Gostekhizdat, 1952. [A German translation is in preparation.]

[37] MARDEN, M., *The geometry of the zeros of a polynomial in a complex variable* (Mathematical Surveys, No. 3). New York: Amer. Math. Society, 1949.

*[38] MARKOV, A. A., *Collected works.* Moscow, 1948.

*[39] MEĬMAN, N. N., *Some problems in the disposition of roots of polynomials.* Uspehi Mat. Nauk, vol. 4 (1949).

[40] MIRSKY, L., *An Introduction to Linear Algebra.* Oxford: Oxford University Press, 1955.

*[41] NAĬMARK, Y. I., *Stability of linearized systems.* Leningrad: Leningrad Aeronautical Engineering Academy, 1949.

[42] PARODI, M., *Sur quelques propriétés des valeurs caractéristiques des matrices carrées* (Mémorial des Sciences Mathématiques, vol. 118), Paris: Gauthiers-Villars, 1952.

[43] PERLIS, S., *Theory of Matrices.* Cambridge. (Mass.): Addison-Wesley, 1952.

[44] PICKERT, G., *Normalformen von Matrizen* (Enz. Math. Wiss., Band I, Teil B. Heft 3, Teil I). Leipzig: Teubner, 1953.

*[45] POTAPOV, V. P., *The multiplicative structure of J-inextensible matrix functions.* Trudy Moscow Mat. Soc., vol. 4 (1955).

*[46] ROMANOVSKIĬ, V. I., *Discrete Markov chains.* Moscow: Gostekhizdat, 1948.

[47] ROUTH, E. J., *A treatise on the stability of a given state of motion.* London: Macmillan, 1877.

[48] ——— *The advanced part of a Treatise on the Dynamics of a Rigid Body.* 6th ed., London: Macmillan, 1905; repr., New York: Dover, 1959.

[49] SCHLESINGER, L., *Vorlesungen über lineare Differentialgleichungen.* Berlin, 1908.

[50] ——— *Einführung in die Theorie der gewöhnlichen Differentialgleichungen auf funktionentheoretischer Grundlage.* Berlin, 1922.

[51] SCHMEIDLER, W., *Vortrage über Determinanten und Matrizen mit Anwendungen in Physik und Technik.* Berlin: Akademie-Verlag, 1949.

[52] SCHREIER, O. and SPERNER, E., *Vorlesungen über Matrizen.* Leipzig: Teubner, 1932. [A slightly revised version of this book appears as Chapter V of [53].]

[53] ——— *Introduction to Modern Algebra and Matrix Theory.* New York: Chelsea, 1958.

[54] SCHWERDTFEGER, H., *Introduction to Linear Algebra and the Theory of Matrices.* Groningen: Noordhoff, 1950.

[55] SHOHAT, J. A. and TAMARKIN, J. D., *The problem of moments* (Mathematical Surveys, No. 1). New York: Amer. Math. Society, 1943.

[56] SMIRNOW, W. I. (Smirnov, V. I.), *Lehrgang der höheren Mathematik,* Vol. III. Berlin, 1956. [This is a translation of the 13th Russian edition.]

[57] SPECHT, W., *Algebraische Gleichungen mit reellen oder komplexen Koeffizienten* (Enz. Math. Wiss., Band I, Teil B, Heft 3, Teil II). Stuttgart: Teubner, 1958.

[58] STIELTJES, T. J., *Oeuvres Completes.* 2 vols., Groningen: Noordhoff.

[59] STOLL, R. R., *Linear Algebra and Matrix Theory.* New York: McGraw-Hill, 1952.

[60] THRALL, R. M. and TORNHEIM, L., *Vector spaces and matrices.* New York: Wiley, 1957.

[61] TURNBULL, H. W., *The Theory of Determinants, Matrices and Invariants.* London: Blackie, 1950.

[62] TURNBULL, H. W. and AITKEN, A. C., *An Introduction to the Theory of Canonical Matrices.* London: Blackie, 1932.

[63] VOLTERRA, V. et HOSTINSKY, B., *Opérations infinitésimales linéaires.* Paris: Gauthiers-Villars, 1938.

[64] WEDDERBURN, J. H. M., *Lectures on matrices.* New York: Amer. Math. Society, 1934.

[65] WEYL, H., *Mathematische Analyse des Raumproblems.* Berlin, 1923. [A reprint is in preparation: Chelsea, 1960.]

[66] WINTNER, A., *Spektraltheorie der unendlichen Matrizen.* Leipzig, 1929.

[67] ZURMÜHL, R., *Matrizen.* Berlin, 1950.

PART B. Papers

[101] AFRIAT, S., *Composite matrices,* Quart. J. Math. vol. 5, pp. 81-89 (1954).

*[102] AIZERMAN (Aisermann), M. A., *On the computation of non-linear functions of several variables in the investigation of the stability of an automatic regulating system,* Avtomat. i Telemeh. vol. 8 (1947).

[103] AISERMANN, M. A. and F. R. GANTMACHER, *Determination of stability by linear approximation of a periodic solution of a system of differential equations with discontinuous right-hand sides,* Quart. J. Mech. Appl. Math. vol. 11, pp. 385-98 (1958).

[104] AITKEN, A. C., *Studies in practical mathematics. The evaluation, with applications, of a certain triple product matrix.* Proc. Roy. Soc. Edinburgh vol. 57, (1936-37).

[105] AMIR MOÉZ ALI, R., *Extreme properties of eigenvalues of a hermitian transformation and singular values of the sum and product of linear transformations,* Duke Math. J. vol. 23, pp. 463-76 (1956).

*[106] ARTASHENKOV, P. V., *Determination of the arbitrariness in the choice of a matrix reducing a system of linear differential equations to a system with constant coefficients.* Vestnik Leningrad. Univ., Ser. Mat., Phys. i Chim., vol. 2, pp. 17-29 (1953).

*[107] ARZHANYCH, I. S., *Extension of Krylov's method to polynomial matrices,* Dokl. Akad. Nauk SSSR, Vol. 81, pp. 749-52 (1951).

*[108] AZBELEV, N. and R. VINOGRAD, *The process of successive approximations for the computation of eigenvalues and eigenvectors,* Dokl. Akad. Nauk., vol. 83, pp. 173-74 (1952).

[109] BAKER, H. F., *On the integration of linear differential equations,* Proc. London Math. Soc., vol. 35, pp. 333-78 (1903).

[110] BARANKIN, E. W., *Bounds for characteristic roots of a matrix,* Bull. Amer. Math. Soc., vol. 51, pp. 767-70 (1945).

[111] BARTSCH, H., *Abschätzungen für die Kleinste charakteristische Zahl einer positiv-definiten hermitschen Matrix,* Z. Angew. Math. Mech., vol. 34, pp. 72-74 (1954).

[112] BELLMAN, R., *Notes on matrix theory,* Amer. Math. Monthly, vol. 60, pp. 173-75, (1953); vol. 62, pp. 172-73, 571-72, 647-48 (1955); vol. 64, pp. 189-91 (1957).

[113] BELLMAN, R. and A. HOFFMAN, *On a theorem of Ostrowski,* Arch. Math., vol. 5, pp. 123-27 (1954).

[114] BENDAT, J. and S. SILVERMAN, *Monotone and convex operator functions,* Trans. Amer. Math. Soc., vol. 79, pp. 58-71 (1955).

[115] BERGE, C., *Sur une propriété des matrices doublement stochastiques,* C. R. Acad. Sci. Paris, vol. 241, pp. 269-71 (1955).

[116] BIRKHOFF, G., *On product integration,* J. Math. Phys., vol. 16, pp. 104-32 (1937).

[117] BIRKHOFF, G. D., *Equivalent singular points of ordinary linear differential equations,* Math. Ann., vol. 74, pp. 134-39 (1913).

[118] BOTT, R. and R. DUFFIN, *On the algebra of networks,* Trans. Amer. Math. Soc., vol. 74, pp. 99-109 (1953).

[119] BRAUER, A., *Limits for the characteristic roots of a matrix,* Duke Math. J., vol. 13, pp. 387-95 (1946); vol. 14, pp. 21-26 (1947); vol. 15, pp. 871-77 (1948); vol. 19, pp. 73-91, 553-62 (1952); vol. 22, pp. 387-95 (1955).

[120] ——— *Über die Lage der charakteristischen Wurzeln einer Matrix,* J. Reine Angew. Math., vol. 192, pp. 113-16 (1953).

[121] ——— *Bounds for the ratios of the coordinates of the characteristic vectors of a matrix,* Proc. Nat. Acad. Sci. U.S.A., vol. 41, pp. 162-64 (1955).

[122] ——— *The theorems of Ledermann and Ostrowski on positive matrices,* Duke Math. J., vol. 24, pp. 265-74 (1957).

[123] BRENNER, J., *Bounds for determinants,* Proc. Nat. Acad. Sci. U.S.A., vol. 40, pp. 452-54 (1954); Proc. Amer. Math. Soc., vol. 5, pp. 631-34 (1954); vol. 8, pp. 532-34 (1957); C. R. Acad. Sci. Paris, vol. 238, pp. 555-56 (1954).

[124] BRUIJN, N., *Inequalities concerning minors and eigenvalues,* Nieuw Arch. Wisk., vol. 4, pp. 18-35 (1956).

[125] BRUIJN, N. and G. SZEKERES, *On some exponential and polar representatives of matrices,* Nieuw Arch. Wisk., vol. 3, pp. 20-32 (1955).

*[126] BULGAKOV, B. V., *The splitting of rectangular matrices,* Dokl. Akad. Nauk SSSR, vol. 85, pp. 21-24 (1952).

[127] CAYLEY, A., *A memoir on the theory of matrices,* Phil. Trans. London, vol. 148, pp. 17-37 (1857); Coll. Works, vol. 2, pp. 475-96.

[128] COLLATZ, L., *Einschliessungssatz für die charakteristischen Zahlen von Matrizen,* Math. Z., vol. 48, pp. 221-26 (1942).

[129] ——— *Über monotone systeme linearen Ungleichungen,* J. Reine Angew. Math., vol. 194, pp. 193-94 (1955).

[130] CREMER, L., *Die Verringerung der Zahl der Stabilitätskriterien bei Voraussetzung positiven koeffizienten der charakteristischen Gleichung,* Z. Angew. Math. Mech., vol. 33, pp. 222-27 (1953).

*[131] DANILEVSKIĬ, A. M., *On the numerical solution of the secular equation,* Mat. Sb., vol. 2, pp. 169-72 (1937).

[132] DILIBERTO, S., *On systems of ordinary differential operations.* In: *Contributions to the Theory of Non-linear Oscillations,* vol. I, edited by S. Lefschetz (Annals of Mathematics Studies, No. 20). Princeton: Princeton Univ. Press (1950), pp. 1-38.

*[133] DMITRIEV, N. A. and E. B. DYNKIN, *On the characteristic roots of stochastic matrices,* Dokl. Akad. Nauk SSSR, vol. 49, pp. 159-62 (1945).

*[133a] ——— *Characteristic roots of Stochastic Matrices,* Izv. Akad. Nauk, Ser. Fiz-Mat., vol. 10, pp. 167-94 (1946).

[134] DOBSCH, O., *Matrixfunktionen beschränkter Schwankung,* Math. Z., vol. 43, pp. 353-88 (1937).

*[135] DONSKAYA, L. I., *Construction of the solution of a linear system in the neighborhood of a regular singularity in special cases,* Vestnik Leningrad. Univ., vol. 6 (1952).

*[136] ——— *On the structure of the solution of a system of linear differential equations in the neighbourhood of a regular singularity,* Vestnik Leningrad. Univ., vol. 8, pp. 55-64 (1954).

*[137] DUBNOV, Y. S., *On simultaneous invariants of a system of affinors,* Trans. Math. Congress in Moscow 1927, pp. 236-37.

*[138] ——— *On doubly symmetric orthogonal matrices,* Bull. Ass. Inst. Univ. Moscow, pp. 33-35 (1927).

*[139] ——— *On Dirac's matrices,* Uč. zap. Univ. Moscow, vol. 2, pp. 2, 43-48 (1934).

*[140] DUBNOV, Y. S. and V. K. IVANOV, On the reduction of the degree of affinor *polynomials,* Dokl. Akad. Nauk SSSR, vol. 41, pp. 99-102 (1943).

[141] DUNCAN, W., *Reciprocation of triply-partitioned matrices,* J. Roy. Aero. Soc., vol. 60, pp. 131-32 (1956).

[142] EGERVÁRY, E., *On a lemma of Stieltjes on matrices,* Acta. Sci. Math., vol. 15, pp. 99-103 (1954).

[143] ——— *On hypermatrices whose blocks are commutable in pairs and their application in lattice-dynamics,* Acta Sci. Math., vol. 15, pp. 211-22 (1954).

[144] EPSTEIN, M. and H. FLANDERS, *On the reduction of a matrix to diagonal form,* Amer. Math. Monthly, vol. 62, pp. 168-71 (1955).

*[145] ERSHOV, A. P., *On a method of inverting matrices,* Dokl. Akad. Nauk SSSR, vol. 100, pp. 209-11 (1955).

[146] ERUGIN, N. P., *Sur la substitution exposante pour quelques systèmes irreguliers,* Mat. Sb., vol. 42, pp. 745-53 (1935).

*[147] ——— *Exponential substitutions of an irregular system of linear differential equations,* Dokl. Akad. Nauk SSSR, vol. 17, pp. 235-36 (1935).

*[148] ——— *On Riemann's problem for a Gaussian system,* Uč. Zap. Ped. Inst., vol. 28, pp. 293-304 (1939).

*[149] FADDEEV, D. K., *On the transformation of the secular equation of a matrix,* Trans. Inst. Eng. Constr., vol. 4, pp. 78-86 (1937).

[150] FAEDO, S., *Un nuove problema di stabilità per le equationi algebriche a coefficienti reali,* Ann. Scuola Norm. Sup. Pisa, vol. 7, pp. 53-63 (1953).

*[151] FAGE, M. K., *Generalization of Hadamard's determinant inequality,* Dokl. Akad. Nauk SSSR, vol. 54, pp. 765-68 (1946).

*[152] ——— *On symmetrizable matrices,* Uspehi Mat. Nauk, vol. 6, no. 3, pp. 153-56 (1951).

[153] FAN, K., *On a theorem of Weyl concerning eigenvalues of linear transformations,* Proc. Nat. Acad. Sci. U.S.A., vol. 35, pp. 652-55 (1949); vol. 36, pp. 31-35 (1950).

[154] ——— *Maximum properties and inequalities for the eigenvalues of completely continuous operators,* Proc. Nat. Acad. Sci. U.S.A., vol. 37, pp. 760-66 (1951).

[155] ——— *A comparison theorem for eigenvalues of normal matrices,* Pacific J. Math., vol. 5, pp. 911-13 (1955).

[156] ——— *Some inequalities concerning positive-definite Hermitian matrices,* Proc. Cambridge Philos. Soc., vol. 51, pp. 414-21 (1955).

[157] ——— *Topological proofs for certain theorems on matrices with non-negative elements,* Monatsh. Math., vol. 62, pp. 219-37 (1958).

[158] FAN, K. and A. HOFFMAN, *Some metric inequalities in the space of matrices,* Proc. Amer. Math. Soc., vol. 6, pp. 111-16 (1958).

[159] FAN, K. and G. PALL, *Imbedding conditions for Hermitian and normal matrices,* Canad. J. Math., vol. 9, pp. 298-304 (1957).

[160] FAN, K. and J. TODD, *A determinantal inequality,* J. London Math. Soc., vol. 30, pp. 58-64 (1955).

[161] FROBENIUS, G., *Über lineare substitutionen und bilineare Formen,* J. Reine Angew. Math., vol. 84, pp. 1-63 (1877).

[162] ——— *Über das Trägheitsgesetz der quadratischen Formen,* S.-B. Deutsch. Akad. Wiss. Berlin. Math.-Nat. Kl., 1894, pp. 241-56, 407-31.

[163] ——— *Über die cogredienten transformationen der bilinearer Formen,* S.-B. Deutsch. Akad. Wiss. Berlin. Math.-Nat. Kl., 1896, pp. 7-16.

[164] ——— *Über die vertauschbaren Matrizen,* S.-B. Deutsch. Akad. Wiss. Berlin. Math.-Nat. Kl., 1896, pp. 601-614.

[165] ——— *Über Matrizen aus positiven Elementen,* S.-B. Deutsch. Akad. Wiss. Berlin. Math-Nat. Kl. 1908, pp. 471-76; 1909, pp. 514-18.

[166] ——— *Über Matrizen aus nicht negativen Elementen,* S.-B. Deutsch. Akad. Wiss. Berlin Math.-Nat. Kl., 1912, pp. 456-77.

*[167] GANTMACHER, F. R., *Geometric theory of elementary divisors after Krull,* Trudy Odessa Gos. Univ. Mat., vol. 1, pp. 89-108 (1935).

*[168] ——— *On the algebraic analysis of Krylov's method of transforming the secular equation,* Trans. Second Math. Congress, vol. II, pp. 45-48 (1934).

[169] ——— *On the classification of real simple Lie groups,* Mat. Sb., vol. 5, pp. 217-50 (1939).

*[170] GANTMACHER, F. R. and M. G. KREĬN, *On the structure of an orthogonal matrix,* Trans. Ukrain, Acad. Sci. Phys.-Mat. Kiev (Trudy fiz.-mat. otdela VUAN, Kiev), 1929, pp. 1-8.

*[171] ——— *Normal operators in a hermitian space,* Bull. Phys-Mat. Soc. Univ. Kasan (Izvestiya fiz.-mat. ob-va pri Kazanskom universitete), IV, vol. 1, ser. 3, pp. 71-84 (1929-30).

*[172] ——— *On a special class of determinants connected with Kellogg's integral kernels,* Mat. Sb., vol. 42, pp. 501-8 (1935).

[173] ——— *Sur les matrices oscillatoires et completement non-négatives,* Compositio Math., vol. 4, pp. 445-76 (1937).

[174] GANTSCHI, W., *Bounds of matrices with regard to an hermitian metric,* Compositio Math., vol. 12, pp. 1-16 (1954).

*[175] GELFAND, I. M. and V. B. LIDSKIĬ, *On the structure of the domains of stability of linear canonical systems of differential equations with periodic coefficients,* Uspehi. Mat. Nauk, vol. 10, no. 1, pp. 3-40 (1955).

[176] GERSHGORIN, S. A., *Über die Abgrenzung der Eigenwerte einer Matrix,* Izv. Akad. Nauk SSSR, Ser. Fiz.-Mat., vol. 6, pp. 749-54 (1931).

[177] GODDARD, L., *An extension of a matrix theorem of A. Brauer,* Proc. Int. Cong. Math. Amsterdam, 1954, vol. 2, pp. 22-23.

[178] GOHEEN, H. E., *On a lemma of Stieltjes on matrices,* Amer. Math. Monthly, vol. 56, pp. 328-29 (1949).

*[179] GOLUBCHIKOV, A. F., *On the structure of the automorphisms of the complex simple Lie groups,* Dokl. Akad. Nauk SSSR, vol. 27, pp. 7-9 (1951).

*[180] GRAVE, D. A., *Small oscillations and some propositions in algebra,* Izv. Akad. Nauk SSSR, Ser. Fiz.-Mat., vol. 2, pp. 563-70 (1929).

*[181] GROSSMAN, D. P., *On the problem of a numerical solution of systems of simultaneous linear algebraic equations,* Uspehi Mat. Nauk, vol. 5, no. 3, pp. 87-103 (1950).

[182] HAHN, W., *Eine Bemerkung zur zweiten Methode von Lyapunov,* Math. Nachr., vol. 14, pp. 349-54 (1956).

[183] ——— *Über die Anwendung der Methode von Lyapunov auf Differenzengleichungen,* Math. Ann., vol. 136, pp. 430-41 (1958).

[184] HAYNSWORTH, E., *Bounds for determinants with dominant main diagonal,* Duke Math. J., vol. 20, pp. 199-209 (1953).

[185] ——— *Note on bounds for certain determinants,* Duke Math. J., vol. 24, pp. 313-19 (1957).

[186] HELLMANN, O., *Die Anwendung der Matrizanten bei Eigenwertaufgaben,* Z. Angew. Math. Mech., vol. 35, pp. 300-15 (1955).

[187] HERMITE, C., *Sur le nombre des racines d'une équation algébrique comprise entre des limites données,* J. Reine Angew. Math., vol. 52, pp. 39-51 (1856).

[188] HJELMSLER, J., *Introduction à la théorie des suites monotones,* Kgl. Danske Vid. Selbsk. Forh. 1914, pp. 1-74.

[189] HOFFMAN, A. and O. TAUSSKY, *A characterization of normal matrices,* J. Res. Nat. Bur. Standards, vol. 52, pp. 17-19 (1954).

[190] HOFFMAN, A. and H. WIELANDT, *The variation of the spectrum of a normal matrix,* Duke Math. J., vol. 20, pp. 37-39 (1953).

[191] HORN, A., *On the eigenvalues of a matrix with prescribed singular values,* Proc. Amer. Math. Soc., vol. 5, pp. 4-7 (1954).

[192] HOTELLING, H., *Some new methods in matrix calculation,* Ann. Math. Statist., vol. 14, pp. 1-34 (1943).

[193] HOUSEHOLDER, A. S., *On matrices with non-negative elements,* Monatsh. Math., vol. 62, pp. 238-49 (1958).

[194] HOUSEHOLDER, A. S. and F. L. BAUER, *On certain methods for expanding the characteristic polynomial,* Numer. Math., vol. 1, pp. 29-35 (1959).

[195] HSU, P. L., *On symmetric, orthogonal, and skew-symmetric matrices,* Proc. Edinburgh Math. Soc., vol. 10, pp. 37-44 (1953).

[196] ——— *On a kind of transformation of matrices,* Acta Math. Sinica, vol. 5, pp. 333-47 (1955).

[197] HUA, L.-K., *On the theory of automorphic functions of a matrix variable,* Amer. J. Math., vol. 66, pp. 470-88; 531-63 (1944).

[198] ——— *Geometries of matrices,* Trans. Amer. Math. Soc., vol. 57, pp. 441-90 (1945).

[199] ——— *Orthogonal classification of Hermitian matrices,* Trans. Amer. Math. Soc., vol. 59, pp. 508-23 (1946).

*[200] ——— *Geometries of symmetric matrices over the real field,* Dokl. Akad. Nauk SSSR, vol. 53, pp. 95-98; 195-96 (1946).

*[201] ——— *Automorphisms of the real symplectic group,* Dokl. Akad. Nauk SSSR, vol. 53, pp. 303-306 (1946).

[202] ——— *Inequalities involving determinants,* Acta Math. Sinica, vol. 5, pp. 463-70 (1955).

*[203] HUA, L.-K. and B. A. ROSENFELD, *The geometry of rectangular matrices and their application to the real projective and non-euclidean geometries,* Izv. Higher Ed. SSSR, Matematika, vol. 1, pp. 233-46 (1957).

[204] HURWITZ, A., *Über die Bedingungen, unter welchen eine Gleichung nur Wurzeln mit negativen reellen Teilen besitzt,* Math. Ann., vol. 46, pp. 273-84 (1895).

[205] INGRAHAM, M. H., *On the reduction of a matrix to its rational canonical form,* Bull. Amer. Math. Soc., vol. 39, pp. 379-82 (1933).

[206] IONESCU, D., *O identitate importantă si descompunere a unei forme bilineare into sumă de produse,* Gaz. Mat. Ser. Fiz. A. 7, vol. 7, pp. 303-312 (1955).

[207] ISHAK, M., *Sur les spectres des matrices,* Sém. P. Dubreil et Ch. Pisot, Fac. Sci. Paris, vol. 9, pp. 1-14 (1955/56).

*[208] KAGAN, V. F., *On some number systems arising from Lorentz transformations,* Izv. Ass. Inst. Moscow Univ. 1927, pp. 3-31.

*[209] KARPELEVICH, F. I., *On the eigenvalues of a matrix with non-negative elements,* Izv. Akad. Nauk SSSR Ser. Mat., vol. 15, pp. 361-83 (1951).

[210] KHAN, N. A., *The characteristic roots of a product of matrices,* Quart. J. Math., vol. 7, pp. 138-43 (1956).

*[211] KHLODOVSKIĬ, I. N., *On the theory of the general case of Krylov's transformation of the secular equation,* Izv. Akad. Nauk, Ser. Fiz.-Mat., vol. 7, pp. 1076-1102 (1933).

*[212] KOLMOGOROV, A. N., *Markov chains with countably many possible states,* Bull. Univ. Moscow (A), vol. 1:3 (1937).

*[213] KOTELYANSKIĬ, D. M., *On monotonic matrix functions of order n,* Trans. Univ. Odessa, vol. 3, pp. 103-114 (1941).

*[214] ——— *On the theory of non-negative and oscillatory matrices,* Ukrain. Mat. Z., vol. 2, pp. 94-101 (1950).

*[215] ——— *On some properties of matrices with positive elements,* Mat. Sb., vol. 31, pp. 497-506 (1952).

*[216] ——— *On a property of matrices of symmetric signs,* Uspehi Mat. Nauk, vol. 8, no. 4, pp. 163-67 (1953).

*[217] ——— *On some sufficient conditions for the spectrum of a matrix to be real and simple,* Mat. Sb., vol. 36, pp. 163-68 (1955).

*[218] ——— *On the influence of Gauss' transformation on the spectra of matrices,* Uspehi Mat. Nauk, vol. 9, no. 3, pp. 117-21 (1954).

*[219] ——— *On the distribution of points on a matrix spectrum,* Ukrain. Mat. Z., vol. 7, pp. 131-33 (1955).

*[220] ——— *Estimates for determinants of matrices with dominant main diagonal*, Izv. Akad. Nauk SSSR, Ser. Mat., vol. 20, pp. 137-44 (1956).

*[221] KOVALENKO, K. R. and M. G. KREĬN, *On some investigations of Lyapunov concerning differential equations with periodic coefficients*, Dokl. Akad. Nauk SSSR, vol. 75, pp. 495-99 (1950).

[222] KOWALEWSKI, G., *Natürliche Normalformen linearer Transformationen*, Leipz. Ber., vol. 69, pp. 325-35 (1917).

*[223] KRASOVSKIĬ, N. N., *On the stability after the first approximation*, Prikl. Mat. Meh., vol. 19, pp. 516-30 (1955).

*[224] KRASNOSEL'SKIĬ, M. A. and M. G. KREĬN, *An iteration process with minimal deviations*, Mat. Sb., vol. 31, pp. 315-34 (1952).

[225] KRAUS, F., *Über konvexe Matrixfunktionen*, Math. Z., vol. 41, pp. 18-42 (1936).

*[226] KRAVCHUK, M. F., *On the general theory of bilinear forms*, Izv. Polyt. Inst. Kiev, vol. 19, pp. 17-18 (1924).

*[227] ——— *On the theory of permutable matrices*, Zap. Akad. Nauk Kiev, Ser. Fiz.-Mat., vol. 1:2, pp. 28-33 (1924).

*[228] ——— *On a transformation of quadratic forms*, Zap. Akad. Nauk Kiev, Ser. Fiz.-Mat., vol. 1:2, pp. 87-90 (1924).

*[229] ——— *On quadratic forms and linear transformations*, Zap. Akad. Nauk Kiev, Ser. Fiz.-Mat., vol. 1:3, pp. 1-89 (1924).

*[230] ——— *Permutable sets of linear transformations*, Zap. Agr. Inst. Kiev, vol. 1, pp. 25-58 (1926).

[231] ——— *Über vertauschbare Matrizen*, Rend. Circ. Mat. Palermo, vol. 51, pp. 126-30 (1927).

*[232] ——— *On the structure of permutable groups of matrices*, Trans. Second. Mat. Congress 1934, vol. 2, pp. 11-12.

*[233] KRAVCHUK, M. F. and Y. S. GOL'DBAUM, *On groups of commuting matrices*, Trans. Av. Inst. Kiev, 1929, pp. 73-98; 1936, pp. 12-23.

*[234] ——— *On the equivalence of singular pencils of matrices*, Trans. Av. Inst. Kiev, 1936, pp. 5-27.

*[235] KREĬN, M. G., *Addendum to the paper 'On the structure of an orthogonal matrix,'* Trans. Fiz.-Mat. Class. Akad. Nauk Kiev, 1931, pp. 103-7.

*[236] ——— *On the spectrum of a Jacobian form in connection with the theory of torsion oscillations of drums*, Mat. Sb., vol. 40, pp. 455-66 (1933).

*[237] ——— *On a new class of hermitian forms*, Izv. Akad. Nauk SSSR, Ser. Fiz.-Mat., vol. 9, pp. 1259-75 (1933).

*[238] ——— *On the nodes of harmonic oscillations of mechanical systems of a special type*, Mat. Sb., vol. 41, pp. 339-48 (1934).

[239] ——— *Sur quelques applications des noyaux de Kellog aux problèmes d'oscillations*, Proc. Charkov Mat. Soc. (4), vol. 11, pp. 3-19 (1935).

[240] ——— *Sur les vibrations propres des tiges dont l'une des extrémités est encastrée et l'autre libre*, Proc. Charkov. Mat. Soc. (4), vol. 12, pp. 3-11 (1935).

*[241] ——— *Generalization of some results of Lyapunov on linear differential equations with periodic coefficients*, Dokl. Akad. Nauk SSSR, vol. 73, pp. 445-48 (1950).

*[242] ——— *On an application of the fixed-point principle in the theory of linear transformations of spaces with indefinite metric*, Uspehi Mat. Nauk, vol. 5, no. 2, pp. 180-90 (1950).

*[243] ——— *On an application of an algebraic proposition in the theory of monodromy matrices*, Uspehi Mat. Nauk, vol. 6, no. 1, pp. 171-77 (1951).

*[244] ——— *On some problems concerning Lyapunov's ideas in the theory of stability*, Uspehi Mat. Nauk, vol. 3, no. 3, pp. 166-69 (1948).

*[245] ——— *On the theory of integral matrix functions of exponential type*, Ukrain. Mat. Z., vol. 3, pp. 164-73 (1951).

*[246] ——— *On some problems in the theory of oscillations of Sturm systems*, Prikl. Mat. Meh., vol. 16, pp. 555-68 (1952).

*[247] KREĬN, M. G. and M. A. NAĬMARK (Neumark), *On a transformation of the Bézoutian leading to Sturm's theorem*, Proc. Charkov Mat. Soc., (4), vol. 10, pp. 33-40 (1933).

*[248] ——— *On the application of the Bézoutian to problems of the separation of roots of algebraic equations*, Trudy Odessa Gos. Univ. Mat., vol. 1, pp. 51-69 (1935).

[249] KRONECKER, L., *Algebraische Reduction der Schaaren bilinearer Formen*, S.-B. Akad. Berlin 1890, pp. 763-76.

[250] KRULL, W., *Theorie und Anwendung der verallgemeinerten Abelschen Gruppen*, S.-B. Akad. Heidelberg 1926, p. 1.

*[251] KRYLOV, A. N., *On the numerical solution of the equation by which the frequency of small oscillations is determined in technical problems*, Izv. Akad. Nauk SSSR Ser. Fiz.-Mat., vol. 4, pp. 491-539 (1931).

[252] LAPPO-DANILEVSKIĬ, I. A., *Résolution algorithmique des problèmes réguliers de Poincaré et de Riemann*, J. Phys. Mat. Soc. Leningrad, vols. 2:1, pp. 94-120; 121-54 (1928).

[253] ——— *Théorie des matrices satisfaisantes à des systèmes des équations differentielles linéaires a coefficients rationnels arbitraires*, J. Phys. Mat. Soc. Leningrad, vols. 2:2, pp. 41-80 (1928).

*[254] ——— *Fundamental problems in the theory of systems of linear differential equations with arbitrary rational coefficients*, Trans. First Math. Congr., ONTI, 1936, pp. 254-62.

[255] LEDERMANN, W., *Reduction of singular pencils of matrices*, Proc. Edinburgh Math. Soc., vol. 4, pp. 92-105 (1935).

[256] ——— *Bounds for the greatest latent root of a positive matrix*, J. London Math. Soc., vol. 25, pp. 265-68 (1950).

*[257] LIDSKIĬ, V. B., *On the characteristic roots of a sum and a product of symmetric matrices*, Dokl. Akad. Nauk SSSR, vol. 75, pp. 769-72 (1950).

*[258] ——— *Oscillation theorems for canonical systems of differential equations*, Dokl. Akad. Nauk SSSR, vol. 102, pp. 111-17 (1955).

[259] LIÉNARD, and CHIPART, *Sur la signe de la partie réelle des racines d'une équation algébrique*, J. Math. Pures Appl. (6), vol. 10, pp. 291-346 (1914).

*[260] LIPIN, N. V., *On regular matrices*, Trans. Inst. Eng 8. Transport, vol. 9, p. 105 (1934).

*[261] LIVSHITZ, M. S. and V. P. POTAPOV, *The multiplication theorem for characteristic matrix functions*, Dokl. Akad. Nauk SSSR, vol. 72, pp. 164-73 (1950).

*[262] LOPSHITZ, A. M., *Vector solution of a problem on doubly symmetric matrices*, Trans. Math. Congress Moscow, 1927, pp. 186-87.

*[263] ——— *The characteristic equation of lowest degree for affinors and its application to the integration of differential equations*, Trans. Sem. Vectors and Tensors, vols. 2/3 (1935).

*[264] ——— *A numerical method of determining the characteristic roots and characteristic planes of a linear operator*, Trans. Sem. Vectors and Tensors, vol. 7, pp. 233-59 (1947).

*[265] ——— *An extremal theorem for a hyper-ellipsoid and its application to the solution of a system of linear algebraic equations*, Trans. Sem. Vectors and Tensors, vol. 9, pp. 183-97 (1952).

[266] LÖWNER, K., *Über monotone Matrixfunktionen*, Math. Z., vol. 38, pp. 177-216 (1933); vol. 41, pp. 18-42 (1936).

[267] ——— *Some classes of functions defined by difference on differential inequalities*, Bull. Amer. Math. Soc., vol. 56, pp. 308-19 (1950).

*[268] LUSIN, N. N., *On Krylov's method of forming the secular equation*, Izv. Akad. Nauk SSSR, Ser. Fiz.-Mat., vol. 7, pp. 903-958 (1931).

*[269] ——— *On some properties of the displacement factor in Krylov's method*, Izv. Akad. Nauk SSSR, Ser. Fiz.-Mat., vol. 8, pp. 596-638; 735-62; 1065-1102 (1932).

*[270] ——— *On the matrix theory of differential equations*, Avtomat. i Telemeh, vol. 5, pp. 3-66 (1940).

*[271] LYUSTERNIK, L. A., *The determination of eigenvalues of functions by an electric scheme*, Electričestvo, vol. 11, pp. 67-8 (1946).

*[272] ——— *On electric models of symmetric matrices*, Uspehi Mat. Nauk, vol. 4, no. 2, pp. 198-200 (1949).

*[273] LYUSTERNIK, L. A. and A. M. PROKHOROV, *Determination of eigenvalues and functions of certain operators by means of an electrical network*, Dokl. Akad Nauk SSSR, vol. 55, pp. 579-82; Izv. Akad. Nauk SSSR, Ser. Mat., vol. 11, pp. 141-45 (1947).

[274] MARCUS, M., *A remark on a norm inequality for square matrices*, Proc. Amer. Math. Soc., vol. 6, pp. 117-19 (1955).

[275] ——— *An eigenvalue inequality for the product of normal matrices*, Amer. Math. Monthly, vol. 63, pp. 173-74 (1956).

[276] ——— *A determinantal inequality of H. P. Robertson*, II, J. Washington Acad. Sci., vol. 47, pp. 264-66 (1957).

[277] ——— *Convex functions of quadratic forms*, Duke Math. J., vol. 24, pp. 321-26 (1957).

[278] MARCUS, M. and J. L. MCGREGOR, *Extremal properties of Hermitian matrices*, Canad. J. Math., vol. 8, pp. 524-31 (1956).

[279] MARCUS, M. and B. N. MOYLS, *On the maximum principle of Ky Fan*, Canad. J. Math., vol. 9, pp. 313-20 (1957).

[280] ——— *Maximum and minimum values for the elementary symmetric functions of Hermitian forms*, J. London Math. Soc., vol. 32, pp. 374-77 (1957).

*[281] MAYANTS, L. S., *A method for the exact determination of the roots of secular equations of high degree and a numerical analysis of their dependence on the parameters of the corresponding matrices*, Dokl. Akad. Nauk SSSR, vol. 50, pp. 121-24 (1945).

[282] MIRSKY, L., *An inequality for positive-definite matrices*, Amer. Math. Monthly, vol. 62, pp. 428-30 (1955).

[283] ——— *The norm of adjugate and inverse matrices*, Arch. Math., vol. 7, pp. 276-77 (1956).

[284] ——— *The spread of a matrix*, Mathematika, vol. 3, pp. 127-30 (1956).

[285] ——— *Inequalities for normal and Hermitian matrices*, Duke Math. J., vol. 24, pp. 591-99 (1957).

[286] MITROVIĆ, D., *Conditions graphiques pour que toutes les racines d'une équation algébrique soient à parties réelles négatives*, C. R. Acad. Sci. Paris, vol. 240, pp. 1177-79 (1955).

[287] MORGENSTERN, D., *Eine Verschärfung der Ostrowskischen Determinantenabschätzung*, Math. Z., vol. 66, pp. 143-46 (1956).

II new

[288] MOTZKIN, T. and O. TAUSSKY, *Pairs of matrices with property L.*, Trans. Amer. Math. Soc., vol. 73, pp. 108-14 (1952); vol. 80, pp. 387-401 (1954).

*[289] NEĬGAUS (Neuhaus), M. G. and V. B. LIDSKIĬ, *On the boundedness of the solutions of linear systems of differential equations with periodic coefficients*, Dokl. Akad. Nauk SSSR, vol. 77, pp. 183-93 (1951).

[290] NEUMANN, J., *Approximative of matrices of high order*, Portugal. Math., vol. 3, pp. 1-62 (1942).

*[291] NUDEL'MAN, A. A. and P. A. SHVARTSMAN, *On the spectrum of the product of unitary matrices*, Uspehi Mat. Nauk, vol. 13, no. 6, pp. 111-17 (1958).

[292] OKAMOTO, M., *On a certain type of matrices with an application to experimental design*, Osaka Math. J., vol. 6, pp. 73-82 (1954).

[293] OPPENHEIM, A., *Inequalities connected with definite Hermitian forms*, Amer. Math. Monthly, vol. 61, pp. 463-66 (1954).

[294] ORLANDO, L., *Sul problema di Hurwitz relativo alle parti reali delle radici di un' equatione algebrica*, Math. Ann., vol. 71, pp. 233-45 (1911).

[295] OSTROWSKI, A., *Bounds for the greatest latent root of a positive matrix*, J. London Math. Soc., vol. 27, pp. 253-56 (1952).

[296] ——— *Sur quelques applications des fonctions convexes et concaves au sens de I. Schur*, J. Math. Pures Appl., vol. 31, pp. 253-92 (1952).

[297] ——— *On nearly triangular matrices*, J. Res. Nat. Bur. Standards, vol. 52, pp. 344-45 (1954).

[298] ——— *On the spectrum of a one-parametric family of matrices*, J. Reine Angew. Math., vol. 193, pp. 143-60 (1954).

[299] ——— *Sur les déterminants à diagonale dominante*, Bul. Soc. Math. Belg., vol. 7, pp. 46-51 (1955).

[300] ——— *Note on bounds for some determinants*, Duke Math. J., vol. 22, pp. 95-102 (1955).

[301] ——— *Über Normen von Matrizen*, Math. Z., vol. 63, pp. 2-18 (1955).

[302] ——— *Über die Stetigkeit von charakteristischen Wurzeln in Abhängigkeit von den Matrizenelementen*, Jber. Deutsch. Math. Verein., vol. 60, pp. 40-42 (1957).

*[303] PAPKOVICH, P. F., *On a method of computing the roots of a characteristic determinant*, Prikl. Mat. Meh., vol. 1, pp. 314-18 (1933).

[304] PAPULIS, A., *Limits on the zeros of a network determinant*, Quart. Appl. Math., vol. 15, pp. 193-94 (1957).

[305] PARODI, M., *Remarques sur la stabilité*, C. R. Acad. Sci. Paris, vol. 228, pp. 51-2; 807-8; 1198-1200 (1949).

[306] ——— *Sur une propriété des racines d'une équation qui intervient en mécanique*, C. R. Acad. Sci. Paris, vol. 241, pp. 1019-21 (1955).

[307] ——— *Sur la localisation des valeurs caractéristiques des matrices dans le plan complexe*, C. R. Acad. Sci. Paris, vol. 242, pp. 2617-18 (1956).

[308] PEANO, G., *Intégration par series des équations différentielles linéaires*, Math. Ann., vol. 32, pp. 450-56 (1888).

[309] PENROSE, R., *A generalized inverse for matrices*, Proc. Cambridge Philos. Soc., vol. 51, pp. 406-13 (1955).

[310] ——— *On best approximate solutions of linear matrix equations*, Proc. Cambridge Philos. Soc., vol. 52, pp. 17-19 (1956).

[311] PERFECT, H., *On matrices with positive elements*, Quart. J. Math., vol. 2, pp. 286-90 (1951).

[312] ——— *On positive stochastic matrices with real characteristic roots*, Proc. Cambridge Philos. Soc., vol. 48, pp. 271-76 (1952).

[313] ——— *Methods of constructing certain stochastic matrices*, Duke Math. J., vol. 20, pp. 395-404 (1953); vol. 22, pp. 305-11 (1955).

[314] ——— *A lower bound for the diagonal elements of a non-negative matrix*, J. London Math. Soc., vol. 31, pp. 491-93 (1956).

[315] PERRON, O., *Jacobischer Kettenbruchalgorithmus*, Math. Ann., vol. 64, pp. 1-76 (1907).

[316] ——— *Über Matrizen*, Math. Ann., vol. 64, pp. 248-63 (1907).

[317] ——— *Über Stabilität und asymptotisches Verhalten der Lösungen eines Systems endlicher Differenzengleichungen*, J. Reine Angew. Math., vol. 161, pp. 41-64 (1929).

[318] PHILLIPS, H. B., *Functions of matrices*, Amer. J. Math., vol. 41, pp. 266-78 (1919).

*[319] PONTRYAGIN, L. S., *Hermitian operators in a space with indefinite metric*, Izv. Akad. Nauk SSSR, Ser. Mat., vol. 8, pp. 243-80 (1944).

*[320] POTAPOV, V. P., *On holomorphic matrix functions bounded in the unit circle*, Dokl. Akad. Nauk SSSR, vol. 72, pp. 849-53 (1950).

[321] RASCH, G., *Zur Theorie und Anwendung der Produktintegrals*, J. Reine Angew. Math., vol. 171, pp. 65-119 (19534).

[322] REICHARDT, H., *Einfache Herleitung der Jordanschen Normalform*, Wiss. Z. Humboldt-Univ. Berlin. Math.-Nat. Reihe, vol. 6, pp. 445-47 (1953/54).

*[323] RECHTMAN-OL'SHANSKAYA, P. G., *On a theorem of Markov*, Uspehi Mat. Nauk, vol. 12, no. 3, pp. 181-87 (1957).

[324] RHAM, G. DE, *Sur un théorème de Stieltjes relatif à certains matrices*, Acad. Serbe Sci. Publ. Inst. Math., vol. 4, pp. 133-54 (1952).

[325] RICHTER, H., *Über Matrixfunktionen*, Math. Ann., vol. 122, pp. 16-34 (1950).

[326] ——— *Bemerkung zur Norm der Inversen einer Matrix*, Arch. Math., vol. 5, pp. 447-48 (1954).

[327] ——— *Zur Abschätzung von Matrizennormen*, Math. Nachr., vol. 18, pp. 178-87 (1958).

[328] ROMANOVSKIĬ, V. I., *Un théorème sur les zéros des matrices non-négatives*, Bull. Soc. Math. France, vol. 61, pp. 213-19 (1933).

[329] ——— *Recherches sur les chaînes de Markoff*, Acta Math., vol. 66, pp. 147-251 (1935).

[330] ROTH, W., *On the characteristic polynomial of the product of two matrices*; Proc. Amer. Math. Soc., vol. 5, pp. 1-3 (1954).

[331] ——— *On the characteristic polynomial of the product of several matrices*, Proc. Amer. Math. Soc., vol. 7, pp. 578-82 (1956).

[332] ROY, S., *A useful theorem in matrix theory*, Proc. Amer. Math. Soc., vol. 5, pp. 635-38 (1954).

*[333] SAKHNOVICH, L. A., *On the limits of multiplicative integrals*, Uspehi Mat. Nauk, vol. 12 no. 3, pp. 205-11 (1957).

*[334] SARYMSAKOV, T. A., *On sequences of stochastic matrices*, Dokl. Akad. Nauk, vol. 47, pp. 331-33 (1945).

[335] SCHNEIDER, H., *An inequality for latent roots applied to determinants with dominant principal diagonal*, J. London Math. Soc., vol. 28, pp. 8-20 (1953).

[336] ——— *A pair of matrices with property P*, J. Amer. Math. Monthly, vol. 62, pp. 247-49 (1955).

[337] ——— *A matrix problem concerning projections*, Proc. Edinburgh Math. Soc., vol. 10, pp. 129-30 (1956).

[338] ——— *The elementary divisors, associated with* 0, *of a singular M-matrix*, Proc. Edinburgh Math. Soc., vol. 10, pp. 108-22 (1956).

[339] SCHOENBERG, J., *Über variationsvermindernde lineare Transformationen*, Math. Z., vol. 32, pp. 321-28 (1930).

[340] ——— *Zur Abzählung der reellen Wurzeln algebraischer Gleichungen*, Math. Z., vol. 38, p. 546 (1933).

[341] SCHOENBERG, I. J., and A. WHITNEY, *A theorem on polygons in n dimensions with application to variation diminishing linear transformations*, Compositio Math., vol. 9, pp. 141-60 (1951).

[342] SCHUR, I., *Über die charakteristischen Wurzeln einer linearen Substitution mit einer Anwendung auf die theorie der Integralgleichungen*, Math. Ann., vol. 66, pp. 488-510 (1909).

*[343] SEMENDYAEV, K. A., *On the determination of the eigenvalues and invariant manifolds of matrices by means of iteration*, Prikl. Matem. Meh., vol. 3, pp. 193-221 (1943).

*[344] SEVAST'YANOV, B. A., *The theory of branching random processes*, Uspehi Mat. Nauk, vol. 6, no. 6, pp. 46-99 (1951).

*[345] SHIFFNER, L. M., *The development of the integral of a system of differential equations with regular singularities in series of powers of the elements of the differential substitution*, Trudy Mat. Inst. Steklov. vol. 9, pp. 235-66 (1935).

*[346] ——— *On the powers of matrices*, Mat. Sb., vol. 42, pp. 385-94 (1935).

[347] SHODA, K., *Über mit einer Matrix vertauschbare Matrizen*, Math. Z., vol. 29, pp. 696-712 (1929).

*[348] SHOSTAK, P. Y., *On a criterion for the conditional definiteness of quadratic forms in n linearly independent variables and on a sufficient condition for a conditional extremum of a function of n variables*, Uspehi Mat. Nauk, vol. 8, no. 4, pp. 199-206 (1954).

*[349] SHREĬDER, Y. A., *A solution of systems of linear algebraic equations*, Dokl. Akad. Nauk, vol. 76, pp. 651-55 (1950).

*[350] SHTAERMAN (Steiermann), I. Y., *A new method for the solution of certain algebraic equations which have application to mathematical physics*, Z. Mat., Kiev, vol. 1, pp. 83-89 (1934); vol. 4, pp. 9-20 (1934).

*[351] SHTAERMAN (Steiermann), I. Y. and N. I. AKHIESER (Achieser), *On the theory of quadratic forms*, Izv. Polyteh., Kiev, vol. 19, pp. 116-23 (1934).

*[352] SHURA-BURA, M. R., *An estimate of error in the numerical computation of matrices of high order*, Uspehi Mat. Nauk, vol. 6, no. 4, pp. 121-50 (1951).

*[353] SHVARTSMAN (Schwarzmann), A. P., *On Green's matrices of self-adjoint differential operators*, Proc. Odessa Univ. Matematika, vol. 3, pp. 35-77 (1941).

[354] SIEGEL, C. L., *Symplectic Geometry*, Amer. J. Math., vol. 65, pp. 1-86 (1943).

*[355] SKAL'KINA, M. A., *On the preservation of asymptotic stability on transition from differential equations to the corresponding difference equations*, Dokl. Akad. Nauk SSSR, vol. 104, pp. 505-8, (1955).

*[356] SMOGORZHEVSKIĬ, A. S., *Sur les matrices unitaires du type de circulants*, J. Mat. Circle Akad. Nauk Kiev, vol. 1, pp. 89-91 (1932).

*[356a] SMOGORZHEVSKIĬ, A. S. and M. F. KRAVCHUK, *On orthogonal transformations*, Zap. Inst. Ed. Kiev, vol. 2, pp. 151-56 (1927).

[357] STENZEL, H., *Über die Darstellbarkeit einer Matrix als Produkt von zwei symmetrischen Matrizen*, Math. Z., vol. 15, pp. 1-25 (1922).

[358] STÖHR, A., *Oszillationstheoreme für die Eigenvektoren spezieller Matrizen*, J. Reine Angew. Math., vol. 185, pp. 129-43 (1943).

*[359] SULEĬMANOVA, K. R., *Stochastic matrices with real characteristic values*, Dokl. Akad. Nauk SSSR, vol. 66, pp. 343-45 (1949).

*[360] ——— *On the characteristic values of stochastic matrices,* Uč. Zap. Moscow Ped. Inst., Ser. 71, Math., vol. 1, pp. 167-97 (1953).

*[361] SULTANOV, R. M., *Some properties of matrices with elements in a non-commutative ring,* Trudy Mat. Sectora Akad. Nauk Baku, vol. 2, pp. 11-17 (1946).

*[362] SUSHKEVICH, A. K., *On matrices of a special type,* Uč. Zap. Univ. Charkov, vol. 10, pp. 1-16 (1937).

[363] SZ-NAGY, B., *Remark on S. N. Roy's paper 'A useful theorem in matrix theory,'* Proc. Amer. Math. Soc., vol. 7, p. 1 (1956).

[364] TA LI, *Die Stabilitätsfrage bei Differenzengleichungen,* Acta Math., vol. 63, pp. 99-141 (1934).

[365] TAUSSKY, O., *Bounds for characteristic roots of matrices,* Duke Math. J., vol. 15, pp. 1043-44 (1948).

[366] ——— *A determinantal inequality of H. P. Robertson,* I, J. Washington Acad. Sci., vol. 47, pp. 263-64 (1957).

[367] ——— *Commutativity in finite matrices,* Amer. Math. Monthly, vol. 64, pp. 229-35 (1957).

[368] TOEPLITZ, O., *Das algebraische Analogon zu einem Satz von Fejér,* Math. Z., vol. 2, pp. 187-97 (1918).

[369] TURNBULL, H. W., *On the reduction of singular matrix pencils,* Proc. Edinburgh Math. Soc., vol. 4, pp. 67-76 (1935).

*[370] TURCHANINOV, A. S., *On some applications of matrix calculus to linear differential equations,* Uč. Zap. Univ. Odessa, vol. 1, pp. 41-48 (1921).

*[371] VERZHBITSKIĬ, B. D., *Some problems in the theory of series compounded from several matrices,* Mat. Sb., vol. 5, pp. 505-12 (1939).

*[372] VILENKIN, N. Y., *On an estimate for the maximal eigenvalue of a matrix,* Uč. Zap. Moscow Ped. Inst., vol. 108, pp. 55-57 (1957).

[373] VIVIER, M., *Note sur les structures unitaires et paraunitaires,* C. R. Acad. Sci. Paris, vol. 240, pp. 1039-41 (1955).

[374] VOLTERRA, V., *Sui fondamenti della teoria delle equazioni differenziali lineari,* Mem. Soc. Ital. Sci. (3), vol. 6, pp. 1-104 (1887); vol. 12, pp. 3-68 (1902).

[375] WALKER, A. and J. WESTON, *Inclusion theorems for the eigenvalues of a normal matrix,* J. London Math. Soc., vol. 24, pp. 28-31 (1949).

[376] WAYLAND, H., *Expansions of determinantal equations into polynomial form,* Quart. Appl. Math., vol. 2, pp. 277-306 (1945).

[377] WEIERSTRASS, K., *Zur theorie der bilinearen und quadratischen Formen,* Monatsh. Akad. Wiss. Berlin, 1867, pp. 310-38.

[378] WELLSTEIN, J., *Über symmetrische, alternierende und orthogonale Normalformen von Matrizen,* J. Reine Angew. Math., vol. 163, pp. 166-82 (1930).

[379] WEYL, H., *Inequalities between the two kinds of eigenvalues of a linear transformation,* Proc. Nat. Acad. Sci., vol. 35, pp. 408-11 (1949).

[380] WEYR, E., *Zur Theorie der bilinearen Formen,* Monatsh. f. Math. und Physik, vol. 1, pp. 163-236 (1890).

[381] WHITNEY, A., *A reduction theorem for totally positive matrices,* J. Analyse Math., vol. 2, pp. 88-92 (1952).

[382] WIELANDT, H., *Ein Einschliessungssatz für charakteristische Wurzeln normaler Matrizen,* Arch. Math., vol. 1, pp. 348-52 (1948/49).

[383] ——— *Die Einschliessung von Eigenwerten normaler Matrizen,* Math. Ann. vol. 121, pp. 234-41 (1949).

[384] ——— *Unzerlegbare nicht-negative Matrizen,* Math. Z., vol. 52, pp. 642-48 (1950).

[385] ——— *Lineare Scharen von Matrizen mit reellen Eigenwerten*, Math. Z., vol. 53, pp. 219-25 (1950).

[386] ——— *Pairs of normal matrices with property L*, J. Res. Nat. Bur. Standards, vol. 51, pp. 89-90 (1953).

[387] ——— *Inclusion theorems for eigenvalues*, Nat. Bur. Standards, Appl. Math. Sci., vol. 29, pp. 75-78 (1953).

[388] ——— *An extremum property of sums of eigenvalues*, Proc. Amer. Math. Soc., vol. 6, pp. 106-110 (1955).

[389] ——— *On eigenvalues of sums of normal matrices*, Pacific J. Math., vol. 5, pp. 633-38 (1955).

[390] WINTNER, A., *On criteria for linear stability*, J. Math. Mech., vol. 6, pp. 301-9 (1957).

[391] WONG, Y., *An inequality for Minkowski matrices*, Proc. Amer. Math. Soc., vol. 4, pp. 137-41 (1953).

[392] ——— *On non-negative valued matrices*, Proc. Nat. Acad. Sci. U.S.A., vol. 40, pp. 121-24 (1954).

*[393] YAGLOM, I. M., *Quadratic and skew-symmetric bilinear forms in a real symplectic space*, Trudy Sem. Vect. Tens. Anal. Moscow, vol. 8, pp. 364-81 (1950).

*[394] YAKUBOVICH, V. A., *Some criteria for the reducibility of a system of differential equations*, Dokl. Akad. Nauk SSSR, vol. 66, pp. 577-80 (1949).

*[395] ZEITLIN (Tseĭtlin), M. L., *Application of the matrix calculus to the synthesis of relay-contact schemes*, Dokl. Akad. Nauk SSSR, vol. 86, pp. 525-28 (1952).

*[396] ZIMMERMANN (Tsimmerman), G. K., *Decomposition of the norm of a matrix into products of norms of its rows*, Nauč. Zap. Ped. Inst. Nikolaevsk, vol. 4, pp. 130-35 (1953).

INDEX

INDEX

[Numbers in italics refer to Volume Two]

APPLIED MATHEMATICS AND PHYSICS

(See also: Biography, Collected Papers)

CORSON, Edward M.: Introduction to Tensors, Spinors, and Relativistic Wave Equations. 2nd (unaltered) ed. 222 pp. 6 1/2 x 9 3/4. ISBN -0315-5,

CRABTREE, Harold: Spinning Tops and Gyroscopic Motion, 2nd ed. 203 pp. 6 x 9. ISBN -0204-3,

EDDINGTON, Arthur S.: The Mathematical Theory of Relativity, 3rd (unaltered) ed. ix + 270 pp. 6 x 9. CIP. ISBN -0278-8,

EVANS, Griffith C.: Logarithmic Potential, 2nd ed.; & BLISS, G. A. Fundamental Existence Theorems, & KASNER, E. Differential-Geometric Aspects of Dynamics. 3 vols in 1. 399 pp. 5-3/8 x 8. ISBN -0305-8,

GANTMACHER, Felix R.: Lectures on Analytical Mechanics, 2nd ed. in prep.

GREENHILL, George: Gyroscopic Theory. vi + 277 pp. + fold-out plates. 6½ x 10¾. ISBN -0205-1,

HEAVISIDE, Oliver: Electrical Papers, 2 vols. 1,183 pp. 5 3/8 x 8. ISBN -0235-3, the set.

———Electromagnetic Theory, including an account of Heaviside's Unpublished Notes, 3rd ed. 3 vols. 1,791 pp. 5¼ x 8. ISBN -0237-X, the set.

HOHENEMSER, K.: Elastokinetik. (Germ.) 89 pp. 5½ x 8½. ISBN -0055-5,

IGNATOWSKY, W. von: Physikalisch-Mathematische Monographien, 3 vols. in 1. (Germ., Engl.) xvi + 232 pp. 6 x 9. ISBN -0201-9,

JACOBI, Carl G. J.: Gesammelte Werke, 2nd ed. (Germ., French) Vols. 1-7, 4,032 pp. 6½ x 9¼. ISBN -0226-4; Vol. 8, (Germ.) 308 pp. 6½ x 9¼. ISBN -0227-2, the 8-vol. set.

———Vorlesungen ueber Dynamik (= vol. 8 of Gesammelte Werke), (Germ.) 308 pp. 6½ x 9¼. ISBN -0227-2,

KASNER, Edward: Differential-Geometric Aspects of Dynamics. See EVANS, G.C.

KLEIN, Felix: The Mathematical Theory of the Top. Included in SIERPINSKI, Waclaw: Congruence of Sets. (General).

LAPLACE, Pierre S.: Celestial Mechanics, Vols. 1-4. xxiv + 136 + 746 pp.; xvii + 990 pp.; xxx + 910 + 117 pp. of tables; xxvi + 1,018 pp. ISBN -0194-2, Vol. 5 (in French), ix + 508 pp. 6 x 9. ISBN -0214-0,

SCHROEDINGER, Erwin: Collected Papers on Wave Mechanics, including Four Lectures on Wave Mechanics, 3rd (augmented) ed. xii + 212 pp. 6 x 9. ISBN -1302-9,

TOLMAN, Richard C.: Relativity, Thermodynamics and Cosmology. 2nd (unaltered) ed. 501 pp. 6 x 9, In prep.

WEYL, Hermann, et al: Das Kontinuum und Andere Monographien, 4 vols. in 1. *Includes:* Das Kontinuum, by H. WEYL; Mathematische Analyse des Raumproblems, by H. WEYL; Neuere Funkltionentheorie, by E. LANDAU; Hypothesen, by B. RIEMANN. (Germ.) 368 pp. 5 3/8 x 8. ISBN 0134-9,

HISTORY

BOCHENSKI, Innocenty M.: History of Formal Logic, 2nd ed. xx + 567 pp. 5 3/8 x 8. ISBN -0238-8,

CAJORI, Florian: History of Mathematics. 3rd (corr.) ed. xi + 518 pp. 6 x 9. CIP. ISBN -0303-1,

———History of the Slide Rule. Included in: BALL, W. R. String Figures. (General Math.)

DICKSON, Leonard Eugene: History of the Theory of Numbers, 3 vols. xii + 468 pp.; xxv + 803 pp.; V + 313 pp. 5 3/8 x 8. ISBN -0086-5, the set.

GOW, James: Short History of Greek Mathematics. xii + 325 pp. 5 3/8 x 8. ISBN -0218-3,

HAWKINS, Thomas: Lebesgue's Theory of Integration: Its Origins and Development, 3rd ed. xv + 227 pp. 6 x 9. ISBN -0282-5,

KLEIN, Felix: Entwicklung der Mathematik im 19. Jahrhundert, 2 vols. in 1. (Germ.) 619 pp. 5¼ x 8. ISBN -0074-1,

MIKAMI, Yashio: The Development of Mathematics in China and Japan. 2nd ed. x + 383 pp. 5 3/8 x 8. ISBN -0149-7,

SCOTT, Joseph Frederick: The Mathematical Work of John Wallis. 2nd (unaltered) ed. xii + 240 pp. 6 x 9. ISBN -0314-7,

TODHUNTER, Isaac: History of the Calculus of Variations in the 19th Century. xii + 532 pp. 5 3/8 x 8. ISBN -0164-0,

———History of the Mathematical Theory of Probability. xvi + 624 pp. 5 3/8 x 8. ISBN -0057-1,

WOODHOUSE, Robert: History of the Calculus of Variations in the 18th Century. ix + x + 154 pp. 5 x 8¼. ISBN -0177-2,

LINEAR ALGEBRA AND MATRIX THEORY

FORDER, Henry G.: The Calculus of Extension, xvi + 490 pp. 5 3/8 x 8. ISBN -0135-7,

GANTMACHER, Felix R.: The Theory of Matrices. Vol. I, x + 374 pp. 6 x 9. ISBN -0131-4, Vol. II, x + 277 pp. 6 x 9. ISBN - 0133-0,

GRASSMANN, Hermann G.: Die Ausdehnungslehre von 1878, 4th ed. (Germ.) xii + 435 pp. 6 x 9. ISBN -0222-1,

KAPLANSKY, Irving: Linear Algebra and Geometry, 2nd ed. 151 pp. 6 x 9. CIP. ISBN -0279-5,

MacDUFFEE, Cyrus C.: The Theory of Matrices. 2nd ed. v + 110 pp. 6 x 9. ISBN -0028-8,

NOMIZU, Katsumi: Fundamentals of Linear Algebra, 2nd ed. (with answers). xii + 361 pp. 6 x 9. CIP. ISBN -0276-0,

PEDOE, Daniel: A Geometric Introduction to Linear Algebra, 2nd ed. xi + 224 pp. 6 x 9. CIP. ISBN -0286-8,